Nicole Webw

D0072889

Nature and Culture

Nature and Culture

Rebuilding Lost Connections

Edited by
Sarah Pilgrim and Jules Pretty

publishing for a sustainable future

London • New York

First published in 2010 by Earthscan

Copyright © Sarah Pilgrim and Jules Pretty, 2010

All rights reserved. No part of this publication may be reproduced, stored in a retrieval system, or transmitted, in any form or by any means, electronic, mechanical, photo-copying, recording or otherwise, except as expressly permitted by law, without the prior, written permission of the publisher.

Earthscan
2 Park Square, Milton Park, Abingdon, Oxfordshire OX14 4RN
Simultaneously published in the USA and Canada by Earthscan
711 Third Avenue, New York, NY 10017
Earthscan is an imprint of the Taylor & Francis Group, an informa business

Earthscan publishes in association with the International Institute for Environment and Development

ISBN: 978-1-84407-821-9 hardback

Typeset by Composition and Design Services
Cover design by Susanne Harris

A catalogue record for this book is available from the British Library

Library of Congress Cataloging-in-Publication Data

Nature and culture: rebuilding lost connections/edited by Sarah Pilgrim and Jules Pretty.
    p. cm.
  Includes bibliographical references and index.
  ISBN 978-1-84407-821-9 (hardback)
  1. Nature–Effect of human beings on. 2. Human beings–Effect of environment on.
3. Human ecology. 4. Biodiversity. 5. Cultural pluralism. 6. Nature and nurture.
I. Pilgrim, Sarah. II. Pretty, Jules N.
  GF75.N34 2010
  304.2–dc22   2010017152

At Earthscan we strive to minimize our environmental impacts and carbon footprint through reducing waste, recycling and offsetting our CO_2 emissions, including those created through publication of this book.

Contents

Contributors

W. M. Adams

William M. Adams is Moran Professor of Conservation and Development at the University of Cambridge, where he has taught since 2004. His research concerns social dimensions of conservation and the evolution of conservation policy and strategy, especially in the UK and Africa. His books include *Future Nature: A Vision for Conservation* (Earthscan, 2003, 2nd edition), *Against Extinction: The Story of Conservation* (Earthscan, 2004), *Green Development: Environment and Sustainability in a Developing World* (Routledge, 2008, 3rd edition) and *Decolonizing Nature: Strategies for Conservation in a Post-colonial Era* (co-edited with Martin Mulligan, Earthscan, 2002). He has also edited a four-volume collection *Conservation* (Earthscan, 2008).

Glenn Albrecht

Glenn Albrecht is Professor of Sustainability at Murdoch University in Perth, Western Australia. He has a PhD in philosophy and works within the transdisciplinary space at the intersection of ecosystem and human health. With co-authors he published *Health Social Science: A Transdisciplinary and Complexity Perspective* (Oxford University Press, 2003) and he has published many book chapters and articles on topics related to ecohealth, human and animal ethics. He is currently engaged in funded research on the mental health impacts of environmental change, including climate change.

E. N. Anderson

Eugene N. Anderson is Professor Emeritus of Anthropology at the University of California, Riverside. He has done fieldwork in Hong Kong, Malaysia, Mexico, British Columbia, California and (for shorter periods) several other areas. He is primarily interested in cultural and political ecology, ethnobiology, and the development of ideas and representations of the environment (especially plants and animals). He has also done research in medical and nutritional anthropology, branching out from a human-ecology focus. He has written several books, including *The Food of China* (Yale University Press, 1988), *Ecologies of the Heart: Emotion, Belief, and the Environment* (Oxford University Press, 1996), *Everyone Eats: Understanding Food and Culture* (New York University Press, 2005) and *Floating World Lost: A Hong Kong Fishing Community* (University Press of the South, 2007). He is currently working on the relationship between ideology, resource management and cultural representations of the environment.

Fikret Berkes

Fikret Berkes is Distinguished Professor at the University of Manitoba and Canada Research Chair in Community-Based Resource Management. He works in the area of commons theory and interrelations between societies and their resources. His scholarly publications include the books, *Sacred Ecology* (Routledge, 2008), *Adaptive Co-management: Collaboration, Learning, and Multi-level Government* (University of British Columbia Press, 2007) co-edited with Derek Armitage and Nancy Doubleday, and *Navigating Social-Ecological Systems: Building Resilience for Complexity and Change* (Cambridge University Press, 2003) co-edited with Johan Colding and Carl Folke.

Tirso Gonzales

Tirso Gonzales, PhD, completed his doctorate in rural sociology at the University of Wisconsin, Madison. His work as a scholar, international consultant and activist has allowed him to work closely with indigenous peoples in the Americas. Currently he is an Assistant Professor of Indigenous Studies at the University of British Columbia, Okanagan, Canada. His current work explores the use of indigenous and non-indigenous research methodologies and techniques on issues central to self-determined indigenous development, indigenous and local histories, indigenous strategic vision, and local management of natural resources. Dr Gonzales has published several articles and chapters on indigenous agriculture and knowledge. He is committed to supporting the agenda of indigenous peoples as well as processes related to indigenous ecological knowledge, cultural affirmation and decolonization. His current research focuses on Latin American Andean indigenous intellectuals, and decolonizing Latin American indigenous studies.

Maria Gonzalez

Maria Gonzalez is a doctoral candidate in the Department of Comparative Literature at the University of Michigan completing a dissertation on Quechua Oral Tradition in Peru and Bolivia. She worked with the non-governmental organization (NGO), Taller de Historia Oral Andina (Oral Andean History Workshop, THOA), in Bolivia between 1999 and 2001 during her fieldwork. She has also worked with Proyecto Andino de Tecnologías Campesinas (Andean Project of Peasant Technologies, PRATEC) on multiple occasions throughout the last 20 years. She has studied the Quechua language for seven years.

David Harmon

David Harmon is Executive Director of the George Wright Society, an association of researchers, managers and other professionals who work in protected areas, and is a cofounder of the NGO Terralingua, which focuses on biocultural diversity. He is the author of *In Light of Our Differences: How Diversity in Nature and Culture Makes Us Human* (Smithsonian Institution Press, 2002) and co-editor of several conservation volumes, including *The Full Value of Parks: From Economics to the Intangible* (Rowman & Littlefield, 2003) and *The Antiquities Act: A Century of American Archaeology, Historic Preservation, and Nature Conservation* (University of Arizona Press, 2006).

Patricia L. Howard

Patricia L. Howard is Professor in the Department of Social Sciences at Wageningen University in The Netherlands as well as Honorary Professor in Biocultural Diversity Studies at the University of Kent. She was lead author of the Scientific Conceptual Framework for the Food and Agriculture Organization's (FAO's) Globally Important Agriculture Heritage Systems (GIAHS) programme. She has published widely on agriculture and sustainability and on human-biodiversity relationships, including many articles and edited books.

Jonathan Loh

Jonathan Loh works on biological and cultural diversity conservation, and developed indicators such as the Living Planet Index and the Index of Linguistic Diversity. He conceived and edits the World Wide Fund for Nature (WWF) Living Planet Report and is an Honorary Research Associate at the Zoological Society of London.

Luisa Maffi

Luisa Maffi, PhD, is one of the pioneers of the field of biocultural diversity. She is co-founder and director of Terralingua, an international NGO devoted to sustaining the biocultural diversity of life through research, education and action. Among her books is the edited volume *On Biocultural Diversity: Linking Language, Knowledge and the Environment* (Smithsonian Institution Press, 2001) and the co-authored volume *Biocultural Diversity Conservation: A Global Sourcebook* (Earthscan, 2010).

Garry Marvin

Garry Marvin is Professor of Human-Animal Studies at Roehampton University, London, UK. He is the author of *Bullfight* (University of Illinois Press, 1994) and the co-author of *Zoo Culture* (University of Illinois Press, 1999) and *Killing Animals* (University of Illinois Press, 2006). He has also published articles on hunting (particularly English foxhunting), hunting taxidermy and zoos. His present research focuses on the experiences of hunters and he is concluding a study of human–wolf relations.

Helen Newing

Helen Newing is a Lecturer in Conservation Social Science at the Durrell Institute of Conservation and Ecology (DICE) and Centre for Biocultural Diversity (CBCD) in the School of Anthropology and Conservation at the University of Kent, UK. Her research focuses on the intersection between conservation and rural development, especially concerning collaborative management, Community Conserved Areas, indigenous rights and biocultural diversity, and interdisciplinary approaches to research. She is the author of a forthcoming book on *Conducting Research in Conservation: Social Science Methods and Practice*, to be published by Routledge in 2010.

Sarah Pilgrim

Sarah Pilgrim, PhD, is a Research Fellow at the University of Essex. Originally trained as an ecologist, she has worked with indigenous and marginal peoples, particularly in Asia, since 2002. She has published extensively on the interaction between biological and cultural diversity and the impacts of cultural revitalization projects on reconnecting people to the land. She was invited to contribute to the European Academies Science Advisory Council report on *Ecosystem Services: Assessment of Current Status of Priority Ecosystem Services in Europe* (2008), and has recently been asked to contribute to the UK government's Foresight Project on Global Food and Farming Futures (2010). Her current research focuses on the resilience of inherited and intentional ecocultures to the impacts of climate change in different parts of the world.

Jules Pretty

Jules Pretty is Professor of Environment and Society in the Department of Biological Sciences at the University of Essex. His books include *This Luminous Coast* (Full Circle, 2010), *The Earth Only Endures: On Reconnecting with Nature and Our Place in it* (Earthscan, 2007), *Agri-Culture: Reconnecting People, Land and Nature* (Earthscan, 2002), and *Regenerating Agriculture: Policies and Practice for Sustainability and Self-reliance* (Earthscan, 1995). He is a Fellow of the Society of Biology and the Royal Society of Arts, former Deputy-Chair of the government's Advisory Committee on Releases to the Environment (ACRE), and has served on advisory committees for a number of government departments. He is a regular speaker, contributor to media, and presenter of the 1999 BBC Radio 4 series *Ploughing Eden*, a contributor and writer for the 2001 BBC TV Correspondent programme *The Magic Bean*, and a panellist in 2007 for Radio 4's *The Moral Maze*. He received an OBE in 2006 for services to sustainable agriculture, and an honorary degree from Ohio State University in 2009. His website is www.julespretty.com.

David Rapport

David Rapport is currently Principal of EcoHealth Consulting and Co-Professor at the Institute for Applied Ecology, Chinese Academy of Sciences, Shenyang.

He has served as founding President of the International Society for Ecosystem Health and as Editor-in-Chief of the international journal *Ecosystem Health*. He held the Tri-council Eco-Research Chair in Ecosystem Health at the University of Guelph. He is a Fellow of the Linnean Society of London. He is lead editor of *Managing for Healthy Ecosystems* (CRC Press, 2003).

James P. Robson

James P. Robson is a PhD candidate in the Natural Resources Institute, University of Manitoba, Canada. His research interests revolve around people-resource dynamics in mountain environments. His current work focuses on the impact of rural out-migration among Zapotec and Chinantec indigenous communities in the high-biodiversity forest landscapes of northern Oaxaca, Mexico. He was formerly at the Institute of Social Research in the Universidad Nacional Autónoma de México (National Autonomous University of Mexico, UNAM).

Colin Samson

Colin Samson is a Senior Lecturer in the Department of Sociology at the University of Essex. He has been working with the Innu peoples of the Labrador-Quebec peninsula in Canada since 1994. His associations with them led to co-authoring the widely-cited report *Canada's Tibet: The Killing of the Innu* which won the Italian Pio Manzo peace prize in 2000. His book *A Way of Life that Does Not Exist: Canada and the Extinguishment of the Innu* (Verso Books, 2003) won the Pierre Savard Award given by the International Council for Canadian Studies in 2006. Colin is currently writing a book on social and cultural 'reversibility', examining the potentials of indigenous peoples to use their own histories, experiences, ideas and practices as ways to reverse the catastrophic effects of colonialism and state domination. He has written a number of recent book chapters and articles on indigenous peoples' human rights and the role of anthropology in indigenous rights conflicts. During 2009–2010 he was a Visiting Scholar at the University of Arizona.

Martina Tyrrell

Martina Tyrrell is a social anthropologist and independent scholar. She was previously a British Academy post-doctoral fellow at the Scott Polar Research Institute, Cambridge University, and Lecturer in Nature, Society and Development at the Department of Geography, University of Reading, UK. Her research explores Inuit environmental knowledge, human-animal relations, and the human dimensions of wildlife conservation. Among her publications are two chapters in *Inuit, Polar Bears and Sustainable Use* (Freeman and Foote, 2009).

Ellen Woodley

Ellen Woodley, PhD, works as an Ecological Consultant in environment and development, specializing in issues of local and traditional ecological knowledge. She is based in Ontario, Canada. She has published chapters with the Millennium Ecosystem Assessment and the Global Environment Outlook (GEO-4) on issues related to culture and biodiversity. She is currently working with communities on adaptation to climate change in Nigeria.

Preface

The network of relationships linking the human race to itself and to the rest of the biosphere is so complex that all aspects affect all others to an extraordinary degree. Someone should be studying the whole system, however crudely that has to be done, because no gluing together of partial studies of a complex nonlinear system can give a good idea of the behaviour of the whole. (Murray Gell-Mann, 1994)

This book is a compilation of writings by forward-thinking conservationists situated in both the natural and social sciences. It addresses two difficult terms – nature and culture – and their relation to one another. Each is used in many differing circumstances, often with different intent. Both, though, are important to all of us and shape our everyday lives. Contributors to this volume differ in their academic backgrounds, conceptualizations of the world, geographical expertise and languages, but they all converge on the importance of understanding the depth and breadth of the multiple interconnections between nature and culture. It is commonly held that these connections have diminished in modern and industrialized societies, but the authors indicate that the rebuilding of these connections is an essential prerequisite to addressing the many global economic, ecological, social and cultural challenges facing humankind today. Many have observed the consequences of this connection being shattered, in some locations forcibly, and the devastating consequences to both people and ecosystems. However, contributors to this volume clearly believe that these connections can be rebuilt for the long-term health of biological and social systems in years to come. All is not lost, but the task is hard.

The chapters in this book centre on a number of common themes. These include the urgent need for interdisciplinarity that crosses natural and social science boundaries. If natural and social systems are intimately interwoven in the real world, then they should be so connected in science. This book attempts to overcome some of the boundaries that exist and look beyond disciplinary divides, calling for a less reductionist approach to conservation in the future, and a more inclusive or holistic way to talk about and protect global diversity. Another theme is ideas about the wild or wilderness, and how humans view and act in nature. These views and ideas inevitably condition our actions towards nature and in some cases promote further disconnection. A common theme centres on the need to build resilience – not just among ecological and cultural systems in parallel, but among whole ecocultural systems.

Here we speculate that resilience can be at its strongest when connections are maintained or rebuilt, and when human and biological systems act together. It is thus an urgent task to consider strategies that focus on rebuilding these connections across all types of culture.

This book has three main objectives:

- to describe the complexity of interconnections that exist between humans and the natural environment in different cultural contexts, and in different regions of the world, and to examine how modern science is adjusting to this emergent paradigm;
- to examine the implications of weakened or even broken connections, particularly caused by rapid socio-economic or ecological shifts, in terms of the health of biological and social systems, and resilience of ecocultural systems as a whole;
- to identify divergent pathways into the future that endeavour to reconnect cultures, in a variety of different ecological and political contexts, to natural landscapes in an attempt to help communities forge their own sustainable pathways into the future.

By discussing the multiple interconnections, we hope that these chapters facilitate the shift towards a new conservation paradigm in which scientists and policy makers start to consider biological and cultural diversity as an interdependent whole, and all actors realize the need for a new integrative conservation approach to global biodiversity. We also hope that this collection of writings can provide guidance to communities currently suffering the effects of disconnection, by providing them with success stories where groups of people have revived their local connections to the land for the long-term health of both human and ecological systems. Most importantly, perhaps, we set out a series of journeys through the lands and minds of hunters, foragers, fishers, farmers and scholars who describe their own connections with nature in many different ways. At times it may seem that there is little hope, that bridges are beyond repair, and that the consequences of development are inevitable. Yet it is in ecocultures across the world that hope begins and spreads, and in some way new futures are created.

Sarah Pilgrim and Jules Pretty
University of Essex
December 2009

Reference

Gell-Mann, M. (1994) *The Quark and the Jaguar: Adventures in the Simple and Complex*, W. H. Freeman, New York

Acronyms

ACDP	Alaska Chukotka Development Program (Chukotka, Russia)
ACRE	Advisory Committee on Releases to the Environment (UK government)
AGRUCO	Agroecología Universidad Cochabamba
AU	agricultural units
CAMPFIRE	Communal Areas Management Program for Indigenous Resources
CBCD	Centre for Biocultural Diversity (University of Kent)
CBD	Convention on Biological Diversity
CC	Comunidad Campesina
CEESP	Commission of Environmental, Economic and Social Policy (IUCN)
CIFOR	Centre for International Forestry Research
CONABIO	Nacional para el Conocimiento y Uso de la Biodiversidad (National Commission for Use and Knowledge of Biodiversity, Mexico)
CONAFOR	Comisión Nacional Forestral (National Forestry Commission, Mexico)
CONAMAQ	Consejo de Ayllus y Markas del Collasuyu (Bolivia)
CONANP	Comisión Nacional de Areas Naturales Protegidas (National Natural Protected Areas Commission, Mexico)
CORENCHI	Regional Committee for Chinantla Alta Natural Resources
COSEWIC	Committee on the Status of Endangered Wildlife in Canada
CRC	Canada Research Chairs
CRRMH	Centre for Rural and Remote Mental Health (New South Wales, Australia)
CWS	Canadian Wildlife Service
DFO	Department of Fisheries and Oceans (Canada)
DICE	Durrell Institute of Conservation and Ecology (University of Kent)
DPSIR	driving force-pressure-state-impact-response (diversity indicator model)
ELF	Ethnolinguistic Fractionalization Index
EPP	Ecocultural Protection Plan
FAO	Food and Agriculture Organization (of the United Nations)
FRLHT	Foundation for Revitalisation of Local Health Traditions (Bangalore)

GIAHS	Globally Important Agriculture Heritage Systems (FAO programme)
HANPP	human appropriation of net terrestrial primary production
HTO	hunters and trappers organization
IBCD	Index of Biocultural Diversity
ICCAs	Indigenous and Community Conserved Areas
IIFB	International Indigenous Forum on Biodiversity
IITC	International Indian Treaty Council
ILD	Index of Linguistic Diversity
ILO	International Labour Organization
IPA	Indigenous Protected Area
IPO	indigenous peoples' organization
IPCC	Intergovernmental Panel on Climate Change
ISE	International Society of Ethnobiology
IUCN	International Union for Conservation of Nature
LGEEPA	Ley General de Equilibrio Ecológico y la Protección al Ambiente (General Law of Ecological Balance and Environmental Protection, Mexico)
LPI	Living Planet Index
LVT	Linguistic Vitality Test
MDGs	Millennium Development Goals
MEA	Millennium Ecosystem Assessment
NACA	Nucelos de Afirmación Cultural Andina (Nuclei of Andean Cultural Affirmation, Peru)
NGO	non-governmental organization
PA	protected area
PES	payments for environmental services
PRATEC	Proyecto Andino de Technologías Campesinas (Andean Project of Peasant Technologies, Peru)
PSR	pressure-state-response (diversity indicator model)
RLI	Red List Index
SARD	Sustainable Agriculture and Rural Development (FAO initiative)
SINAP	Sistema Nacional de Áreas Protegidas (National System of Protected Areas, Nicaragua)
SEI	Site of Ecocultural Importance
TEK	traditional ecological knowledge
THOA	Taller de Historia Oral Andina (Oral Andean History Workshop, Bolivia)
UNAM	Universidad Nacional Autónoma de México (National Autonomous University of Mexico)
UNDP	United Nations Development Programme
UNEP	United Nations Environment Programme
UNESCO	United Nations Educational, Scientific and Cultural Organization
UNICEF	United Nations Children's Fund
UNPFII	United Nations Permanent Forum on Indigenous Issues

VCA   Voluntary Conservation Area
VITEK   Vitality Index of Traditional Ecological Knowledge
WHO   World Health Organization
WWF   World Wide Fund for Nature (World Wildlife Fund in US and Canada)

1
Nature and Culture: An Introduction

Sarah Pilgrim and Jules Pretty

The State-of-the-Art

There is a widespread recognition across cultures that the diversity of life involves both the living forms (biological diversity) and the worldviews and cosmologies of what life means (cultural diversity) (Posey, 1999; Berkes et al, 2000; Maffi, 2001; Harmon, 2002). Thomas (2009) stated, 'the most valuable assets of any traditional community are its lands and its culture'. What has become clear is that these assets are so inextricably linked that one cannot exist without the other, and indeed differentiating between the two, particularly in the context of traditional societies, is a somewhat arbitrary activity. Many would consider this distinction to be a social construct in itself. Even when considered as a dichotomy, it is clear that nature and culture converge on many levels that span belief systems, social and institutional organizations, norms, stories, knowledge, behaviours and languages. As a result, there exists a mutual feedback between cultural systems and the environment, with shifts in one commonly leading to changes in the other. Thus the division commonly made between nature and culture is not universal and, in many cases, is a product of modern industrialized thought shaped by the need to control and manage nature (Berkes, 2008).

Though this combined concept has been slow to emerge in many industrial contexts, it represents the majority view in most resource-dependent communities, many of which perceive biological and cultural diversity as part of the same interconnected whole. Reflecting upon this, Berkes and Folke (2002) suggest that distinctions between social and natural systems are somewhat artificial. Traditional societies have, after all, interacted with biodiversity through adaptive and co-evolutionary processes for thousands of generations (Balée, 1994; Norgaard, 1994; Denevan, 2001; Maffi, 2001; Toledo, 2001;

Gunderson and Holling, 2002; Harmon, 2002; Heckenberger et al, 2007). Berkes and Folke (2002) suggest that the term 'social-ecological system' helpfully refers to this integrated concept of humans and nature.

Conceptualizations of the relationship between human societies and nature have historically shaped the way in which we see the world and our actions towards it. A variety of social science sub-disciplines have now developed new terms to describe branches of research relating to environmental conservation, and include environmental or ecological anthropology, environmental politics, ecological economics and environmental history (Rapport, 2006). The growth of these sub-disciplines has led professional societies to establish formal working groups on conservation issues, and some disciplines even have separate conservation-oriented professional societies. Between them, these recently established interdisciplinary fields investigate a range of relevant conservation research questions at a variety of scales and using many methods.

Some of these emergent disciplines help to explore bridges between the natural and social sciences. But although many have the potential to contribute to understanding of the interactions between nature and culture, at present there are no generally accepted and recognized conceptual or methodological approaches for achieving this. By being fragmented in this way, these new sub-disciplines can appear uncoordinated and disconnected when it comes to the advancement of scientific knowledge, local implementation, and the development of national and international policies. In this collection, one aim is to map how to go beyond divisive definitions by investigating the bridges linking nature with culture, and the far-reaching community efforts that have been initiated to rebuild these links.

The Biodiversity and Cultural Diversity Complex

Biodiversity is defined as the variation of life at the level of gene, species and ecosystem (CBD, 1992). Much has been written on its importance in terms of intrinsic value, anthropocentric uses, and role in today's economic markets and in providing subsistence livelihood options for resource-dependent communities worldwide (Constanza et al, 1997; Gunderson and Holling, 2002; MEA, 2005). Biodiversity represents the product of thousands of years of evolution. At the same time, it serves as an absorptive barrier, providing protection from, and thus resilience against, environmental perturbations. Resilience theory emphasizes that all systems have limits of change (tipping points). Within these limits, systems can tolerate and adapt to perturbations while still sustaining normal function. Going beyond these thresholds, however, results in the destabilization of the system (Rotarangi and Russell, 2009). Biodiversity is now a recognized prerequisite to ecosystem health and resilience, as well as an essential precondition to sustainable livelihoods, human health and many other social objectives, as reflected in the Millennium Development Goals (MDGs) (MEA, 2005; Rapport, 2006).

Culture can be defined as a combination of sets of practices, networks of institutions and systems of meanings. Cultural systems code for the knowledge, practices, beliefs, worldviews, values, norms, identities, livelihoods and social organizations of human societies. Different cultures value nature in different ways and thus have different connections with their natural environments. The maintenance of cultural diversity into the future, and the knowledge, innova-tions and outlooks it contains, increases the capacity of human systems to adapt and cope with change (Gunderson and Holling, 2002; Harmon, 2002). Therefore, in the same way that biological diversity increases the resilience of natural systems, cultural diversity has the capacity to increase the resilience of social systems. Rotarangi and Russell (2009) suggest that 'the maintenance and evolution of identity and culture of indigenous people and communities is premised on such resilience'.

As will be discussed throughout this book, nature and culture converge on many levels from values, beliefs and norms to practices, livelihoods, knowl-edge and languages. As a result, there exists a mutual feedback between cultural systems and the environment (Maffi and Woodley, 2007). This book investigates this concept by the examination of case studies which analyse how different cultures interact with biodiversity and how nature, in turn, has shaped their worldviews, knowledge and practices, particularly in light of current climate change. As a result, human and ecosystem health is predicated on a concomitant effort to sustain this inextricable connection.

Beliefs, Cosmologies and Worldviews: Our Place in Nature

Cultural systems are broadly based upon the way in which people interpret the world around them (Geertz, 1973). Human meanings and interpretations are perhaps the most diverse in their linkage to the natural world, based on dependence and daily interactions, values, knowledge, perceptions and belief systems, and how strongly these centre upon nature. Reflecting this, it has been suggested that the difference in cultural worldviews and cosmologies of nature between industrialized and resource-dependent (or subsistence-oriented) communities stems from a difference in need and purpose (Milton, 1998; Berkes, 2004).

Many industrial cultures perceive nature and culture as two separate entities, thus the prevailing modernist view tends to be of a nature-culture dichotomy, whereby humans are seeking to assert their dominance over nature. However, some cultures hold a more inclusive view perceiving humans as inter-dependent components of nature. In this case, nature is regarded as a force that manages human existence. In practice, the worldviews of human commu-nities form a spectrum between these two extremes. What is more, perceptions of nature are dynamic and with the coming challenges of climate change and peak oil, it is conceivable that those communities whose livelihoods appear (on the surface) to be resource-independent, may have to undergo substantial changes in their perceptions and practices in the near future.

Milton (1998, 1999) has considered human communities' relationship with nature in some depth, and has suggested that some feel an acute (strong) sense of oneness with nature. These communities do not recognize a distinction between nature and culture. Instead, they view themselves as part of the same continuous system as the lands to which they belong. Although relationships/kinships with non-human entities (such as plants, animals, spirits and gods) are easily observable, the relationship with nature as a whole is often more intrinsic and subtle, so that it goes unspoken and unrecognized. Thus to have a strong sense of oneness with nature is to not recognize a distinction between nature and culture. On the other hand, communities with a weak sense tend to perceive humans as separate from nature. They do, nevertheless, tend to acknowledge a reciprocal relationship based upon respect.

This inclusive view of nature, or sense of oneness (strong or weak), is not universal, and many human communities instead hold an exclusive and reductionist view of nature. Some have even gone beyond viewing nature and culture as separate entities, and instead perceive them as extreme opposing entities, whose interaction generally leads to one or the other being damaged in some way (hence the establishment of people-free protected areas and exclusion zoning). This diversity of perspectives, in itself, contributes to cultural diversity. Ellen (1996) recognized this diversity and proposed that three definitions of nature exist in the modern industrialized cosmos: nature as a category of 'things'; nature as space that is not human; and nature as inner essence (Milton, 1998). E.O. Wilson conjectured that all humans, no matter their culture, have an innate connection with nature based on our common histories as hunter-gatherers. He termed this innate bond 'biophilia' (a love of nature) (Kellert and Wilson, 1993). This theory is supported by evidence that many modern people living in urban areas still acknowledge a spiritual or affective relationship with nature and the outdoors (Milton, 1999; Pretty, 2004, 2007; Pretty et al, 2007, 2008).

Goodin's green theory of value (1992) suggests that all humans want to see some sense and pattern to their lives, and nature provides the backdrop against which this can occur and in which cultural processes, activities and belief systems can develop. Thus it enables human lives to be set in a larger context and explains why non-human nature is often thought of as sacred (Milton, 1999; Berkes, 2004). Similar to Wilson's biophilia hypothesis, Goodin's theory reflects a cultural belief in the value of nature, which is related to people's dependence upon the local environment and, subsequently, is reflected in peoples' actions and behaviours towards it.

Livelihoods and Resource Management Practices: Human Dominion and Nature

Human cultures and their associated behaviours shape biodiversity through the direct selection of plants and animals and the reworking of whole landscapes (Sauer, 1965). Such landscapes can be described as anthropogenic nature; their composition, be it of introduced species, agricultural monocultures or genetically modified crops, a reflection of local culture and a product of human history, including the context in which individuals and groups live their lives (Milton, 1999). Food is one example of how human cultures shape and determine the composition of ecological landscapes. Food plays a role above and beyond nutrition in human societies; it helps to define our identity as individuals, societies and distinct cultures. Food can act as a social marker, representing social structure and politics, and during religious and spiritual ceremonies. Food also epitomizes how a culture uses, classifies and thinks about its natural resources, as such the diversity of diets today reflects the diversity of cultures that exist. Diets originally evolved from the resources available on the local landscape, and thus sustaining traditional diets, dishes and foods, acts to retain connections to both ancestors and the landscape.

Adams (1996) describes nature as a 'cultural archive, a record of human endeavour and husbandry'. Even ecologies previously thought to be natural and pristine are now known to be the result of long-term cultural interactions (e.g. resource-dependent livelihood practices) according to recent archaeological and ethnographic evidence (Stephenson and Moller, 2009), negating the term and concept of wilderness (Callicott and Nelson, 1998). For this reason, many anthropologists perceive landscapes to be a partial social construct, formed from the connection and interaction between people and place (Adams, 2010, this volume). Thus few landscapes are considered non-human today, except for the extremes of the poles or the depths of the oceans, although global climate change is bringing this assertion into question, acknowledged with the naming of this era as the 'anthropocene'.

Traditional human cultures often have what may be considered as a subtler, yet just as significant, ecological footprint, which is nonetheless critical in moulding the local/regional landscape. This is most likely a product of their continuing natural resource dependence. Unlike industrial countries, where human communities have, in many cases, shaped and manipulated the landscape without restraint (urbanization being one product of this dominance), many indigenous and traditional cultures have developed livelihood practices that inevitably alter the landscape, but do so with a level of respect and restraint, so as to ensure natural resource security for future generations. The survival of these communities and their landscapes through to the present day is testimony to the success of many of these strategies (Callicott and Nelson, 1998). Recognizing this, many scientists and policy makers now acknowledge the contribution that traditional cultural practices can make to biodiversity conservation both now and in the future, particularly in little-known ecosystems or where state-imposed management schemes have failed (CBD, 1992; Veitayaki, 1997).

Although natural resource-based practices and knowledge bases vary greatly between human cultures, and even between communities within the same culture, sustainable management practices often derive from systems whereby resource harvesting is coupled with environmental management (Western and Wright, 1994; Turner and Berkes, 2006). Community-based conservation is the process by which biodiversity is protected by and with the local community using their local knowledge and practices. That is not to say that all livelihood activities developed within resource-dependent communities lead to biodiverse outcomes, but that within many traditional cultures there exist practices, skills and knowledge, developed from worldviews, belief systems and livelihood dependencies, that sustainably manage ecological integrity more successfully than modern industrial societies have managed (Nepstad et al, 2006; Turner and Berkes, 2006). This book considers such communities and their resource management practices in more detail (Robson and Berkes, 2010, this volume).

Local Resource Knowledge and its Transmission

Mā te, ka marama;
Mā te marama, ka mātau;
Mā te mātau, ka ora

Through knowledge comes understanding;
Through understanding comes wisdom;
Through wisdom comes life

(from the Māori, New Zealand, see Williams, 2009)

Berkes (2001) indicates that 'knowledge-belief-practice' complexes are key to linking nature with culture. Local knowledge of nature (termed variously traditional knowledge, indigenous knowledge, local ecological knowledge or ecoliteracy) is accumulated within a society and transferred through cultural modes of transmission, such as stories and narratives, as people travel over the land, spatially and temporally (Pilgrim et al, 2007, 2008; Singh et al, 2010). It comprises a compilation of observations and understandings contained within social memory that try to make sense of the way the world behaves. Societies then use this collective knowledge to guide their actions towards the natural world. As a body of knowledge, it is rarely written down, enabling this cultural resource to remain dynamic and current, adapting with the ecosystem upon which it is based (Berkes, 2001; Turner and Berkes, 2006).

One reason for scientists' tendency to overlook, or even dismiss, local knowledge is that it is rarely capable of being generalized (Jacobson and Stephens, 2009). It tends to be locally distinct, place-based, set within a cultural context, and inclusive of all of the inter-related components of the human-environment complex in that area. The importance of this knowledge is becoming more widely recognized by scientists and scientific institutions around the world. Stephenson and Moller (2009), in discussing the interrelations between local knowledge and modern science, emphasize the value and need to integrate both forms of knowledge capital, providing that both are taken within their respective cultural, spiritual and social contexts. They argue that we need to go beyond the dualism (local knowledge versus science) which emphasizes a superiority of one form and inferiority of the other, and towards an understanding of the role that both knowledge bases can play in the future of conservation.

Likewise, Jacobson and Stephens (2009) state that any unchallenged dichotomy, such as that placed on local knowledge and science, can undermine the value of one component. However, they also warn of the risks associated with 'value-free' science, and suggest that we need to understand the continuities and values of both sides of the dichotomy within different contexts, without compromising the distinctiveness or integrity of either. Both types of knowledge, for instance, are embedded within their respective belief systems, and employ different modes of enquiry as a result. By opening up to the multiple legitimate voices that exist, conservation research will become integrative, working both with and for local indigenous and marginal groups, in order to begin to understand the complex human–ecological interactions that exist. Such partnerships are critical if these systems are to be better understood. Thus there is a need in future conservation research for both modern science, which emphasizes knowledge seeking, and local knowledge, which emphasizes knowledge holding (Stephenson and Moller, 2009).

Berkes (2009) also considers the relationship between modern science and traditional knowledge with respect to the future for conservation. He argues that a key difference between both forms of knowledge is that the latter focuses on process rather than content. Evidence for this and the up-to-date nature of local knowledge is explicit when considering indigenous peoples' understandings of current climate change. Elders are unable to transmit knowledge on the impacts of climate change, as most are relatively recent; they can, however, teach processes of knowledge acquisition and development. This is leading to an in-depth understanding of local ecosystem dynamics relating to global climate change within traditional resource-dependent societies. Furthermore, local knowledge recognizes and appreciates the multiple levels of interconnections that exist between nature and culture, which modern day science is striving so hard to come to terms with.

On this basis Berkes, too, argues for the integration of science and local knowledge. He suggests that these two knowledge frameworks no longer need to exist in opposition, and instead we should work on building dialogue and partnerships to link them. Perhaps the most significant aspect of local knowledge is that it derives from frequent interactions with the land, which would be impossible if communities were to become disconnected (either physically or psychologically) from their homelands. Local knowledge is based on being able to read the signs and signals of the land, and then make sense of these observations. These cultural understandings of the environment not only give rise to sustainable management practices, but also to knowledge of species requirements, ecosystem dynamics, sustainable harvesting levels and ecological interactions, to name but a few (Pilgrim et al, 2007, 2008; Singh et al, 2010). If sustained through stories, ceremonies and discourse, this culturally ingrained knowledge can enable its holders to live within the constraints of the local environment, without the need for catastrophic learning in the event of major resource depletion (Turner and Berkes, 2006). Thus, it can be perceived as a form of cultural insurance for the future, providing a source of creativity and innovation, as well as a range of solutions for coping with future challenges.

By going unrecorded, the knowledge of resource-dependent communities is often contained solely within the local and often threatened language. Languages encode cultural knowledge bases in a way that is often non-translatable into other languages, but nonetheless ties its speakers to their landscape inextricably.

Their stories, proverbs and names can lose meaning outside of the physical context of the local environment. In this way, languages can be described as a resource for nature (Maffi, 1998) and, realizing this, a growing body of literature now exists on the multiple interconnections between linguistic, cultural and biological diversity (Maffi, 1998, 2001). However, diverse languages and knowledge bases are threatened today by the same drivers that lead to the erosion of both biological and cultural diversity.

As described by Berkes (2009), one of the biggest challenges to integrating local knowledge with modern science is persuading researchers and scientists to accept that there are, in fact, different ways of knowing, based on culture, semiotics and values, and all have an integrity and distinctiveness that makes them invaluable to the future of conservation. Instead of trying to blend these different knowledge bases, we should be able to appreciate and respect their different epistemologies and cultural contexts, in order to form cross-cultural partnerships for the benefit of human and ecological systems as a whole (Berkes, 2009). Thus the challenge is to move beyond researching local knowledge and to start integrating local knowledge into research.

Socio-Cultural Institutions and their Role in Shaping Landscapes

There is widespread acknowledgement that culturally created landscapes are worthy of protection. Sites that have been set aside for cultural reasons, and have subsequently maintained high natural value, are often designated as internationally recognized Protected Areas under International Union for the Conservation of Nature (IUCN) category V. Therefore conservation does not only derive from an intention to conserve. It can derive from complex belief systems that comprise human religions and are embodied in a diversity of social institutions. In fact, the great majority of non-industrial societies who have succeeded in protecting the productivity of their ecosystems over time have done so primarily through the use of local cultural institutions. Key to their success is the manifestation of objects of nature as spiritual, culturally powerful symbols that command a sense of respect, and are in some cases revered by society. Despite the diversity of cultures that exist globally, many have independently evolved informal regulations, norms and social taboos pertaining to the respectful treatment of nature, which evolve into a form of environmental ethics.

It has long been thought that biodiversity exists outside agriculture. However agrobiodiversity is a key contributor to biological diversity across the world and is thus central to the resilience of many human and ecological systems. Most agrobiodiversity exists where traditional cultural institutions, such as kinship, still play a significant role. Cultural (non-market based) institutions co-evolve with specific ecosystems over time and act to define locally acceptable practices and behaviours, and in some cases, have a greater influence than external market signals. Thus when considering agri-cultural systems (Pretty, 2002), it is important to understand the interactions between culture and agrobiodiversity in terms of identity, cosmology and religion, ecological knowledge, language and aesthetics, social position and status, and common property rights and regimes.

Humans have a long history of developing regimes and rules to protect and preserve natural places in a steady state. These diverse and location-specific rule systems form informal institutional frameworks within communities, legitimated by shared values. Often termed tenure systems, these frameworks have regulated the use of private and common property throughout history, for instance, by defining access rights and appropriate behaviours (Ostrom, 1990; Turner and Berkes, 2006). Where these systems are robust, they can maintain the productivity and diversity of the natural environment without the need for formal legal enforcement sanctions. Compliance derives from shared values and informal internally derived community sanctions, such as moral influence from elders. In some places, formalized payment mechanisms (e.g. Payments for Ecosystem Services) have been put in place to reinforce these norms and reward traditional societies for the diversity of environmental services their ways of life maintain, promoting protection for intellectual property and ownership of knowledge.

Socially embedded norms and institutions therefore arise from a combination of local knowledge bases, cultural belief systems and distinct worldviews. These contextual systems of collective action are intimately linked to the land upon which they are based and, subsequently, are enormously diverse. They govern the use of resources across a wide range of contexts, from forests to fisheries, demonstrating remarkable diversity and flexibility. How humans know the world, therefore, governs behaviour and practices that in turn shape landscapes, which form a cultural archive of human endeavours (Adams, 1996). Amidst a diversity of cultures comes a diversity of meanings, leading to a diversity of actions, and providing an array of biodiversity outcomes. This nature-culture continuum or interconnection has existed through the past and into the present, and is therefore likely to be sustained in the future.

Common Drivers of Diversity Loss and System Degradation

A healthy system is able to maintain full functionality in times of stress, i.e. one that is resilient to incremental changes and perturbations. The diversity of a system is frequently used as a proxy for health, since a diverse system has more adaptive capacity and is therefore more likely to cope with change. However there have been unparalleled losses in biological and cultural diversity in recent decades. As a consequence, both human and ecological systems are becoming less stable (e.g. through the disruption of livelihoods, governance, resource pools and cultural traditions).

It is now understood that many causes of biodiversity loss are also responsible for the loss of cultural diversity. Despite this, the loss of biodiversity is often considered as a separate policy issue to that of cultural diversity (e.g. through language loss or assimilation). Both have undergone an unprecedented rate of decline in recent decades, shifting towards monocultures of the land, people and mind. Common drivers of erosion include a shift in consumption patterns (even in traditional societies who interact with the capitalist economy), the globalization of food systems (Berkes, 2001), and the commodification of natural resources. These drivers are reinforced by pressures of assimilation (attempting to integrate minority cultures into dominant society) and urbanization, and are at their most damaging when they lead to rapid and unanticipated periods of socio-economic change, jeopardizing local system resilience.

Furthermore, resource dependent societies are frequently being suppressed by culturally inappropriate education systems, based around a globalized model of education that fails to take into account cultural differences. This leads to a loss of linguistic diversity and local knowledge. Increased deforestation, unsustainable agricultural production and externally imposed land tenure arrangements resulting from market interests are significant drivers of change, threatening or altogether dismissing culturally embedded ownership and management practices (Tyrrell, 2010, this volume). Limited market opportunities are causing diversification away from resource-based livelihoods and towards environmentally disconnected activities and cultures. This has the capacity to create a deviation from traditional resource management systems and local communities' stake in their natural environment. Moreover, the dominance of modern healthcare systems, at the expense of local knowledge and traditional healthcare systems and practices, is threatening the long-term interdependencies between nature and culture in many societies.

Extreme natural events comprise one of the most rapid drivers of change, particularly when coupled with anthropogenic stressors (Rapport and Whitford, 1999). Tools commonly used in externally imposed resource management also create common drivers and threats, such as exclusive policies (e.g. some nature reserves or state-imposed management systems). A lack of transboundary cooperation and geopolitical instability threaten global diversity, as do weak institutions and a lack of resources, particularly when developing resource management strategies in non-Organisation for Economic Co-operation and Development (OECD) countries. Amplifying this is the widespread encroachment and reclamation of traditional lands in search of rapid economic returns.

The combination of social, economic and political drivers has led to global climate change and other environmental threats including overexploitation and habitat destruction, which, in turn, has led to unprecedented rates of species extinctions. This is eroding the resilience of human and ecological systems, particularly in resource-dependent societies. Furthermore, the degradation of ecosystems with its attendant issues of food security, spread of human pathogens, newly emerging and resurging infectious diseases, and the creation of psychological ills, is a major cause of ill health today (Rapport et al, 1998; Rapoort and Lee, 2003; Rapport and Mergler, 2004). Thus an unprecedented combination of pressures is emerging to threaten the health of human and ecological systems across the world, by forcing communities towards or over critical thresholds, leading to vulnerability and decline. These threats are paving the way for the homogenization of cultures and landscapes as demonstrated by assessments of the state of global and sub-global environments and cultural systems (Maffi, 2001; MEA, 2005; Rapport, 2006; Pretty et al, 2007, 2008).

Ecocultures: Paving the Way towards Resilience brightspots

It is evident that human and environment systems are intimately linked in ways that we are only just beginning to appreciate (Pretty et al, 2007), and certain cultural and ecological components are necessary to ensure system resilience, whereby systems can absorb and cope with changes without losing critical functioning (Holling, 1973; Costanza et al, 1997). However, due to recent and intense periods of diversity loss (both biological and cultural), there is now a growing recognition that human and ecological systems are more vulnerable than formerly predicted. Thus the challenge that lies before us is immense. This book seeks to find possible solutions to this challenge, and looks toward cultures and societies that have successfully maintained their links to nature, termed here 'Ecocultures', to provide possible guidance in creating novel, diverse and sustainable paths into the future. guyana

Ecocultures comprise human cultures that have retained, or strive to regain, their connection with the local environment, and in doing so, are improving their own resilience in light of the multitude of pressures they face, including global climate change. The term ecoculture represents the inextricable links and interplay that can be observed between ecological and cultural systems. This term is not being used as a replacement for the widely accepted socio-ecological system concept, but more an advancement of this notion, whereby ecocultural systems not only comprise the social institutions and distinct frameworks of a community, but also the worldviews, identity, values, distinct cultural practices and behaviours that make a community or group culturally distinct. Thus the phrase 'ecocultural resilience' can be used to emphasize the need to adopt a holistic approach to resilience-building as a consequence of the interconnected complexity of human and ecological systems. Rotarangi and Russell (2009) argue that 'social-ecological resilience has so far mostly been discussed in the absence of critical cultural dimensions and holistic concepts which define indigenous communities (e.g. culturally specific local dynamics, connections to place, language and social relationships)'. Here, we try to highlight the importance of cultural dimensions which define and shape human interactions and relationships with the natural environment.

This volume sets out to consider the depth and complexity of interconnections that exist between nature and culture, both conceptually and within actual communities. It looks at how modern science-based disciplines are having to adapt and converge to deal with the challenges these interrelations represent. But perhaps most importantly, in understanding the complexity of these interconnections, we seek to understand possible solutions to the loss of biological and cultural diversity. That is, to reconnect nature with culture where disconnection has occurred, and to strengthen connections where they have persevered but are now threatened, and therefore develop plans of action from community through to international policy level. It is no longer sufficient just to understand these interconnections and to discuss their prioritization in the international conservation arena. We have to find exemplars of communities, cultures and even nation states that are succeeding in strengthening these connections, and in so doing, should take guidance from these cases when considering how to deal with ecocultural ill health resulting from environmental-cultural disconnection.

Contributions to this Volume: Mapping the Way

This volume begins by considering how modern science and its artificially constructed scientific disciplines are dealing with the challenges brought about by ecocultural systems. Helen Newing discusses the various sub-disciplines that have emerged as a consequence of the need for interdisciplinary approaches, focusing on the evolution of two in particular; that is ethnobiology and conservation biology. Newing considers how these two sub-disciplines have worked to bridge disciplinary divides and their success in sharing methodological tools and concepts. The author then reflects on the challenges faced by interdisciplinary scientists working at the interface, both in the field and in training, including the need to understand the theoretical basis of different disciplines. Newing reflects on cross-disciplinary power relations and the unheard voices of multi-disciplinary teams, but acknowledges the indispensible role that truly interdisciplinary scientists will play in the future in dealing with forthcoming global challenges.

In the next chapter, David Harmon, Ellen Woodley and Jonathan Loh assess the emergence of the concept of biocultural diversity in the modern scientific framework. They discuss the definitions of this concept, and the difficulties associated with describing and quantifying this construct. A key tool for any scientist interested in the methodological basis of assessing biocultural diversity, Harmon and colleagues discuss the challenges faced by scientists attempting to develop measurement indicators and applying the driving force-pressure-state-impact-response (DPSIR) framework to such a holistic concept. The authors describe the current measures that exist for assessing biological diversity, such as the Convention on Biological Diversity (CBD) 2010 indicators and the IUCN Red List. They then go on to discuss the less well developed cultural indicators, including the use of linguistic diversity as recorded in the Ethnologue. Finally, the authors discuss the development of indices for the assessment and monitoring of biocultural diversity. An integrated index is needed to represent something meaningful about the interactions between biological and cultural diversity, which also represents trajectories over time, and can be used to guide policy decisions. However, the authors acknowledge that the outlook is dim, with biocultural indicators being developed as diversity is being lost. With projected losses for the future, the authors suggest that the selection and management of biocultural indicators should be in the hands of those who have most at stake, namely indigenous and marginal communities.

In Chapter 4, W. M. Adams takes a conceptual look at the relationship between humans and nature, and the evolution of landscapes, not just as an ecological construction, but also as a cultural construction. Adams observes that 'ideas from nature are profoundly cultural', and that even the meaning of the word 'nature' is not a given, but instead it diverges between different communities and cultural groups. This is most visible in the projected ideas of landscape within different cultural groups. The author considers the relationship between literary scholars and the landscapes of Britain, including the concept of wild lands versus privatized enclosures. Adams considers the place of the Lake District National Park in the literary, social and cultural history of Britain, and the role of Wordsworth in particular in securing the hills of the Lake District in the minds and imaginations of his readers. He then goes on to discuss early wilderness concepts and their role in laying the foundations for the social construct that is wilderness and the conservation ideas we maintain today. He considers different views of wilderness and how wilderness has come to be seen within the imagination of different cultural groups, such as the Kikuyu of Kenya. Thus Adams speculates that landscape itself is a social construct which differs greatly between different cultural groups, and leads us to question, if landscapes are indeed social constructs, what do we want to construct for the future?

Chapter 5 by Tirso Gonzales and Maria Gonzalez considers the interrelations between nature and culture as seen through the 'Andean cosmovision of ever'. This forms the basis of the way of life for Andean peasant communities. It comprises a holistic worldview that humans, nature and the gods or deities are all intertwined as part of one interconnected system, and within that system, all are equal and should be treated with respect. However, as the authors point out, this does not coincide with the dominant colonial viewpoint, and this has led to the Andean cosmovision being marginalized for the last 500 years. The authors look at the establishment of Andean cultural places, *ayllu*, and their survival, in spite of dominant ever-expanding monocultures of the mind, land and spirit. They consider the representation of this worldview within Andean languages, beliefs, ethics and practices, and the Andean concept of sustainability that preceded the term sustainability in the modern development paradigm. The chapter then considers the differences in Andean knowledge and knowledge creation compared with the reductionist tendencies of modern day science, including the emphasis placed by the former on place-based knowledge and the process of knowing. Finally, the authors suggest how to integrate the ecocentric, integrative nature of the Andean cosmovision into modern scientific discourse for the future of sustainable development.

In Chapter 6, David Rapport and Luisa Maffi explore further the idea of ecocultures and the concept of Ecocultural Distress Syndrome, including the implications of diversity loss for these complex systems. The authors consider what comprises a healthy ecocultural system in terms of structure, function and resilience, based on ecological, socio-cultural, economic and governance dimensions. The authors then discuss the implications of a decline in ecocultural health through the loss of diversity using three case studies; the degradation of satoyama (rural landscapes) of Japan, the desertification of the Mesopotamian marshlands traversing Iraq and Iran, and the damage to the Inner Mongolia grasslands. Rapport and Maffi then consider these case studies in light of Ecocultural Distress Syndrome, which indicates the loss of organization, vitality and resilience of the system. In doing so, the authors paint a dim picture for the future of ecocultures and their continuing demise, and emphasize the need for a global strategy to repair ecocultural health around the world. They claim that science on its own is not enough, and political will is needed, not just in terms of policy development but also in terms of implementation, using the CBD 2010 targets as an example. They highlight the need for ecocultural conservation to become a global priority in order to stabilize the health of human and ecological systems and abate the decline of diversity into the future.

Chapter 7 considers natural resource management, not in terms of environmental management but in terms of the management of human behaviours. Martina Tyrrell assesses Western conservation perspectives in comparison with local conservation perspectives in the Canadian Arctic, focusing on the Western need to dominate landscapes and the impacts of externally imposed management ideas that overlook local cultural values and practices. Tyrrell considers Inuit place in today's globalized society, and the challenges of trying to live culturally sustaining and meaningful lives in the face of political, economic, social and cultural processes of globalization. She goes on to describe the reciprocal relationship that traditionally exists between Inuit and animals, including Inuit views on hunting as a reciprocal act of engagement. However the Inuit worldview has long been overlooked by Canadian conservation policy, including the reciprocal relations upon which Inuit hunting practices are founded. Where external hunting regulations have been imposed, Tyrrell describes irreparable changes in Inuit-animal relations; from a relationship of reciprocity and respect has emerged the view that hunting is little more than an Inuit 'right'. This has had devastating effects for some wildlife populations living within the Canadian Arctic. Tyrrell powerfully describes the shift in the role of two Arctic species, the beluga whale and the polar bear, from sentient beings to basic resources and, thus, documents the damaging effects that can result from culturally uninformed wildlife conservation policies, and the impact these can have on ecocultural systems.

In the following chapter, Garry Marvin further considers the construction of hunting, but this time within the context of modern recreational hunters. Focusing on what he calls the 'nature hunter', Marvin analyses the ethics of hunting, allowing us to explore the thoughts and minds of the hunter during different stages of the hunt. Through Marvin's account, we join the hunter on his journey, through preparation, stalking, to the kill or indeed miss, and the after-hunt stories. Marvin argues that the act of hunting is less about the actual kill, and rather about the process of the hunt. He argues that we need to understand this process in order to understand the actual motives of the hunter. He notes that even hunters condemn hunting in certain situations, and the specific circumstances by which an animal is engaged preceding the kill are vital to the ethics of the hunter. For nature hunters in particular, the kill is not the aim of the hunt, but rather the hunt is the aim of the kill. Marvin sets out a journey into the hunting complex from the eyes of the hunters he accompanied. In this way, it is possible to learn about the mindset, views and ethics of the nature hunter, such as the self-imposed technology limits and the emphasis on 'fair chase'. In the view of the hunter, hunting enables a person to move beyond being *in* nature to being *of* nature, thus enabling the shift from onlooker to actor. Marvin concludes that understanding the hunting complex offers us insight into the spectrum of human-animal relations that exist within modern day societies.

In Chapter 9, Patricia L. Howard deals with the interrelations between agrobiodiversity and culture, including the role of human cultures in sustaining practices that lead to agrobiodiversity. In doing so, Howard considers the dimensions of cultural evolution in relation to the ecological adaptation of landscapes, focusing on five key dimensions and their embeddedness within social institutions; namely cultural identity, cosmology and religious beliefs, traditional knowledge, social position and status, and natural resource tenure regimes. The author considers the role that agrobiodiversity plays within a culture, and the development of cultural needs and preferences based on this diversity. Furthermore, Howard considers the role that traditional belief systems play in agrobiodiversity management, and the socially contextualized knowledge upon which these systems are based. This includes the ways in which local knowledge is accrued and the modes and routes of transmission in relation to social structure and status. Delving further into the social context, the author describes the links between agrobiodiversity and social status, and the cultural roles of identity, belief systems, knowledge, social systems and tenure regimes in social institutions, and what this means for ecocultural resilience in the future.

Eugene Anderson takes this debate further in Chapter 10 by considering how humans relate to food and food production systems. Anderson describes the mainstreaming of food products that has occurred throughout recent history, at the expense of biological diversity, and what this has meant for food and human cultures around the world today. The author also considers instances where this is not the case, and cultures have sustained their local food systems and thus connections to the land through the use of intricate management practices. This tends to occur where modern management techniques have failed due to severe environmental conditions, but still urbanization and modernization threaten to encroach. By examining case studies from around the world, Anderson outlines the diversity of local management systems developed in the context of human cultures and adapted to specific ecological conditions, such as the burning regimes of the US, the terraced landscapes of Peru, and the vertical farming systems of Europe. Finally, Anderson considers the future of food systems and modern food movements such as the slow food and locavore movements, and their roles in reconnecting human societies with their local landscapes.

The final section of the book is concerned with rebuilding the connections between human cultures and their landscapes. It considers success stories and examples that can be taken forward and incorporated into policy frameworks, and action plans intended to alleviate the damage caused by disconnection. James P. Robson and Fikret Berkes begin this section by discussing Indigenous and Community Conserved Areas (ICCAs) and their associated livelihoods, cultures and belief systems that centre on nature. These constitute systems of voluntary conservation which are based on spiritual or livelihood complexes, or indeed both, and which are only just being recognized as an integrated conservation strategy. They are effective in conserving both biological and cultural diversity (despite the latter being a relatively recent construct), under the realization that the future of conservation lies at least partly in the past (Kothari, 2009). The authors explore the emerging role of ICCAs within communities of Oaxaca, in southern Mexico. Conservation in these communities is often far removed from modern conservation discourse or the sustainability paradigm, instead it stems from a combination of respect and dependence on the natural environment, and the services it provides. The authors then go on to compare ICCAs with alternative Protected Area designations in a range of geographic contexts. Robson and Berkes suggest that ICCAs may provide a pre-existing framework for the integrative conservation of biological and cultural diversity on which future models can be based. In doing so, we stand to enter into a new conservation paradigm, termed 'next generation' conservation, that comprises holistic and socially appropriate objectives whereby cultural and livelihood needs are not just combined with but integrated into conservation objectives.

In Chapter 12, Glenn Albrecht considers the human health implications of environmental disconnection stemming from ecological degradation. Cultures all over the world have concepts in their language that relate psychological stress to the state of the environment. This, in itself, is not a new area. However, this strong interdependency has only recently been recognized within Western discourse. Albrecht discusses this oversight and the term 'solastalgia', which he uses to describe the feelings of devastation and disassociation a person feels when their home environment is negatively transformed. Albrecht discusses two new diagnostic categories that link environmental and human health, namely 'psychoterratic' (mental health) and 'somaterratic' (physical health) distress. Central to these new terms and concepts is their place-based nature and the integral link to the landscapes upon which they are based. Albrecht explores two case studies to illustrate psychoterratic distress, in the US Appalachia and amongst communities living in Hunter Valley, Australia. The author goes on to introduce the concept of 'soliphilia' (the love of life) as a means to combat feelings of solastalgia, and the potential of creating new ways of living in terms of sustainability and reconnecting with the land. He argues that by reconnecting nature with culture and understanding the importance of this connection for ecocultural health, we will be able to drive the development of sustainable solutions to local, regional and global health challenges.

In Chapter 13, Sarah Pilgrim, Colin Samson and Jules Pretty consider the use of 'revitalization projects' in reconnecting nature with culture. Revitalization projects are a highly diverse group of projects with a similar objective in mind: to maintain or reclaim the culture of local peoples and reconnect them to the land for long-term individual and societal health. To understand these projects, including their objectives and outcomes, six broad categories of revitalization project have been developed. Many revitalization efforts span several of these categories, but all strive to meet the objectives of at least one. The authors discuss these different project categories in turn with examples of local to national efforts from around the world. They then consider the impacts that these projects have had for both biological and cultural diversity, in terms of knowledge revival, natural resource management, economic stability, cultural revival of practices and strengthening community bonds. Pilgrim et al discuss the potential of revitalization projects as an integrative framework for the protection of biological and cultural diversity in the future, both in industrial and non-industrial contexts. The authors conclude that revitalization projects offer a new community-centred approach to dealing with the problems of disconnection, and by being locally driven, projects are more likely to encourage long-term support and participation.

In the concluding chapter, Jules Pretty and Sarah Pilgrim debate the future of ecocultural systems. In doing so, the authors consider the novel combination of mounting pressures that exist and threaten ecocultural systems today. They also discuss the dominant model of convergence of lifestyles and resource use patterns, and the implications this has had on consumption levels globally. Finally, they discuss the need for the divergence of pathways if a sustainable future is to be achieved. The authors consider the need for effective policy to address ecocultural health, the current policy options that exist, the need to prioritize biological and cultural diversity conservation in an integrated approach, and the consequences of overlooking the links between nature and culture in future policy targets.

The realization of the multiple interconnections between nature and culture has led to the unification of natural and social science fields that were previously isolated. One component of this book is to develop an understanding of how the academic sector is reforming to deal with the convergence of these two previously isolated concepts. By cutting across a range of disciplines, this book aims to provide a systems overview of the interconnections between nature and culture, both conceptually and physically. In doing so, we seek to understand the implications of diversity loss for the resilience of ecocultural systems, and also to find examples of where these connections have been sustained or rebuilt to the benefit of human and environmental health. Reflecting this, a dominant theme in environmental discourse today is to find a new holistic way of speaking about and conserving nature and culture without fragmentation. This book aims to take a step towards achieving this by offering a new forward-thinking approach to the conservation and strengthening of ecocultural systems.

By being at the forefront of the current paradigm shift in conservation, the contributors to this volume strive not only to offer novel perspectives to try to guide 'next generation' conservation forward (Robson and Berkes, 2010, this volume), but offer positive solutions and frameworks by which this can be achieved. Where communities have succeeded in sustaining ecocultural health, the authors of this book have tried to learn from these communities and identify key tools that can be utilized by other communities (human and ecological) around the world suffering the consequences of disconnection. Furthermore, the need to overcome and bridge pre-existing boundaries is a key theme, and authors of this volume not only endeavour to describe current boundaries (e.g. between disciplines, understandings, cultures, paradigms, worldviews, languages and institutional frameworks) but to offer novel tools and mechanisms that can be used to overcome these divides.

The interaction between nature and culture is gradually becoming accepted in scientific and conservation discourse as our understanding of the existing web of complex interconnections grows. However, it is as our understanding progresses, that we are seeing more and more broken connections. Thus this book, written in the light of sustainability science, aims not just to document these interconnections, but to offer positive solutions and success stories to policy makers, governmental and non-governmental researchers, and indeed communities suffering from the pressures of disconnection. In offering a range of solutions and policy options, we hope to provide diverse pathways that may lead to divergence, but that will facilitate communities in forging a pathway towards sustainable futures.

References

Adams, W. M. (1996) *Future Nature: A Vision for Conservation*, Earthscan, London

Balée, W. (1994) *Footprints of the Forest: Ka' apor Ethnobotany*, Columbia University Press, New York

Berkes, F. (2001) 'Religious traditions and biodiversity', *Encyclopaedia of Biodiversity*, vol 5, pp109–120

Berkes, F. (2004) 'Rethinking community-based conservation', *Conservation Biology*, vol 18, no 3, pp621–630

Berkes, F. (2008) *Sacred Ecology (2nd edition)*, Routledge, New York

Berkes, F. (2009) 'Indigenous ways of knowing and the study of environmental change', *Journal of the Royal Society of New Zealand*, vol 39, no 4, pp151–156

Berkes, F., Colding, J. and Folke, C. (2000) 'Rediscovery of traditional ecological knowledge as adaptive management', *Ecological Applications*, vol 10, no 5, pp1251–1262

Berkes, F. and Folke, C. (2002) 'Back to the future: Ecosystem dynamics and local knowledge', in L. H. Gunderson and C. S. Holling (eds) *Panarchy: Understanding Transformations in Human and Natural Systems,* Island Press, Washington, DC, pp121–146

Callicott, J. B. and Nelson, M. P. (eds) (1998) *The Great New Wilderness Debate,* University of Georgia Press, Athens

CBD (1992) 'Convention on Biological Diversity', UNEP, www.cbd.int/convention/

Costanza, R., d'Arge, R., de Groot, R., Farber, S., Grasso, M., Hannon, B., Limburg, L., Naeem, S., O'Neil, R. V., Paruelo, J., Raskin, R. G., Sutton, P. and van den Belt, M. (1997) 'The value of the world's ecosystem services and natural capital', *Nature*, vol 387, pp253–260

Denevan, W. M. (2001) *Cultivated Landscapes of Native Amazonia and the Andes*, Oxford University Press, Oxford, UK

Ellen, R. F. (1996) 'The cognitive geometry of nature: A contextual approach', in G. Palsson and P. Descola (eds) *Nature and Society: Anthropological Perspectives*, Routledge, London and New York

Geertz, C. (1973) *The Interpretation of Cultures,* Basic Books, New York

Goodin, R. E. (1992) *Green Political Theory,* Polity Press, Cambridge

Gunderson, L. H. and Holling, C. S. (eds) (2002) *Panarchy: Understanding Transformations in Human and Natural Systems,* Island Press, Washington, DC

Harmon, D. (2002) *In Light of Our Differences,* Smithsonian Institution Press, Washington, DC

Heckenberger, M. J., Russell, J. C., Toney J. R. and Schmidt, M. J. (2007) 'The legacy of cultural landscapes in the Brazilian Amazon: Implications for biodiversity', *Philosophical Transactions of the Royal Society B*, vol 362, pp197–208

Holling, C. S. (1973) 'Resilience and stability of ecological systems', *Annual Review of Ecology and Systematics*, vol 4, pp1–23

Jacobson, C. and Stephens, A. (2009) 'Cross-cultural approaches to environmental research and management: A response to the dualisms inherent in Western science?' *Journal of the Royal Society of New Zealand*, vol 39, no 4, pp159–162

Kellert, S. R. and Wilson, E. O. (eds) (1993) *The Biophilia Hypothesis*, Island Press, Washington, DC

Kothari, A. (2009) 'Protected areas and people: The future of the past', *Parks*, vol 17, no 2, pp23–34

Maffi, L. (1998) 'Language: A resource for nature', *Nature and Resources, UNESCO Journal on the Environment and Natural Resources Research*, vol 34, no 4, pp12–21

Maffi, L. (ed.). (2001) *On Biocultural Diversity: Linking Language, Knowledge and the Environment*, Smithsonian Institution Press, Washington, DC

Maffi, L. and Woodley, E. (2007) 'Biodiversity and culture', UNEP's 4th Global Environment Outlook Report, UNEP, Nairobi

MEA (2005) *Ecosystems and Human Well-being: Current State and Trends*, Millennium Ecosystem Assessment, Island Press, Washington, DC

Milton, K. (1998) 'Nature and the environment in indigenous and traditional cultures', in D. E. Cooper and J. A. Palmer (eds) *Spirit of the Environment: Religion, Value and Environmental Concern*, Routledge, London and New York, pp86–99

Milton, K. (1999) 'Nature is already sacred', *Environmental Values*, vol 8, pp437–449

Nepstad, D., Schwartzman, S., Bamberger, B., Santilli, M., Ray, D., Schlesinger, P., Lefebvre, P., Alencar, A., Prinz, E., Fiske, G. and Rolla, A. (2006) 'Inhibition of Amazon deforestation and fire by parks and indigenous lands', *Conservation Biology*, vol 20, pp65–73

Norgaard, R. B. (1994) *Development Betrayed*, Routledge, London

Ostrom, E. (1990) *Governing the Commons: The Evolution of Institutions for Collective Action*, Cambridge University Press, Cambridge

Pilgrim, S., Cullen, L., Smith, D. J. and Pretty, J. (2008) 'Ecological knowledge is lost in wealthier communities and countries', *Environmental Science and Technology*, vol 42, no 4, pp1004–1009

Pilgrim, S., Smith, D. J. and Pretty, J. (2007) 'A cross-regional assessment of the factors affecting ecoliteracy: Implications for policy and practice', *Ecological Applications*, vol 17, no 6, pp1742–1751

Posey, D. A. (ed) (1999) *Cultural and Spiritual Values of Biodiversity*, UNEP and Intermediate Technology Publications, Nairobi

Pretty, J. (2002) *Agri-Culture: Reconnecting People, Land and Nature*, Earthscan, London

Pretty, J. (2004) 'How nature contributes to mental and physical health', *Spirituality and Health International*, vol 5, no 2, pp68–78

Pretty, J. (2007) *The Earth Only Endures: On Reconnecting with Nature and Our Place in it*, Earthscan, London

Pretty, J., Adams, B., Berkes, F., de Athayde, S., Dudley, N., Hunn, E., Maffi, L., Milton, K., Rapport, D., Robbins, P., Samson, C., Sterling, E., Stolton, S., Takeuchi, K., Tsing, A., Vintinner, E. and Pilgrim, S. (2008) 'How do nature and culture intersect?' Plenary paper for Conference *Sustaining Cultural and Biological Diversity In a Rapidly Changing World: Lessons for Global Policy*, organized by AMNH,

IUCN-The World Conservation Union/Theme on Culture and Conservation, and Terralingua, New York, 2–5 April, 2008

Pretty, J., Ball, A. S., Benton, T., Guivant, J., Lee, D., Orr, D., Pfeffer, M. and Ward, H. (eds) (2007) *Sage Handbook on Environment and Society*, Sage, London

Rapport, D. J. (2006) 'Sustainability science: An ecohealth perspective', *Sustainability Science*, vol 2, pp77–84

Rapport, D. J., Costanza, R. and McMichael, A. (1998) 'Assessing ecosystem health: Challenges at the interface of social, natural, and health sciences', *Trends Ecology and Evolution*, vol 13, no 10, pp397–402

Rapport, D. J. and Lee, V. (2003) 'Ecosystem approaches to human health: Some observations on north/south experiences', *Environmental Health*, vol 3, no 2, pp26–39

Rapport, D. J. and Mergler, D. (2004) 'Expanding the practice of ecosystem health', *EcoHealth*, vol 1, no 2, pp4–7

Rapport, D. J. and Whitford, W. G. (1999) 'How ecosystems respond to stress? Common properties of arid and aquatic systems', *Bioscience*, vol 49, no 3, pp193–203

Rotarangi, S. and Russell, D. (2009) 'Social-ecological resilience thinking: Can indigenous culture guide environmental management?' *Journal of the Royal Society of New Zealand*, vol 39, no 4, pp209–213

Sauer, C. O. (1965) 'The morphology of landscape' (originally published in 1925), in J. Leighly (ed) *Land and Life*, University of California Press, Berkeley, pp315–350

Singh, R. K., Pretty, J. and Pilgrim, S. (2010) 'Traditional knowledge and biocultural diversity: Learning from tribal communities for sustainable development in northeast India', *Journal of Environmental Planning and Management*, vol 53, no 4, pp511–533

Stephenson, J. and Moller, H. (2009) 'Cross-cultural environmental research and management: Challenges and progress', *Journal of the Royal Society of New Zealand*, vol 39, no 4, pp139–149

Thomas, W. H. (2009) 'The Forest Stewards Initiative: A new institution for safeguarding traditional ecological knowledge in Papua New Guinea', *Journal of the Royal Society of New Zealand*, vol 39, no 4, pp187–191

Toledo, V. M. (2001) 'Biodiversity and indigenous peoples', *Encyclopaedia of Biodiversity*, vol 3, pp451–463

Turner, N. and Berkes, F. (2006) 'Coming to understanding: Developing conservation through incremental learning in the Pacific Northwest', *Human Ecology*, vol 34, pp495–513

Veitayaki, J. (1997) 'Traditional marine resource management practices used in the Pacific islands: An agenda for change', *Ocean and Coastal Management*, vol 37, pp123–136

Western, D. and Wright, R. M. (eds) (1994) *Natural Connections*, Island Press, Washington, DC

Williams, J. (2009) '"O ye of little faith": Traditional knowledge and Western Science', *Journal of the Royal Society of New Zealand*, vol 39, no 4, pp167–169

Part I
Science in Practice

2
Bridging the Gap: Interdisciplinarity, Biocultural Diversity and Conservation

Helen Newing

Cultural and biological diversity have recently come together in the concept of biocultural diversity, defined as 'the diversity of life in all of its manifestations: biological, cultural, and linguistic, which are inter-related (and likely co-evolved) within a complex socio-ecological adaptive system' (Terralingua, 2008). It is evident from this definition that research on biocultural diversity must span disciplines from both the natural and social sciences. However, bridging disciplinary divides is notoriously difficult, both because of institutional barriers within research systems and also because of deeper theoretical challenges. But there has also been important progress in developing interdisciplinary approaches to biocultural diversity research arising from the contrasting histories of two of the most relevant 'interdisciplinary disciplines' – ethnobiology and conservation biology. Both emerged in the 1980s and focus on the relationship between the natural and the social world, whether this is expressed in terms of 'human dimensions' to biodiversity conservation or the more integrated concept of biocultural diversity. Both have a strong advocacy element – they attempt to address 'real world' problems relating to the conservation of biological and/or cultural diversity. However, their histories and their approaches to interdisciplinarity differ. What, then, can we learn about bridging disciplinary divides related to biocultural diversity and conservation?

What Do We Mean by Bridging the Gap?

Interactions between different disciplines have been classified in many ways. In this chapter, I follow Klein (1990), Moran (2002) and others by using interdisciplinarity as a generic term to refer to 'any form of dialogue or interaction between two or more disciplines' (Moran, 2002; see also Dillon, 2008; Lau and Pasquini, 2008). Where it is useful to distinguish between different kinds or levels of interaction, I use the terms 'multidisciplinarity', 'interdisciplinarity' and 'transdisciplinarity' respectively to refer to the use of two or more disciplines in parallel, more integrated approaches where different disciplinary perspectives inform and intertwine with one another, and transformative approaches involving the creation of novel perspectives that transcend disciplinary boundaries (Klein, 1990; Barry et al, 2008).

A second useful distinction is between theoretical and instrumental concerns. In theoretical terms, the aim of interdisciplinarity is to bridge the gaps between the formal disciplines in order to maintain a unified body of knowledge. However, throughout much of the 20th century, interdisciplinarity has been promoted for far more immediate and applied purposes – to address specific 'real world' problems that go beyond the scope of any single discipline (Klein, 1990). In this sense, interdisciplinarity is 'a means of solving problems and answering questions that cannot be addressed satisfactorily using single methods or approaches' (Marzano et al, 2006) where the problem, not the discipline, defines the tools of study. Building on this distinction, a third, related typology distinguishes between (i) the 'borrowing' of specific methods and models from one discipline by another; (ii) a temporary collaboration for the purpose of practical problem-solving; (iii) the progressive convergence of subjects and methods in different disciplines; or (iv) the emergence of a new 'interdiscipline' in its own right (Klein, 1990; Barry et al, 2008).

Biocultural Diversity: Bridging Disciplinary Divides

The list of disciplines that are relevant to the task of biocultural diversity conservation is formidable. Anthropology, development studies, geography and politics all offer insights into 'community', common pool resource theory, forms of participation and governance, gender relations and broader power relations. Anthropology, environmental psychology and environmental education give further different perspectives on people's attitudes to the environment. Environmental economics has contributed tools and conceptual frameworks related to valuation studies, rational decision making and financial incentives for conservation. History, geography and political ecology provide a temporal perspective, offering tools to identify key events and track trends over time. The natural resource management disciplines have developed participatory methodologies that later became prominent in lines of enquiry in both conservation biology and ethnobiology. More recently, technical advances in mapping have been used in documenting the global and regional distribution of both biological and cultural diversity, in mapping (biological) metapopulations in order to inform the design of protected areas systems, and also in participatory mapping at a local scale. The latter can be used further to understand natural resource use and relations with the landscape, as well as in direct support of land and resource claims. A recent addition to the list of relevant subdisciplines is conservation psychology which, like conservation biology, emphasizes advocacy and action to further conservation goals (Saunders and Myers, 2003; Clayton and Brook, 2005; Saunders et al, 2006).

Clearly, no single new discipline can cover all of these different perspectives in any detail. In the following sections, I will trace the history of ethnobiology and conservation biology respectively in order to discuss how they have approached the need to bridge so many disciplines.

Ethnobiology

Ethnobiology – the disciplinary 'home' of the concept of biocultural diversity – has a long academic history rooted in anthropology (Schultes and von Reis, 1995), but began to consolidate as a sub-discipline in its own right in the 1980s. This process was closely connected to the perceived biodiversity crisis – particularly the destruction of tropical forests and the associated displacement of indigenous and traditional forest peoples. There was a proliferation of ethnobotanical case studies documenting local knowledge of plants and their uses (e.g. Plotkin, 1995; Prance, 1995), which were used to support arguments in international policy for the economic viability of sustainable use of tropical forests instead of their destruction. Anthropological and ethnolinguistic techniques to elicit local knowledge and determine culturally specific cognitive classification systems were combined with survey methods from the field of botany, valuation techniques from environmental economics and phytochemical and pharmacological techniques to isolate bioactive compounds and test their efficacy as medicines. Similar studies were carried out on a wide range of taxa and other features of the natural environment, including insects, mammals, birds, fish, soils and habitat types (Schultes and von Reis, 1995), and the results were compared to the body of knowledge in the corresponding natural science disciplines. Over time, more attention was paid to dynamic aspects of traditional knowledge systems such as knowledge transmission and erosion, especially in terms of the effects of increased integration of traditional societies into the market economy (e.g. Zent, 1999; Pilgrim et al, 2008). Through this work there was a progressive integration of theoretical approaches and methodological tools from different disciplines, resulting in the successful emergence of a new 'transdisciplinary' subdiscipline in its own right.

As the field of ethnobiology advanced, the focus expanded to include not only documentation of knowledge and uses of plants and animals but also 'ethnoecological' perspectives on broader human–environment interactions (Gragson and Blount, 1999; Hunn, 2002). Of particular relevance to conservation and biocultural systems, ethnobiological research on agricultural systems and management of wild plants in the Amazon opened a debate about the extent to which ecosystems that had until then been perceived as wilderness were in fact shaped by humans (Posey and Balée, 1989). This fed into strong advocacy statements concerning the links between cultural and biological diversity (ISE, 1989). Similarly, research in West Africa using methodological tools from anthropology and geography suggested that far from being a cause of deforestation, traditional farming practices had played an important role in increasing forest cover throughout the 20th century (Fairhead and Leach, 1994, 1998). This work highlighted the extent to which traditional practices had shaped and maintained habitats and ecosystems around the world, and led to an increasingly applied approach looking at traditional knowledge as a basis for sustainable resource management. Accordingly, there was an increasing emphasis on participatory rights-based approaches to support indigenous peoples' struggles to retain or regain access to their ancestral lands. Hunn (2002) describes the evolution of ethnobiology in terms of four stages: (i) extractive research on indigenous knowledge of plants and animals; (ii) 'ethnoscience' focusing on different cultural perspectives on the natural world; (iii) traditional knowledge as a basis for sustainable natural resource management; and (iv) an emerging focus on participatory research 'of, by and for indigenous peoples, traditional societies and local communities'.

The concept of biocultural diversity has become a prominent theme in ethnobiology. It was the theme of two of the last five conferences of the International Society of Ethnobiology (ISE) – in 2000 and 2008 – and has also been the subject of several books (e.g. Maffi, 2001a; Stepp et al, 2002; Carlsson and Maffi, 2004). Largely thanks to the work of the international non-governmental organization Terralingua (Maffi, 2001a, www.terralingua.org), it has also gained ground in international conservation policy, particularly the Convention on Biological Diversity (CBD) and the International Union for Conservation of Nature's (IUCN's) Commission of Environmental, Economic and Social Policy (CEESP). Detailed quantitative analyses of the global distribution of biological diversity and cultural diversity, using linguistic diversity as a proxy for the latter, revealed a high degree of correlation between the two and strengthened the evidence for a codependent relationship (Maffi, 2001a, b). Theoretical frameworks and methodological tools were adapted from conservation biology, including the distinction between in-situ and ex-situ conservation (Zent, 1999), the concept of endemism, the use of linguistic distribution maps, the compilation of a Red Book of endangered languages based on the Red Data Book of endangered species (Maffi, 2001a,b), and more recently, precisely defined quantitative biocultural diversity indices (Loh and Harmon, 2005; Harmon et al, 2010, this volume).

Conservation Biology

As ethnobiology moved beyond studies of traditional knowledge to broader studies on and advocacy for biocultural systems as a whole, there was a major thematic convergence with the discipline of conservation biology. Like ethnobiology, conservation biology has a long history as a concept but emerged as an academic discipline in the 1980s, largely in response to the increasingly urgent need to address the perceived biodiversity crisis (Noss, 1999). Many of the early conservation biologists were field biologists whose study sites faced an immediate threat of destruction from human activities, yet their professional training in ecology left them ill-equipped to deal with these threats. At the global level, the establishment of international databases by the World Conservation

Monitoring Centre (WCMC) tracking the status of different species and ecosystems highlighted the extent of the crisis. Conservation biology, then, was, from the beginning, an overtly mission-oriented discipline (Noss, 1999; Meine et al, 2006). It aimed to develop new applied interdisciplinary perspectives and produce a generation of professionals who were better equipped to address the 'human dimensions' of biodiversity conservation in a changing global context (Buscher and Wolmer, 2007).

Accordingly, a substantial body of social science research has emerged, building on a wide range of disciplinary traditions. Emerging thematic strands include financial incentive systems, rational decision-making, valuation studies and common pool resource theory; the relationship between human populations, livelihoods and the state of biodiversity; environmental attitudes and behaviour; environmental law and policy; and a wide range of issues associated with different aspects of community conservation and collaborative natural resource management. The latter field has built on insights from anthropology, geography, development studies and political ecology to unpick the concept of 'community', develop perspectives on traditional ecological knowledge and participatory approaches, analyse institutional aspects and power relations, and, most recently, explore broader issues of governance and decentralization. The leading researchers in these different fields would not regard themselves as conservation biologists, but these different aspects are increasingly included within the scope of the discipline, as indicated by their inclusion in the first two textbooks on social aspects of conservation (Russell and Harshbarger, 2003; Mulder and Coppolillo, 2005). Whilst it is probably true to say that the concept of biocultural diversity is still marginal in the field of conservation biology, there is an extensive body of research on the closely related concepts of socioecological systems and resilience that represents perhaps the closest convergence between ethnobiological and conservation biological perspectives (Berkes and Folke, 2002; Folke, 2004; Butler and Oluoch-Kosura, 2006; Folke and Gunderson, 2006).

Progress, however, in bridging the gap between the social and natural science components of the emerging discipline of conservation biology has been frustratingly slow (Touval and Dietz, 1994; Noss, 1997; Meffe, 1998; Eriksson, 1999; Siebert, 2000; Mascia et al, 2003). There are two reasons for this. Firstly, the massive breadth of the subject area and the shortage of researchers with expertise covering both the natural sciences and the social sciences has meant that, initially, specific projects were carried out through multidisciplinary teamwork. This involved temporary collaborations between established researchers from a wide range of disciplinary backgrounds. Thus, rather than progressing through the cumulative integration of tools and perspectives from different disciplines into a growing body of knowledge, the development of the discipline of conservation biology has remained somewhat fragmented. Secondly, the sense of urgency of the conservation 'mission' meant that the emerging 'meta-discipline' (Meffe, 1998) of conservation biology had a strong emphasis on practical problem-solving rather than on building abstract theoretical frameworks as an end in itself. This applied particularly to the area of 'human dimensions' – defined by Jacobson and McDuff (1998) as 'a variety of people-oriented management considerations and a cross-disciplinary range of inquiry'. For many conservation biologists, biology is and should remain the primary focus of the discipline; human dimensions are of subsidiary interest in order to address immediate threats to biodiversity. This view has been accentuated by continuing – and at times heated – controversy over community approaches to conservation and continued conflict between conservation and indigenous rights (Gray et al, 1998; Terborgh, 1999; Chatty and Colchester, 2002; Colchester, 2003; Chapin, 2004; Dowie, 2005; West et al, 2006). For many 'old school' conservationists, community conservation has proved a distraction to the primary mission of protecting and preserving biodiversity and, therefore, theoretical research into community aspects of conservation is of little interest.

By the early 2000s, debate about the roles of the sciences became so acute that in a landmark article reflecting on the state of the discipline, Mascia and colleagues reemphasised that 'the real question for debate, of course, is not whether to integrate the social sciences into conservation but how to do so' (Mascia et al, 2003). Following publication of this article, a Social Science Working Group (SSWG) was established within the Society for Conservation Biology (SCB) (www.conbio.org/WorkingGroups/SSWG/) 'dedicated to strengthening conservation social science and its application to conservation practice'. However, the need to integrate social science perspectives into mainstream conservation biology, and indeed the extent to which the discipline should attempt to do so, remain matters for debate. Lidicker (1998) goes as far as to suggest that attempts to produce a single metadiscipline are misguided:

> ... *most current discussions of these matters seem to confuse conservation biology with conservation. They are not the same. Conservation needs conservation biologists, for sure, but it also needs conservation sociologists, conservation political scientists, conservation chemists, conservation economists, conservation psychologists, and conservation humanitarians.*

Nonetheless, the need to consolidate the discipline of conservation biology and define a thematic and disciplinary core is widely recognized. Discussions of what should be included in the core material reflect the above debates, with much concern that too broad a thematic scope runs the risk of producing people who are a 'jack of all trades, master of none'. Specifically, some authors are concerned that rigorous training in ecological techniques and the principles of (deductive) research design and statistical analysis may be neglected. Others point to the danger of training idiot savants – 'individuals skilled in certain areas (in this case, the technical biological aspects of conservation) but largely inept in other aspects of the field' (Jacobson and McDuff, 1998). As Harrison et al (2008) assert in relation to similar issues in geography, to produce good science that bridges disciplinary divides 'research must be rigorous on both sides – not academic on one side and "vernacular" on the other'. For this to be so, students must gain a solid grounding in both the social and the natural sciences and begin to develop skills in their integration. However, it has also been suggested that training programmes in conservation biology should be vocational rather than purely academic – they should aim primarily at preparing young people for professional practice rather than training up 'little professors' (Noss, 1997). Thus, coverage of vocational and managerial skills competes with the social sciences and natural sciences for time allocations in training programmes.

There is an extensive literature on training in conservation biology and much of it focuses on these issues. Overall, it suggests that fears that the natural sciences are losing ground are unfounded and in many programmes there is a strong emphasis on vocational training; however, coverage of disciplinary perspectives from the social sciences remains relatively weak. For example, Van Heezik and Seddon (2005) conducted a survey of 54 graduate programmes in wildlife management and conservation biology in the US, the UK and Australia, the 5 most common topics were biostatistics, conservation biology, habitat issues, conservation genetics and remote sensing. Similarly, Bonine et al (2003) reviewed different types of conservation training in the US, UK, Brazil, Peru, Madagascar and Indonesia. The majority of courses targeting non-OECD country students were found to have a 'practical science-based programme that addresses conservation issues at a site-specific or local scale'. Furthermore non-academic programmes run by international conservation organizations tended to focus on biological training and on non-academic skills rather than on the social sciences.

The situation in terms of interdisciplinary coverage is hard to discern because of inconsistent use of the term 'interdisciplinary'. Rather than referring to the relationship between different disciplinary approaches, it has been used variously to mean content relating to the social sciences, to human dimensions (in thematic terms rather than in terms of disciplinary influences), to non-academic vocational skills or to any component that is not strictly based in the natural sciences. Van Heezik and Seddon (2005) define interdisciplinarity as including (i) 'human dimensions and sociopolitical issues (e.g. environmental policy, ethics, planning, regulations, human ecology, public education, and community relations)' and (ii) interpersonal skills, including 'communicating about the environment and conflict resolution'. By these criteria, between 70 and 80 per cent of programmes were interdisciplinary. Niesenbaum and Lewis (2003) identified interdisciplinary content in conservation textbooks by searching for a set of 35 keywords, many of which were associated with thematic coverage (United Nations; World Bank; corporation; media) and applied skills (communication; planning; conflict resolution) rather than any particular disciplinary learning. They found that interdisciplinary aspects tended to be restricted to a few chapters rather than integrated throughout. An analysis of the curricula of 26 training programmes produced a similar conclusion; that almost all programmes included some interdisciplinary content, however, it was often covered only as an add-on in the final session of a natural science based component. They concluded that interdisciplinarity is being 'ghettoized'. Several other authors have also drawn attention to the distinction between learning about individual methods and approaches and learning how to integrate them (Touval and Dietz, 1994; Noss, 1997; Jacobson and McDuff, 1998; Eriksson, 1999), and reflect further on the need to combine academic and practical aspects of training (Saberwal and Kothari, 1996; Inouye and Dietz, 2000; Nordenstam and Smardon, 2000; White et al, 2000; Clark, 2001; Scholte, 2003; Kroll, 2005; Perez, 2005; Lopez et al, 2006; Martinich et al, 2006; Takacs et al, 2006).

Ethnobiology and Conservation Biology Compared

Ethnobiology and conservation biology have taken very different approaches to interdisciplinarity and it is therefore instructive to compare progress. Ethnobiology has evolved from a home within the established discipline of anthropology and, through the successive borrowing and gradual integration of specific tools from different disciplines, it has successfully consolidated as a new 'transdisciplinary' subdiscipline with its own set of unique methodological tools and conceptual frameworks. This has most likely been facilitated by its relatively well-defined thematic focus, centring on traditional ecological knowledge and different cultural perspectives on the environment. However, with the evolution of the concept of biocultural diversity and increased emphasis on actions to further its conservation, the scope of ethnobiology is broadening.

From its establishment as a discipline, conservation biology has been far broader in thematic scope, attempting to address all issues that affect the conservation of biodiversity. Its disciplinary origins in field biology have provided a solid theoretical basis for the development of its own natural science-based tools and conceptual frameworks (Noss, 1999), but no single disciplinary tradition has informed and supported the development of a consolidated social science perspective. Moreover, there has been limited success in bridging the divide between the natural and social sciences.

An additional distinction between conservation biology and ethnobiology is in the relationship between science and advocacy or practice. Both have a strong sense of mission, aiming to contribute to the conservation of global diversity, whether biological, cultural or both. The first meeting of the ISE in 1989 produced a strongly worded action call not only to carry out research, but also to 'guarantee the preservation of vital biological and cultural diversity'. Zent (1999) points to the 'ethnoecological imperative' ('the urgent need to document and conserve traditional or local environmental knowledge and management systems before they are gone forever') and the Code of Ethics of the ISE (ISE, 2008), which affirms the commitment of the society and its members to work collaboratively to support community driven development and 'protect the inextricable linkages between cultural, linguistic and biological diversity'. Similarly the Code of Ethics of the SCB emphasises the professional responsibility of all conservationists to contribute to the 'mission' of conservation (www.conbio.org/aboutus/Ethics/). However, in terms of their disciplinary evolution, conservation biology has focused far more on practical problem-solving than ethnobiology, although the relationship between objective academic enquiry and advocacy continues to be the cause of much debate amongst conservation biologists (e.g. see Mulvey and Lydeard, 2000; Gill, 2001; Meffe, 2007; Scott et al, 2007). This may become more of an issue in ethnobiology as the ideal of participatory research ('of, by and for indigenous peoples, traditional societies and local communities'; Hunn, 2002) develops.

Crossing Disciplinary Boundaries

As Lidicker (1998) states, what is needed to develop interdisciplinary approaches to conservation is 'to educate people to understand a specific intellectual area, to appreciate the contributions of other disciplines, and to be comfortable working in teams'. An examination of some of the common challenges encountered in multidisciplinary teamwork suggests that there is a long way to go in addressing this need. Multidisciplinary teams bringing together natural and social scientists are notorious for communication difficulties, and frequently fail to meet expectations. Marzano et al (2006) interviewed natural and social scientists working on interdisciplinary projects in the UK's Rural Economy and Land Use (RELU) Programme, and concluded 'each side found the other opaque and impenetrable', with 'social scientists being particularly daunted by (the seeming impenetrability of) bodies of scientific knowledge, and natural scientists being more unused to discussions of how knowledge is produced'. Similarly a workshop with natural and social scientists involved in environmental management in Australia highlighted problems related to different conceptual approaches, timescales of the different disciplines and the need to dedicate time to communication about different research paradigms (Strang, 2009; see also Sillitoe, 2004).

Fox et al (2006) carried out an online survey of 360 biological and social scientists that identified some of the barriers to integrating social sciences and conservation. These include the lack of a common vocabulary, a lack of knowledge of conservation among social scientists and frustration amongst natural scientists with the social science literature (particularly with the level of theory and jargon). These and other authors also report an unequal power dynamic between the disciplines. Social scientists are often in a minority in teams working on environmental issues and may not be involved in initial planning stages, thus they often have to fit their research into what they regard as an inappropriate conceptual framework (Campbell, 2005; Fox et al, 2006; Strang, 2009). Hence social scientists can easily become marginalized.

The sense of 'mission' in conservation-related research introduces an additional complicating factor. Campbell (2005) writes of the problems faced by a social scientist working with conservation biologists who place an overriding value on the conservation of biological (but not necessarily cultural) diversity. Where there are potential conflicts between biodiversity conservation and the interests of local people, this can put the social scientist in a difficult ethical position. As Campbell reports, 'it is sometimes exhausting to be the one person speaking a different language or trying to represent a broad suite of social science concerns'.

All of the authors cited above state that, in order to be successful, a multi-disciplinary project must devote considerable time to building relationships between team members and discussing, or even providing training in, different research paradigms (Sillitoe, 2004; Campbell, 2005; Marzano et al, 2006; Strang, 2009). This suggests that many natural science-based disciplines are still failing to prepare people adequately for interdisciplinary collaboration.

Challenges to Interdisciplinary Collaboration

Challenges associated with interdisciplinary collaboration have a long history, dating back at least as far as the establishment of formal academic systems in the late 19th and early 20th centuries, and are by no means unique to attempts to bridge the natural and social sciences. Klein (1990) describes how initiatives were taken repeatedly throughout the 20th century to encourage interdisciplinary collaboration, but outcomes rarely met expectations. For example, a major funding programme for interdisciplinary research in the US in the late 1960s 'was plagued by disciplinary chauvinism and the psychological, social and epistemological problems of working across disciplines' (Klein, 1990). Three factors that have been identified as contributing to problems of this kind are related to turf wars between established disciplines, a lack of knowledge of different disciplinary paradigms amongst individual researchers, and a simple lack of interest in the kinds of questions and issues addressed by different disciplines.

In terms of turf wars, according to Lau and Pasquini (2008) '[disciplinary] experts tend generally to regard fields other than their own with considerable suspicion – spurious at worst, at best irrelevant'. Klein (1990) writes colourfully about disciplines as territories, each of which is 'staked off by its own patrolled boundaries and no trespassing notices'. According to this metaphor, interdisciplinarity is a place where 'researchers, teachers and practitioners cross the "no man's land" between the disciplines, making their way across the "academic demilitarized zone"' (Klein, 1990). Geertz (in Sillitoe, 2004) characterizes academic disciplines as tribal in terms of their allegiances and territorial defence, and Turner (2002, in Youngblood, 2007) describes the history of relations between human and physical geography as 'a history of internecine conflict'. Territorial boundaries are reinforced through the 'indoctrination' of each new generation of adherents to their own disciplinary tribe and through vested interests in the institutional and bureaucratic structures that mark disciplinary divides.

There are barriers at every level of university system to the development of interdisciplinary perspectives. Undergraduate degree programmes and textbooks taking an interdisciplinary approach are scarce, and thus Masters programmes that attempt to bridge disciplinary divides must cater for students from a wide range of disciplinary backgrounds, many of whom are already committed to a particular disciplinary approach. Postgraduate students often have trouble identifying a university department that can provide support across disciplinary boundaries (Jacobson and McDuff, 1998; Nyhus et al, 2002), and once they are accepted, may find that they are constrained by departmental expectations in the disciplinary perspectives that are open to them. At the postdoctoral level, interdisciplinary or applied researchers can find it hard to compete for faculty positions and grants because of the higher prestige given to theoretical and single-discipline research (Lau and Pasquini, 2004, 2008; Fox et al, 2006). Financial and administrative systems, until recently, tended to discourage collaboration across departments within individual universities (e.g. Campbell, 2005; Fox et al, 2006; Kainer et al, 2006), and this is replicated in national funding systems, such that early career academics face further challenges in raising funds for interdisciplinary research.

A similar problem persists in publishing. Interdisciplinary environmental journals are often of lower status than established single-discipline journals (Campbell, 2005), and promotional structures favour single-author publications, discouraging multidisciplinary teamwork (Lidicker, 1998; Campbell, 2005; Fox et al, 2006). Even if a journal editor accepts an interdisciplinary article for review, the individual reviewers may not have the breadth of expertise to evaluate the work adequately, which can result in inappropriate and conflicting requirements for revisions (Campbell, 2005). For these reasons, Marzano et al (2006) state that interdisciplinarity is perceived to be risky in career terms, not only in job prospects but also in relation to both funding and publishing. This is particularly the case in terms of applied research, which tends to have lower prestige value than theoretical, single-discipline research (Fox et al, 2006).

Underpinning these institutional barriers, however, is a lack of basic comprehension between researchers using different disciplinary paradigms. This is probably most acute between quantitative and qualitative approaches to research rather than between the natural and social sciences *per se*. From a quantitative perspective, 'valid' science is defined by rigorous statistical analysis of numerical data representing precisely defined variables that are usually identified in advance of data collection; only thus can conclusions be drawn with any degree of certainty. In qualitative research, on the other hand, the emphasis is on in-depth description and complexity. Furthermore, it often uses inductive techniques rather than working from a specific hypothesis. Standards of validity and rigour are different, with more concern for contextual or ecological validity (how far the situation under which the research is carried out represents 'real life': Newing et al, *in press*) than the internal validity of quantitative, deductive approaches. From a qualitative perspective, quantitative studies often seem simplistic and 'reductionist' in approach, too heavily influenced by the researcher's expectations (as expressed in the hypotheses and variables chosen for study), and of little relevance to the broader picture. Conversely, from a quantitative perspective, qualitative research lacks 'data' (numbers), a clear research design (no clear hypothesis or precisely defined variables), and above all, because of the absence of statistics, valid 'proof'. Qualitative presentations may be perceived as meandering rhetoric – often with much theoretical jargon – peppered with short anecdotes from the field, but little if any 'real' science.

The quantitative–qualitative divide was the subject of extended and heated debates during the latter half of the 20th century. These debates were, to a large extent, resolved at a theoretical level, resulting in the general acceptance that qualitative and quantitative perspectives each had strengths and weaknesses, and are best applied in mixed-method combinations (Tashakkori and Teddlie, 1998). Nonetheless, in practice the quantitative-qualitative divide remains a major barrier to successful interdisciplinary research. Many qualitative researchers do not understand statistics, whereas quantitative scientists have no training in qualitative techniques. More fundamentally, students in many disciplines, especially in the natural sciences, receive no training in science theory except as it applies to their own discipline, and thus remain unaware of alternative epistemological and ontological paradigms. Whilst it is relatively straightforward for a researcher from one discipline to master a specific methodological technique from another, it is far harder to grasp and accept deeper conceptual differences in what constitutes valid knowledge and what assumptions are made about the nature of the world – especially for people who have already been 'indoctrinated' into a single disciplinary perspective.

Returning to the natural science/social science divide, natural scientists are almost invariably schooled in deductive, quantitative, positivistic approaches, whereas there is much more variation in the social sciences. It is not surprising, then, that social scientists in multidisciplinary teams sometimes feel that their research is unappreciated by their natural science colleagues. This is less of a problem for quantitative than qualitative social scientists. Several of the more successful areas of research on environmental issues that span the natural and social sciences involve quantitative social science perspectives and techniques, such as sociological surveys, economic valuation studies and attitudinal surveys from environmental education. As Phillipson and Lowe (2006) point out in relation to the RELU:

> The quantitative inclinations of economics and its commitment (at least in its neoclassical form) to positive (as opposed to normative) knowledge and explanations equip it well for technical collaboration with natural scientists. A key challenge for RELU will be to engage the full breadth of the potential social science contribution.

Research on Biocultural Diversity

Many of the epistemological and ontological barriers to communication are especially important for research into biocultural diversity, because similar challenges are present in exchanges between scientific and traditional knowledge systems (Sillitoe, 2004) and indeed in any cross-cultural research. Traditional knowledge may be based on premises that are completely alien to scientists, especially in terms of worldviews and belief systems. Despite this, the importance of taking traditional knowledge into account in conservation management is now widely acknowledged, although opinions still diverge about its epistemological status in relation to deductive science. Where is the 'proof' of its validity? At a deeper level, traditional knowledge systems are holistic and socially and culturally embedded; they encompass not only factual knowledge but also spiritual beliefs and worldviews (Berkes et al, 2000). Positivistic approaches to research that attempt to break them down into discrete technical facts are therefore inappropriate, and according to Sillitoe (1998), writing in the context of development

> *We need to establish that it is dangerous to do this and demonstrate the importance of understanding environmental interactions and development opportunities within their sociocultural contexts.*

Lastly, the interaction between scientific and traditional knowledge mirrors the perceived inequalities in power and prestige described above for multidisciplinary teams involving natural and social scientists. Participatory approaches in development recognise that highly technical approaches, which cast scientists in the role of 'experts', serve to discourage active participation by overpowering and devaluing local voices. This is an important issue in biocultural diversity research that aims at an equal partnership with local people.

Conclusions

It is apparent that interdisciplinarity research into biocultural diversity and conservation needs to bridge not one kind of gap but several – between different disciplinary traditions, between research and practice, between different thematic areas, and also between different cultures. Ethnobiology and conservation biology are probably the two most important disciplines attempting to do so. Ethnobiology has been more successful in building novel transdisciplinary perspectives and technical tools, and has a greater focus on cultural difference. Conservation biology has advanced further in terms of informing professional practice and has attempted to embrace a much wider thematic scope. Each discipline faces substantial challenges in its further development, and can learn from the strengths and weaknesses of the other. However, in terms of furthering research into biocultural diversity and conservation, what is important is not the creation and consolidation of specific (inter- or meta-) disciplines, but the creation of the conditions in which research can make use of whatever disciplinary tools become pertinent as new issues and perspectives arise. This is not only of academic interest, but it is particularly urgent in view of the escalation of global environmental change related to global warming and sea level rise. Identifying measures to strengthen the resilience of socioecological systems at the local level is essential in order to mitigate the effects of the crisis on biological and cultural diversity. In my view, the most important challenge to doing so is to build a community of scholars who have a basic grounding in science theory and understand the epistemological and ontological premises of the major research paradigms (see also Noss, 1997; Jacobson and McDuff, 1998; Maffi, 2004). I urge qualitative scientists interested in biocultural diversity and conservation to make the effort to gain a basic knowledge of statistics, and quantitative scientists to come to terms with the principles of qualitative research. Whilst this will not of course negate the institutional barriers to interdisciplinarity, it is a fundamental requirement for clear communication and ongoing interdisciplinary exchange.

References

Barry, A., Born, G. and Weszkalnys, G. (2008) 'Logics of interdisciplinarity', *Economy and Society*, vol 37, pp20–49

Berkes, F., Colding, J. and Folke, C. (2000) 'Rediscovery of traditional ecological knowledge as adaptive management', *Ecological Applications*, vol 10, pp1251–1262

Berkes, F. and Folke, C. (2002) 'Back to the future: Ecosystem dynamics and local knowledge', in L. H. Gunderson and C. S. Holling (eds) *Panarchy: Understanding Transformations in Human and Natural Systems*, Island Press, Washington, DC, pp121–146

Bonine, K., Reid, J. and Dalzen, R. (2003) 'Training and education for tropical conservation', *Conservation Biology*, vol 17, pp1209–1218

Buscher, B. and Wolmer, W. (2007) 'Introduction: The politics of engagement between biodiversity conservation and the social sciences', *Conservation and Society*, vol 5, no 1, pp1–21

Butler, C. D. and Oluoch-Kosura, W. (2006) 'Linking future ecosystem services and future human well-being', *Ecology and Society*, vol 11, no 1, p30

Carlsson, T. J. S. and Maffi, L. (eds) (2004) 'Ethnobotany and conservation of biocultural diversity', *Advances in Economic Botany*, vol 15, New York Botanical Garden, New York

Campbell, L. (2005) 'Overcoming obstacles to interdisciplinary research', *Conservation Biology*, vol 19, pp574–577

Chapin, M. (2004) 'A Challenge to conservationists', *Worldwatch Magazine*, Nov/Dec, pp17–31

Chatty, D. and Colchester, M. (eds) (2002) *Conservation and Mobile Indigenous Peoples: Displacement, Forced Settlement, and Sustainable Development*, Berghahn Books, New York and Oxford

Clark, T. (2001) 'Developing policy-oriented curricula for conservation biology: Professional and leadership education in the public interest', *Conservation Biology*, vol 15, pp31–39

Clayton, S. and Brook, A. (2005) 'Can psychology help save the world? A model for conservation psychology', *Analyses of Social Issues and Public Policy*, vol 5, pp87–102

Colchester, M. (2003) *Salvaging Nature: Indigenous Peoples, Protected Areas and Biodiversity Conservation*, World Rainforest Movement, Montevideo

Dillon, P. (2008) 'A pedagogy of connection and boundary crossings: Methodological and epistemological transactions in working across and between disciplines', *Innovations in Education and Teaching International*, vol 45, pp255–262

Dowie, M. (2005) 'Conservation refugees: When protecting nature means kicking people out', *Orion Magazine*, Nov/Dec

Eriksson, L. (1999) 'Graduate conservation education', *Conservation Biology*, vol 13, p955

Fairhead, J. and Leach, M. (1994) 'Contested forests: Modern conservation and historical land-use in Guineas Ziama Reserve', *African Affairs*, vol 93, no 373, pp481–512

Fairhead, J. and Leach, M. (1998) *Reframing Deforestation, Global Analyses and Local Realities: Studies in West Africa*, Routledge, New York and London.

Folke, C. (2004) 'Traditional knowledge in social-ecological systems', *Ecology and Society*, vol 9, no 3, article 7

Folke, C. and Gunderson, L. (2006) 'Facing global change through social-ecological research', *Ecology and Society*, vol 11, no 2, article 43

Fox, H., Christian, C., Cully-Nordby, J., Pergams, P., Peterson, G. and Pyke, C. (2006) 'Perceived barriers to integrating social science and conservation', *Conservation Biology*, vol 20, pp1817–1820

Gill, R. B. (2001) 'Professionalism, advocacy and credibility: A futile cycle?', *Human Dimensions of Wildlife*, vol 6, no 1, pp21–32

Gragson, T. L. and Blount, B. G. (1999) *Ethnoecology: Knowledge, Resources and Rights*, University of Georgia Press, Athens and London

Gray, A., Parellada, A. and Newing, H. (eds) (1998) *From Principles to Practice: Indigenous Peoples and Biodiversity Conservation in Latin America*, IWGIA, Copenhagen

Harrison, S., Massey, D. and Richards, K. (2008) 'Conversations across the divide', *Geoforum*, vol 39, pp549–551

Hunn, E. S. (2002) 'Traditional environmental knowledge: Alienable or inalienable intellectual property', pp3–10 in J. R. Stepp, R. S. Wyndham and R. K. Zarger (eds) *Ethnobiology and Biocultural Diversity*, ISE/University of Georgia Press, Athens, Georgia

Inouye, D. and Dietz, J. (2000) 'Creating academically and practically trained graduate students', *Conservation Biology*, vol 14, pp595–596

ISE (1989) 'Declaration of Belem', http://ise.arts.ubc.ca/global_coalition/declaration.php, accessed 12 July 2009

ISE (2008) 'ISE Code of Ethics (with 2008 additions)', http://ise.arts.ubc.ca/global_coalition/ethics.php, accessed 25 June 2009

Jacobson, S. K. and McDuff, M. D. (1998) 'Training idiot savants: The lack of human dimensions in conservation biology', *Conservation Biology*, vol 12, no2, pp263–267

Kainer, K., Schmink, M., Covert, H., Stepp, J., Bruna, E., Dain, J., Espinosa, S. and Humphries, S. (2006) 'A graduate education framework for tropical conservation and development', *Conservation Biology*, vol 20, pp3–13

Klein, J. T. (1990) *Interdisciplinarity: History, Theory and Practice*, Wayne State University Press, Detroit

Kroll, A. J. (2005) 'Integrating professional skills in wildlife student education', *The Journal of Wildlife Management*, vol 71, no 1, pp226–230

Lau, L. and Pasquini, M. (2004) 'Meeting grounds: Perceiving and defining interdisciplinarity across the arts, social sciences and sciences', *Interdisciplinary Science Reviews*, vol 29, pp49–64

Lau, L. and Pasquini, M. (2008) '"Jack of all trades"? The negotiation of interdisciplinarity within geography', *Geoforum*, vol 39, pp552–560

Lidicker, W. Z. (1998) 'Revisiting the human dimension in conservation biology', *Conservation Biology*, vol 12, no 6, pp1170–1171

Loh, J. and Harmon, D. (2005) 'A global index of biocultural diversity', *Ecological Indicators*, vol 5, pp231–241

Lopez, R. R., Hays, K. B., Wagner, M. W., Locke, S. L., McCleery, R. A. and Silvy, N. J. (2006) 'Integrating land conservation planning in the classroom', *Wildlife Society Bulletin*, vol 34, no 1, pp223–228

Maffi, L. (2001a) *On Biocultural Diversity: Linking Language, Knowledge, and the Environment*, Smithsonian Institute Press, Washington, DC

Maffi, L. (2001b) 'Linking language and environment: A coevolutionary perspective', pp24–48 in C. L. Crumley (ed), *New Directions in Anthropology and the Environment*, Altamira Press, Walnut Creek

Maffi, L. (2004) 'Maintaining and restoring biocultural diversity: The evolution of a role for ethnobiology', pp9–35 in T. J. S. Carlsson and L. Maffi (eds), 'Ethnobotany and conservation of biocultural diversity', *Advances in Economic Botany*, vol 15, New York Botanical Garden, New York

Martinich, J., Solarz, S. and Lyons, J. (2006) 'Preparing students for conservation careers through project-based learning', *Conservation Biology*, vol 20, pp1579–1583

Marzano, M., Carss, D. N. and Bell, S. (2006) 'Working to make interdisciplinarity work: Investing in communication and interpersonal relationships', *Journal of Agricultural Economics*, vol 57, no 2, pp185–197

Mascia, M., Brosius, J., Dobson, T., Forbes, B., Horowitz, L., McKean, M. and Turner, N. (2003) 'Conservation and the social sciences', *Conservation Biology*, vol 17, pp649–650

Meffe, G. (1998) 'Softening the boundaries', *Conservation Biology*, vol 12, pp259–260

Meffe, G. (2007) 'Conservation focus: Policy advocacy and conservation science', *Conservation Biology*, vol 21, p11

Meine, C., Soule, M. and Noss, R. (2006) '"A mission-driven discipline": The growth of conservation biology', *Conservation Biology*, vol 20, pp631–651

Moran, J. (2002) *Interdisciplinarity*, Routledge, London

Mulder, M. B. and Coppolillo, P. (2005) *Conservation: Linking Ecology, Economics and Culture*, Princetown University Press, New Jersey

Mulvey, M. and Lydeard, C. (2000) 'Let's not abandon science for advocacy: Reply to Berg and Berg', *Conservation Biology*, vol 14, pp1924–1925

Newing, H., Eagle, C., Puri, R. and Watson, C. (in press) *Conducting Research in Conservation: A Social Science Perspective*, Routledge, New York and London.

Niesenbaum, R. and Lewis, T. (2003) 'Ghettoization in conservation biology: How interdisciplinary is our teaching?', *Conservation Biology*, vol 17, pp6–10

Nordenstam, B, and Smardon, R. (2000) 'A perspective of educational needs in environmental science and policy for the next century', *Environmental Science and Policy*, vol 3, pp57–58

Noss, R. (1997) 'The failure of universities to produce conservation biologists', *Conservation Biology*, vol 11, pp1267–1269

Noss, R. (1999) 'Is there a special conservation biology?', *Ecography*, vol 22, pp113–122

Nyhus, P. J., Westley, F. R., Lacy, R. C. and Miller, P. S. (2002) 'A role for natural resource social science in biodiversity risk assessment', *Society and Natural Resources*, vol 15, pp923–932

Perez, H. (2005) 'What students can do to improve graduate education in conservation biology', *Conservation Biology*, vol 19, pp2033–2035

Phillipson, J, and Lowe, P. (2006) 'Special issue guest editorial: The scoping of an interdisciplinary agenda', *Journal of Agricultural Economics*, vol 57, pp163–164

Pilgrim, S., Cullen, L., Smith, D. J. and Pretty, J. (2008) 'Ecological knowledge is lost in wealthier communities and countries', *Environmental Science and Technology*, vol 42, no 4, pp1004–1009

Plotkin, M. (1995) 'The importance of ethnobotany in tropical forest conservation', in R. E. Schultes and S. von Reis (eds) *Ethnobotany: The Evolution of a Discipline*, Dioscorides Press, Portland, Oregon, pp147–156

Posey, D. and Balée, W. (1989) *Resource Management in Amazonia: Indigenous and Folk Strategies*, New York Botanical Garden, New York

Prance, G. (1995) 'Ethnobotany today and in the future', in R.E. Schultes and S. von Reis (eds) *Ethnobotany: The Evolution of a Discipline*, Dioscorides Press, Portland, Oregon, pp60–68

Russell, D. and Harshbarger, C. (2003) *Groundwork for Community-Based Conservation: Strategies for Social Research*, Altamira Press, Walnut Creek

Saunders, C. D., Brook, A. T. and Myers Jr, O. E. (2006) 'Using psychology to save biodiversity and human well-being', *Conservation Biology*, vol 20, no 3, pp702–705

Saunders, C. and Myers Jr., O. E. (2003) 'Exploring the potential of conservation psychology', *Human Ecology Review*, vol 10, no 2, ppiii–v

Saberwal, V. K. and Kothari, A. (1996) 'The human dimension in conservation biology curricula in developing countries', *Conservation Biology*, vol 10, pp1328–1331

Scholte, P. (2003) 'Curriculum development at the African Regional Wildlife Colleges, with special reference to the Ecole de Faune, Cameroon', *Environmental Conservation*, vol 30, no 3, pp249–258

Schultes, R. E. and von Reis, S. (eds) (1995) *Ethnobotany: The Evolution of a Discipline*, Dioscorides Press, Portland, Oregon

Scott, J., Rachlow, J., Lackey, R., Pidgorna, A., Aycrigg, J., Feldman, G., Svancara, L., Rupp, D., Stanish, D. and Steinhorst, R. (2007) 'Policy advocacy in science: Prevalence, perspectives and implications for conservation biologists', *Conservation Biology*, vol 21, pp29–35

Siebert, S. (2000) 'Creating academically and practically training graduate students', *Conservation Biology*, vol 14, pp595–596

Sillitoe, P. (1998) 'The development of indigenous knowledge', *Current Anthropology*, vol 39, no 2, pp223–252

Sillitoe, P. (2004) 'Interdisciplinary experiences: Working with indigenous knowledge in development', *Interdisciplinary Science Reviews*, vol 29, no 1, pp6–23

Stepp, J. R., Wyndham, R. S. and Zarger, R. K. (eds) (2002) *Ethnobiology and Biocultural Diversity*, International Society of Ethnobiology/University of Georgia Press, Athens, Georgia

Strang, V. (2009) 'Integrating the social and natural sciences in environmental research: A discussion paper', *Environment Development and Sustainability*, vol 11, pp1–18

Takacs, D., Shapiro, D. and Head, W. (2006) 'From is to should: Helping students translate conservation biology into conservation policy', *Conservation Biology*, vol 20, pp1342–1348

Tashakkori, A. and Teddlie, C. (1998) *Mixed Methodology: Combining Qualitative and Quantitative Approaches (Applied Social Research Methods Series, vol 46)*, Sage Publications, Thousand Oaks and London

Terborgh, J. (1999) *Requiem for Nature*, Island Press, Washington

Terralingua (2008) 'Unity in biocultural diversity', www.terralingua.org/html/home.html, accessed 30 July 2008

Touval, J. and Dietz, J. (1994) 'The problem of teaching conservation problem solving', *Conservation Biology*, vol 8, pp902–904

Van Heezik, Y. and Seddon, P. (2005) 'Structure and content of graduate wildlife management and conservation biology programs: An international perspective', *Conservation Biology*, vol 19, pp7–14

West, P., Igoe, J. and Brockington, D. (2006) 'Parks and peoples: The social impact of protected areas', *Annual Review of Anthropology*, vol 35, pp251–77

White, R., Fleischner, T. and Trombulak, S. (2000) 'The status of undergraduate educa-tion in conservation biology', www.conbio.org/Resources/education/StatusUnder-gradEducation.pdf, accessed 23 February 2009

Youngblood, D. (2007) 'Interdisciplinary studies and bridging disciplines: A matter of process', *Journal of Research Practice*, vol 3, article M18

Zent, S. (1999) 'The quandary of conserving ethnoecological knowledge: A Piaroa example', pp90–124 in T. L. Gragson and B. G. Blount (eds) *Ethnoecology: Knowl-edge, Resources and Rights*, University of Georgia Press, Athens, Georgia

3
Measuring Status and Trends in Biological and Cultural Diversity

David Harmon, Ellen Woodley and Jonathan Loh

What are Biological and Cultural Diversity?

Biological diversity, or 'biodiversity' for short, can be defined in many ways, but since the emergence of the concept in the late 1980s definitions have converged on the following aim: they all attempt to capture a sense of the sheer breadth of life. The United Nations Environment Programme's *Global Environment Outlook-4* (UNEP GEO-4) puts it this way:

> *Biodiversity is the variety of life on Earth. It includes diversity at the genetic level, such as that between individuals in a population or between plant varieties, the diversity of species, and the diversity of ecosystems and habitats* (UNEP, 2007).

The UNEP definition goes on to state that, '[b]iodiversity also incorporates human cultural diversity, which can be affected by the same drivers as biodiversity, and which has impacts on the diversity of genes, other species and ecosystems' (UNEP, 2007). Most other definitions exclude culturally transmitted information by limiting the term to variation produced through genetic heredity.

Cultural diversity can also be defined in many ways. Harmon (2002) calls it:

> *...The variety of human expression and organization, including that of interactions among groups of people and between these groups and the environment. Thus defined, the term is global in scope, both physically and conceptually. It comprises the entire index of discrete behaviors, ideas, and artifacts as exhibited by the whole of humankind, no matter where they live. It also covers*

the complete range of types of cultural interaction, as well as their outcomes or manifestations.

Through the Convention on the Protection and Promotion of the Diversity of Cultural Expressions, the United Nations Educational, Scientific and Cultural Organization (UNESCO) defines it as follows:

> *'Cultural diversity' refers to the manifold ways in which the cultures of groups and societies find expression. These expressions are passed on within and among groups and societies.*
>
> *Cultural diversity is made manifest not only through the varied ways in which the cultural heritage of humanity is expressed, augmented and transmitted through the variety of cultural expressions, but also through diverse modes of artistic creation, production, dissemination, distribution and enjoyment, whatever the means and technologies used* (UNESCO, 2005a).

Biological diversity is typically classified according to three nested levels of organization: genes, species and ecosystems. Cultural diversity can be classified according to a parallel division: memes (the most basic discrete units of culture), languages or ethnolinguistic groups, and cultures or culture areas.

Diversity Integrated: The Concept of Biocultural Diversity

The term 'biocultural diversity' was coined in the 1990s as a way to express 'the concept of an intimate link between biological, cultural, and linguistic diversity' and as an organizing principle through which the implications of this link could be explored (Maffi, 2005b). In the overview of the development of the concept, Maffi (2005b) defines the term as 'the diversity of life in all its manifestations – biological, cultural, and linguistic – which are interrelated within a complex socio-ecological adaptive system'. The concept is intended to be as broadly inclusive as possible, and stresses the idea that the various realms of diversity are, in many ways, indissoluble. Harmon and Loh have offered the following more detailed definition:

> *Biocultural diversity is the total variety exhibited by the world's natural and cultural systems. It may be thought of as the sum total of the world's differences, no matter what their origin. It includes biological diversity at all its levels, from genes to populations to species to ecosystems; cultural diversity in all its manifestations (including linguistic diversity), ranging from individual ideas to entire cultures, the abiotic or geophysical diversity of the Earth, including that of its landforms and geological processes, meteorology, and all other inorganic components and processes (e.g., chemical regimes) that provide the setting for life; and,*

importantly, the interactions among all of these (Harmon and Loh, 2004; see also Loh and Harmon, 2005; Harmon, 2007).

Beyond this, and of critical importance, biocultural diversity is an evolutionary systems concept that emphasizes adaptive interactions among all three realms of diversity: biological, cultural and abiotic. One way that these interactions find expression is in the form of cultural landscapes (see Brown et al, 2005, for a conservation-oriented overview, and Heckenberger et al, 2007, for an example of how new findings in archaeology and historical ecology are reconceptualizing parts of Amazonia as cultural landscapes).

Although the existence of an 'inextricable link' between biological and cultural diversity had been posited as early as 1988 by an international congress of ethnobotanists, the idea of biocultural diversity really began to take shape in the early 1990s as various scholars independently observed that an extinction crisis affecting linguistic and cultural diversity had begun to unfold, one which not only parallels the species extinction crisis but also seems driven by the same factors. A cascade of research during the rest of the decade and into the first years of the 21st century established several crucial points:

- Distributions of species and languages show remarkable areas of overlap.
- Biological and cultural/linguistic diversity face many threats in common.
- Certain biogeographical conditions promote the development and retention of both biological and cultural diversity.
- Retention of biological and cultural diversity is often associated with indigenous peoples.
- Conventional disciplinary approaches are unable to portray an integrated picture of the state of the diversity of life in all its forms, biotic and abiotic; rather, a new interdisciplinary way of looking at diversity is needed.
- This new approach – based on the biocultural diversity concept – needs to exist not just as theorizing, but as a means to generate practical measures to ensure the perpetuation of the world's diversity.

Initial work to correlate biological and cultural diversity, using linguistic diversity as a proxy, led to cross-mappings of the global geographic distributions of biological and cultural diversity and to analyses of the overlaps in these distributions (Mace and Pagel, 1995; Harmon, 1996). This early work led to further developments, such as taking the World Wide Fund for Nature's (WWF's) representation of the world's ecoregions (and specifically the 'Global 200' conservation-priority ecoregions) and mapping the world's languages onto them (Oviedo et al, 2000; Skutnabb-Kangas et al, 2003). These, as well as later mappings that use different representations of global biodiversity (Stepp et al, 2004; Loh and Harmon, 2005), consistently show that areas of high biodiversity and high cultural diversity tend to coincide, especially in the tropics (Amazon Basin, Central Africa, Indomalaysia/Melanesia). Biologically megadiverse countries also tend to be culturally megadiverse (Harmon, 2002), although there are of course many nuances that need to be accounted for (Moore et al, 2002; Sutherland, 2003).

Although the biocultural diversity concept was inspired by comparisons of megadiverse areas, an important aspect of the approach is that it places value on every component of the world's diversity, no matter where it is found, how abundant it is, or how phylogenetically unique it is. For example, in areas such as the Arctic and the arid tropics, biodiversity is comparatively low, but this does not exclude such areas from being considered important bioculturally. In these regions, cultural features are closely linked with the biophysical environment and, just as in megadiverse regions of the world, are facing similar threats.

Principles of Measuring Status and Trends in Diversity

The basic questions and measurements

An underlying principle common to all diversity measurements is that they are teleological – that is, they are done for a specific purpose. Moreover, different purposes call for different types of measurement, different assumptions behind the measurements, and different sampling techniques. Depending on what they measure, some indicators will be more or less relevant to the questions at hand; some will have data available on their state at levels of disaggregation essential to the analysis, and some won't; some will be more robust than others, and less sensitive to changes in policy or management interventions; and some will be easier to communicate to a target audience than others. All of these factors are crucial to indicator selection because often the purpose of measurement is to establish the groundwork upon which to base a policy or management decision.

Diversity measurement: Challenges in common

Lack of complete global inventories For many units of diversity measurement, complete global inventories do not exist. In terms of species, for example, the worldwide numbers of bird and mammal species are nearly completely known, although a small number of new species continue to be found, but ongoing global inventories of fish, amphibian and reptile species remain far from complete. For many invertebrate taxa, the number of species can only be estimated with wide margins of error. Therefore, biological diversity indicators that incorporate species data usually exclude the latter categories, even though invertebrates account for most of the world's species richness. Likewise, the number of languages is far from settled: each of the 15 successive editions up to 2005 of *Ethnologue,* the leading global handbook, included more languages than the last (although the 2009 edition reports a leveling off).

Insufficient knowledge of status and trends Even where inventories are more or less complete, there is often a dearth of knowledge about the abundance, richness, and evenness of the units being measured. Again using species and languages as examples, we may know a lot about status and trends in the populations of many birds and mammals, and in the demographics of widely spoken or state-supported languages, but almost nothing about the thousands of small, inconspicuous species or locally spoken languages.

Lack of taxonomic capacity For both biological and cultural diversity, there is a lack of specialists trained and able to distinguish between and classify the units of measurements, especially in the diversity-richest areas of the world where such expertise is most needed. Most zoologists are specialists in vertebrate species, yet the vast majority of species are invertebrates. Similarly, relatively few of the world's linguists are trained in the study of small indigenous languages, which form the backbone of the global linguistic corpus. (There are exceptions, e.g. in Australia, where several universities have strong emphases on indigenous linguistics.)

Inexactness of units of measure At a fundamental level, there is serious disagreement among experts over how to define some of the most common units of diversity measurement. There are several competing species concepts, and perennial problems in distinguishing species from subspecies; there is even less consensus on how to define an ecosystem. For cultural diversity, defining the bounds of ethnic groups is difficult, and distinguishing languages from dialects is a problem that resists straightforward solutions (see Harmon, 2002, for a complete discussion).

The DPSIR Framework for Indicators

As noted above, different diversity measurements call for their own unique assumptions and methods. However, one theoretical model has proven especially useful as a framework for diversity indicators: the pressure–state–response (PSR) model (see Rapport and Singh, 2006). This model has been expanded and is now usually seen in a somewhat more elaborate form as the driving force–pressure–state–impact–response (DPSIR) framework (UNEP, 2007).

The DPSIR model should be seen as a cause-and-effect chain which describes how changes come about in any given environment or ecosystem. It is sometimes helpful to think of a 'target' or focal point of a system, the state of which is the primary concern of the conceptual model.

Drivers are the underlying, indirect forces in a society or economy which create *pressures* on a system or 'target'. These drivers may be demographic or economic trends, scientific or technological developments, or shifts in societal preferences or behaviour. *Pressures* are the direct activities or practices which cause changes to the *state* of the system or 'target'; they may also be referred to as threats. The main pressures on biodiversity, for example, include land use or habitat change, over-exploitation of biological resources, invasive species or genes, pollution, disease and climate change. The concepts of 'drivers' and 'pressures' are sometimes blurred, but are distinguished essentially by their distance from the 'target', in time or space, along the cause-effect chain. Also, drivers tend to be socioeconomic while pressures tend to be the resulting ecological agents of change. *State* refers to some quality or qualities of the system or target with which we are concerned. If the target is a forest, for example, *state* would refer to its size, health, productivity, or the diversity of species it supports. *Impacts* are referred to when changes in the *state* of a system affect human well-being (and the state of the humans in question is not itself the target of the model). For example, if the state of a forest is declining it could result in a reduction in the ecological services it provides to local communities, such as hydrological regulation or supply of wild food or medicines. *Responses* are societal activities, policies or interventions designed to resist changes in state by reducing pressures or drivers. For example, the establishment of a protected catchment area could be a response to the loss of water flow regulation (*impact*) caused by a reduction in forest cover (*state*) due to agricultural encroachment (*pressure*) resulting from population growth (*driver*).

In this chapter, we are concerned with indicators or measures of state, and the target with which we are concerned will be the biological, cultural or biocultural diversity of any given system.

The *state* of many terrestrial ecosystems can be measured using remote sensing. Most remote sensing measures only the extent of a particular habitat type or biome, and does not normally provide information about the quality of the habitat. Forests are an example. They are the habitat type most suited to monitoring by satellite, but the first global forest maps did not become available until the late 1990s. Moreover, other measures are needed to assess the health or quality of the habitat and the species within forests, since seemingly healthy looking forests can be undergoing changes in structure and composition if there are changes in the composition of species within them. The extent of habitat loss in other biomes such as grasslands and wetlands are harder to monitor using remote sensing, and global maps do not yet exist. Marine and freshwater ecosystems such as rivers and lakes, generally speaking, do not change in area over time, and yet the biodiversity they contain may disappear completely, so simply measuring habitat extent is not an adequate indicator of the state of the ecosystem. Remote sensing can also provide information relating to ecosystem productivity, but diversity is best measured using more direct measures at ground (or sea) level.

Overview of Biological Diversity Indicators

The abundance of many wild species or species populations has been monitored extensively over the past 40 years or more, especially in the developed world. Measures of population or abundance give a quantifiable indicator of the state of a species, group of species (such as farmland birds), or of the habitat in which they live (such as temperate grasslands). A number of biodiversity indicators have been developed for these purposes, although most population monitoring has been directed towards bird and mammal species. The status of a far greater number of species has been assessed using the International Union for Conservation of Nature (IUCN) Red List categories (extinct, extinct in the wild, critically endangered, endangered, vulnerable, near threatened or least concern), although these assessments are based on a mixture of measures of state, such as abundance or habitat extent, and pressure, such as over-exploitation. With some exceptions, the Red List status of most bird, mammal and amphibian species, and many other vertebrate, invertebrate and plant species, have been assessed at least once, and often more than once, whereas population trend data are available for only a fraction of these. Population trends, on the other hand, can provide information on species at finer geographic scales and over shorter time intervals than Red Lists, so is a more sensitive indicator. Yet some species do not immediately show population changes when habitats are altered, and long-term monitoring may be needed to track the time lags between the onset of external pressures and the responses of these species. In combination, information provided by both indicators based on species abundance – such as the Living Planet Index or the farmland bird index in the UK – and indicators based on Red Lists – the Red List Indices – can provide status and trend information across a large number of taxonomic groups, ecosystems and biomes, and the last decade has seen an effort to gather much more data.

Examples of Biological Diversity Indices

The CBD 2010 headline indicators   At a meeting of the Convention on Biological Diversity (CBD) in 2001, and again at the World Summit on Sustainable Development in 2002, more than 180 of the world's governments made a commitment:

> *To achieve by 2010 a significant reduction of the current rate of biodiversity loss at the global, regional and national level as a contribution to poverty alleviation and to the benefit of all life on Earth.*

To measure progress towards the 2010 target, as it is generally known, the CBD has adopted 22 'headline indicators', grouped under 7 'focal areas'. Around 40 measures are now in the process of development by a number of intergovernmental and non-governmental organizations (NGOs), some of them from scratch, in order to provide information on whether or not the world's governments will have succeeded in meeting the target. These initiatives are coordinated for the purposes of reporting to the CBD under the umbrella of the 2010 Biodiversity Indicators Partnership (see www.twentyten. net) by the UNEP World Conservation Monitoring Centre.

The indicators do not follow the DPSIR framework, and are by no means comprehensive. Some are in a more advanced state of development than others, and for some, as yet, little or no data are available. It is unlikely that all of the headline indicators will be able to deliver trend information by 2010, and because the target is to reduce the *rate* of biodiversity loss, data from at least three points in time will be required to know if the target has been met. There are currently 18 measures which have been selected by the CBD under the focal area of 'Status and trends in the components of biodiversity'. It would be beneficial for policy-making, communicating with the public and deciding future interventions if, over the coming decade, a smaller set of more comprehensive, aggregated indicators were to be used for monitoring biodiversity at the global level (Mace and Baillie, 2007).

The Living Planet Index (LPI) The LPI, which began in 1997, is a WWF-supported project to develop a measure of the changing state of the world's biodiversity over time. The index uses population trends of vertebrate species from around the world to assess the state of global biodiversity. It tracks a total of nearly 10,000 populations of around 2,000 species. Indices for marine, terrestrial and freshwater species are calculated and then averaged to create an aggregated index. Indices can be produced at different biogeographic and taxonomic scales and in this way the LPI is an indicator of the health of ecosystems at different levels. Between 1970 and 2005 the index showed an overall 30 per cent decline. There is no significant change in the direction of the index since its peak in 1976, which suggests that global biodiversity decline shows no sign of abating, and the 2010 target has not yet been met.

Separate indices are produced for terrestrial, marine and freshwater species, so the LPI can also be considered to be a measure of trends in ecosystem health for different biomes or habitats. Although vertebrates represent only a fraction of all known species, it is assumed that their population trends are typical of global biodiversity as a whole. No attempt is made to select species on the basis of geography, ecology or taxonomy, so the LPI dataset contains more population trends from well-researched regions, biomes and species. In compensation, temperate and tropical regions are given equal weight within the terrestrial and freshwater indices, as are ocean basins within the marine index, with equal weight being given to each species within each region or ocean basin. An assumption is made that the available population time series data are representative of vertebrate species in the selected ecosystems or regions, and that vertebrates are a good indicator of overall biodiversity trends (Loh et al, 2005).

Regional LPIs Different regions of the world show varying trends in species populations, reflecting the differing anthropogenic and environmental pressures on biodiversity. The terrestrial index reveals a marked difference in trends between tropical and temperate species. While temperate species populations have remained reasonably stable on average, tropical species have declined by more than 50 per cent. This rapid fall in tropical species populations reflects the rapid conversion of natural habitat to cropland or pasture over the last 50 years, driven ultimately by the growth in human population and increasing world demand for food, fibre and timber. The conversion of natural habitat to farmland in temperate regions, on the other hand, largely occurred long before 1970, and the consequent decline in species populations is not therefore reflected in the temperate index. The LPI does not say that the current state of biodiversity is worse in the tropics than temperate regions, but that the trends over the last three decades have been worse.

Red List Index (RLI) The RLI is a well-developed indicator with some global-level time-series data. The RLI tracks the changes in threat status, as recorded in the IUCN Red List of Threatened Species, across all species in a given taxonomic group. The IUCN Red List places species into threat categories according to an assessment of extinction risk, ranked from least concern through vulnerable, endangered and critically endangered to extinct. The only taxonomic groups which have been comprehensively assessed are birds, mammals and amphibians. So far, the Red List Index has been applied only to birds and amphibians, but will soon be completed for mammals as well. In order to give a better indication of trends in biodiversity as a whole, the RLI is being applied to large, random samples of species from a number of less well-known taxonomic groups, including invertebrates and plants, but these will not be ready until all these groups have been assessed for threat status at least twice. The RLI for birds, which has the most comprehensive data for any taxonomic group, shows that more than 2 per cent of all birds species have become more threatened (moved up to a higher extinction risk category) since 1988 whereas less than 0.5 per cent have become less threatened (moved down a category).

Overview of Cultural Diversity Indicators

With concern about the status of the world's cultural diversity continuing to mount, increasing attention is being paid to indicators of culture.

'Culture' is any set of human activities that creates a distinctive way of life which forms the basis of a people's identity. Maffi (2005a) refers to this as the 'broad anthropological sense' of culture and notes that development of indicators in this sense is a fairly new endeavour. Within this sense of 'culture', cultural indicators are derived from complex phenomena, for example, language, traditional ecological knowledge, religious and spiritual belief and ceremonial behaviour, food systems and ethnic identity.

Fully developed, integrative cultural indicators in this anthropological sense are often lacking in assessments of the state of the planet. Organizations such as the United Nations Development Programme (UNDP), UNEP, IUCN and the CBD are grappling with this issue, as are some individual researchers. This work is still breaking new ground and there are few, if any, widely accepted standards as yet in this emerging area.

At the global level, only a handful of such cultural indicators have global data sets. Thus far, the most frequently used global cultural indicator has been linguistic diversity – or, more precisely, language richness, the number of different languages spoken worldwide, which is used as a proxy for both linguistic diversity and the global variety of cultures. As outlined below, languages and linguistic diversity form a key part of several global diversity indices currently under development.

Another way to divide cultural diversity indicators is by means of transmission. There are two broad categories of units of cultural transmission: individual traits and groups. Traits are definable attributes or elements of culture that can be analysed, such as behaviours, practices and beliefs. Groups are larger-scale entities characterized by commonalities, such as ethnic groups, speech communities or villages (Holden and Shennan, 2005). Cultural indicators could be divided along these lines as well. Within these two broad categories of cultural transmission, we can further distinguish between 'vertical' modes (cultural traits being passed down to the younger generation by parents, elders and other teachers of a previous generation) and 'horizontal' modes (cultural traits being diffused among cultures by imposition or borrowing/ adoption) (Pagel and Mace, 2004).

Examples of Cultural Diversity Indices

Greenberg's linguistic diversity indices In an influential paper published in 1956, the linguist Joseph Greenberg proposed eight indices of linguistic diversity. The indices are based on the probability of two members of a population sharing the same language. (As a comparison, one widely used measure of biological diversity, the Simpson diversity index, represents the probability that two randomly selected individuals in a habitat belong to the same species.) Greenberg identified the indices with the letters $\underline{A}$ through $\underline{H}$. Of these, the $\underline{A}$ Index, which measures the probability that two randomly chosen individuals from a population in a given area will speak the same language, has found the most favour because it is the simplest and most straightforward to determine. It is calculated according to the following formula:

$$\underline{A} = 1 - \Sigma_i (p_i^2)$$

where p_i is the proportion of a population in a given area that speaks language i.

Hence 0 represents the situation in which everyone in a given area speaks the same language, and 1 the opposite situation, in which everyone in a given area speaks a different language (Greenberg, 1956).

Because it can potentially be derived from census data, the $\underline{A}$ Index offers the possibility of tracking changes in linguistic diversity over time.

Canada's indices of linguistic continuity and ability In a quantitative study of vitality and moribundity (Norris, 1998), Statistics Canada used 1996 census responses to calculate an 'index of continuity' and an 'index of ability' for the country's native languages. The index of continuity measures language vitality by comparing the number of people who speak it at home with the number who learned it as their mother tongue of origin. In this index, a 1:1 ratio is scored at 100, and represents a perfect maintenance situation in which every mother-tongue speaker keeps the language as a home language. Any score lower than 100 indicates a decline in the strength of the language. The index of ability compares the number who report being able to speak the language (at a conversational level) with the number of mother-tongue speakers. Here, a score of more than 100 indicates that an increment of people have learned it as a second language, and may suggest some degree of language revival (Norris, 1998).

Florey's Linguistic Vitality Test (LVT) The linguist Margaret Florey has developed an LVT to address shortcomings in the methodology of assessing language endangerment (Florey, 2006). Even though studies of endangered languages have expanded rapidly in the last 20 years, Florey argues that the field's methods have not kept pace: characterizations of language endangerment scenarios and speaker ability are often ad hoc, there is little common use or understanding of terminologies, and the results of studies lack comparability across languages. The LVT is intended to be a standardized tool to:

- gain an informed overview of linguistic vitality in a site based on empirical data;
- learn how linguistic ability varies in and between communities, e.g. according to age, gender, religious affiliation, special roles;
- permit comparisons of linguistic vitality between sites; and
- assess intergenerational transmission of linguistic and other indigenous knowledge.

There are three components of the LVT:

1 Lexical recognition; designed to test receptive ability in the endangered language – that is, the language which is in the process of being replaced by a dominant, non-endangered language. Respondents are asked to identify 5 photo sets and a total of 53 test items. The first three sets test recognition of common nouns, while the fourth and fifth sets test comprehension of simple sentences. Each photo set is displayed in front of the respondent, and the contents of the photos are explained to the respondent in the dominant language. Then a taped description of each of the photos in turn, described in the endangered language, is played for the respondent and the respondent is asked to attempt to select a photo that correctly matches it.
2 Translation sentences; designed to test productive ability in the endangered language. Respondents are asked to translate 3 sets of sentences (75 total sentences) from the dominant language to the endangered language. The first test sentence is read (or a recording played) to the respondent, who then attempts to translate it into the endangered language. Respondents who are reasonably confident with the first translation set move on to translation set two and then on to translation set three.
3 Discourse; designed to test creative ability in the endangered language.

McConvell and Thieberger's analysis of trends in Australia's Aboriginal languages   Drawing on a wide range of studies and precepts (including those described above in Norris, 1998), McConvell and Thieberger (2001) put together a status report on the Aboriginal languages of Australia – arguably the most endangered body of indigenous languages in the world. This report gives a comprehensive treatment of the factors that produce moribundity (and vitality) in small languages as they struggle to co-exist with a large, socio-politically dominant language. They use census and other data to develop age-class analyses of particular Aboriginal languages. From these age-class data, they create an Endangerment Index for these languages, which is the percentage of speakers aged 0–19 divided by the percentage of speakers aged 20–39. The intent is to try to see whether there is a drop-off in speaker percentage among the youngest generation. Languages with an index value of greater than one are considered 'strong'; those with values of less than one, 'endangered'.

McConvell and Thieberger go on to discuss a number of caveats and qualifications, especially concerning languages that had been considered 'endangered' by earlier analysts (using different methods of analysis) but which rated higher than one in their Endangerment Index. These caveats, for example, point to potential problems with how census data on languages are to be interpreted. A key lesson from their study is that close familiarity with the situation on the ground and at relevant local scales is necessary for an accurate picture of cultural diversity.

Ethnolinguistic Fractionalization Indices (ELFs) ELFs are very similar to Greenberg's diversity indices, and use a formula almost identical to his. ELFs measure the probability that two people from a given area, chosen at random, will be from different ethnolinguistic groups.

ELFs have been in use for a long time. For example, using one of the 20th century's landmark compendiums of culture-group data, the *Atlas Narodov Mira*, or *New World Atlas* (State Geological Committee of the Soviet Union, 1964). Taylor and Hudson (1972) included an ELF for 129 countries in the first edition of their *Handbook of Political and Social Indicators*. This ELF has subsequently seen wide use in the fields of political science, econometrics and other fields (Bossert et al, 2005).

Index of Linguistic Diversity (ILD) The ILD is designed to be the first global index of trends in linguistic diversity, as measured by changes in the number of mother-tongue speakers. The objective is to provide quantitative data that will show whether languages (particularly indigenous languages) are losing speakers, and if so, at what pace.

The ILD is drawn from a global database of language demographic information gathered primarily from successive editions of *Ethnologue,* which is generally recognized as the best such source available, as well as data from a number of other publications. The index is based on a random sample of 1500 languages from the 2005 *Ethnologue.*

For the 1500 sampled languages, trends are determined from mother-tongue speaker data taken from nine editions of *Ethnologue:* 2005 (15th edition), 2000 (14th), 1996 (13th), 1992 (12th), 1988 (11th), 1984 (10th), 1978 (9th), 1958 (5th), and 1951 (1st). It should be noted that not all languages have trend data; of the 1500 languages, about 1000 have trend data available, i.e. at least two unique data points. In all, about 2700 unique data points form the basis for the trend data in the ILD.

The ILD was recently completed, and the first results are about to be published (Harmon and Loh, 2010, in press). The main finding is that the world's linguistic diversity declined by 20 per cent over the period 1970–2005. In addition, the diversity of the world's indigenous languages declined slightly more over the same period, although in certain regions (e.g. the Americas), the decrease was much more.

Overview of Biocultural Diversity Indicators

Biocultural diversity indicators are difficult to develop because they must represent something meaningful about this complex concept. Biocultural diversity emphasizes adaptive interactions between and among three wide-ranging systems of diversity – biological, cultural and abiotic – with feedback relationships among factors in each system. Identifying how different factors interact is a fundamental challenge.

A second challenge is exploring the pros and cons of combining indicators of biological, cultural and abiotic diversity into a single index. To simplify the problem, consider the interactions between biological and cultural diversity alone, setting aside abiotic diversity. It can be argued that aggregating biological and cultural diversity loses the residual variation in each system independent of the other. Furthermore, the frameworks of conventional biological diversity indices cannot all be easily applied in the cultural domain – and the significance of this depends on what one's objectives are in constructing an index in the first place. If, for example, one wants an index that goes beyond combining measures of richness and relative abundance, then one must include dissimilarity information for the entities represented in a sample. This involves using distance- and abundance-weighted indices. But each such step requires progressively more and higher-quality information (James Steele, pers comm).

A third challenge is considering whether biocultural diversity indicators can (or even should) be used to develop an integrated conservation strategy. Here, a number of complicated evolutionary interactions need to be plumbed. For example, is the 'hotspot' approach the best to take? If it is, what kind of hotspots should we be looking for? As Steele (pers comm) notes, recent work suggests that hotspots for phylogenetic diversity are not the same as hotspots for endemism. Yet both are important considerations in biological and cultural diversity. Beyond this, the argument for there being an analogy between biological and cultural diversity in terms of creating an optimal conservation strategy is less clear-cut, because humans can switch cultures and languages, while species cannot consciously evolve into something else.

Despite these challenges, it is clear that the interactions between the biological and cultural diversity systems are significant, and that indicators of these interactions are a critical part of understanding how the systems relate to one another. What makes a particular indicator potentially suitable to a biocultural diversity perspective? The best biocultural diversity indicators are those that:

- operate at a parallel level of organization across the biological and cultural diversity realms – meso-scale indicators that combine species and language measurements are the preponderant examples to date, but there is no reason in principle why micro-scale (genetic/memetic) or macro-scale (ecosystem/cultural area) indicators could not be developed;
- reflect interactions between humans and the environment;
- are highly sensitive to changes in status and trends; and
- that are readily understandable to the public and policy-makers.

Indigenous Peoples' Organizations and Indicator Development

While indigenous peoples' cultures constitute about 70–80 per cent of the world's cultural diversity they represent the minority population in most countries (IUCN, 1997). This has prompted concern over how well indigenous peoples' issues are represented by conventional indicators. For many indigenous peoples, there is a close relationship with their traditional lands and territories, which form a core part of their identity and spirituality and are deeply rooted in their culture, language and history. For this reason, indicators of well-being overlap with measures of rights of access to intact and non-degraded land and resources. This overlap reflects interactions between cultures and the environment, as suggested in the above criteria for good biocultural indicators. Many indigenous peoples' organizations (IPOs) have been working to determine indicators that measure aspects of culture such as traditions, ceremonies, food systems and values that define a peoples' or a community's cultural identity. Since these aspects of culture are closely interconnected with the local ecology, many of these indicators are, in fact, integrated and biocultural. They tend to measure both cultural resilience (well-being of indigenous peoples by the extent of use of traditional knowledge, customs, practices and languages that are the basis of cultural identity) as well as ecological integrity. The idea of resilience is implicit in and interdependent with the notion of diversity. It is an underemphasized concept in the biocultural diversity discussion. To ensure resilience among diverse cultures of the world, collaboration is needed to safeguard the ecological integrity of ecosystems upon which cultural traditions depend.

The Working Group on Indicators of the International Indigenous Forum on Biodiversity (IIFB) was formed in 2006 to respond to the immediate need to identify and test indicators relevant to the implementation of the Strategic Plan for the CBD. The Strategic Plan, along with Goal 9 in the framework for monitoring achievement towards the 2010 target, is to maintain the sociocultural diversity of indigenous and local communities.

At the same time the IIFB working group on indicators was being established, the United Nations Permanent Forum on Indigenous Issues (UNPFII) identified a need to develop indicators relevant to indigenous peoples and the Millennium Development Goals (MDGs). This was largely in response to several reports showing that efforts to reach some MDG targets were displacing indigenous communities from their ancestral lands and thereby accelerating the loss of lands and resources critical for indigenous peoples' livelihoods. The UNPFII therefore decided to work on indicator development across its mandated areas of health, human rights, economic and social development, environment, education and culture.

The IIFB Working Group on Indicators is developing a limited number of indicators in the thematic areas of (i) traditional knowledge, innovations and practices; (ii) customary sustainable use of biodiversity; (iii) goods and services from biodiversity; and (iv) effective participation of indigenous and local communities in national, regional and international policy processes. The group identified the 12 global core themes and issues of concern to indigenous peoples, and produced a list of sub-core issues and indicators. From this list, the group drew up a shortlist of indicators relevant to the CBD and to the its monitoring framework of goals, targets and indicators.

The Global Consultation on the Rights to Food and Food Security for Indigenous Peoples, held in Bilwi, Nicaragua, in 2006, was a collaborative effort between the Food and Agriculture Organization's (FAO's) Sustainable Agriculture and Rural Development (SARD) Initiative and the International Indian Treaty Council (IITC). This consultation, together with another meeting held in Guatemala in 2002, and a questionnaire administered by the IITC in 2004/5, resulted in the determination of 11 thematic areas for indicator development. These thematic areas were then further reduced to a list of five to meet FAO requirements for policy directives. The indicators do not appear to be integrated into the list developed at the International Experts Seminar. The indicators developed at the 2nd Global Consultation are strongly biocultural. It should be noted that indigenous peoples strongly support a human rights-based approach to the development of indicators, which distinguishes structural, process and outcome indicators. Outcome indicators alone will not reveal the underlying structural and procedural changes needed for advances towards eradicating poverty and securing the well-being of indigenous peoples.

In 2007/8 the IITC launched the next phase of the indicators project, which focused on disseminating, providing training in and 'field testing' the cultural indicators with UN bodies, development agencies and NGOs, and, in particular, indigenous communities. While still in the training phase, the main lesson learned to date is that the cultural indicators, once they are presented and explained, and examples are provided to community members, work as an effective tool and a methodology which communities can use to discuss, measure, assess, and evaluate both changes and impacts. The identified themes and the innovative three-part format for each set of indicators (structural, process and outcome indicators) assist indigenous communities not only to assess impacts but to look at changes over time, explore root causes and develop responses on several levels.

Indices of Biocultural Diversity

Index of Biocultural Diversity (IBCD)   The IBCD (Loh and Harmon, 2005) is the first attempt to quantify global biocultural diversity by means of a country-level index. Using a combination of five indicators, the IBCD establishes rankings of biocultural diversity for 238 countries and territories. The index uses the number of languages, religions and ethnic groups present within each country as a proxy for its cultural diversity, and the number of bird, mammal species and plant species as a measure of its biological diversity. The IBCD has three parts:

1   A *biocultural diversity richness component* (BCD-RICH), which is a relative measure of a country's 'raw' biocultural diversity using unadjusted counts of the five indicators.

2   An *areal component* (BCD-AREA), which adjusts the indicators for land area and therefore measures a country's biocultural diversity relative to its physical extent.

3   A *population component* (BCD-POP), which adjusts the indicators for human population and therefore measures a country's biocultural diversity relative to its population size.

The IBCD gives equal weight to cultural and biological diversity, so a country's overall biocultural diversity score is calculated as the average of its cultural diversity score (aggregated from the diversity scores for languages, religions and ethnic groups) and its biological diversity score (aggregated from the diversity scores for bird/mammal species and plant species). When values for these indicators are ranked on a global basis, it becomes apparent that biocultural diversity is not evenly distributed. A few countries are megadiverse, with very large values; then the ranking rapidly diminishes to much lower values found in more typical countries. The IBCD therefore uses a common log scale to produce a linear distribution.

To compensate for the fact that large countries tend to have a greater biological and cultural diversity than small ones simply because of their greater area (or greater population), Loh and Harmon calculated two additional diversity values for each country by adjusting first for land area (BCD-AREA) and second for population size (BCD-POP). This was done by measuring how much more or less diverse a country is in comparison with an expected value based on its area or population alone. The method used is a modified version of that used by Groombridge and Jenkins (2002).

The expected diversity was calculated using the standard formula for the species–area relationship $log\ S = c + z\ log\ A$ where S = number of species, A = area, and c and z are constants derived from observation. Examples for bird/mammal species and languages are in Figures 3.1 and 3.2, respectively.

To calculate the deviation of each country from its expected value, the expected log N_i value is subtracted from the observed log N_i value. The index is calibrated such that the world, or maximum, value is set equal to 1.0, the minimum value is set equal to zero and the average or typical value is 0.5 (meaning no more or less diverse than expected given a country's area or population).

By combining the results of BCD-RICH, BCD-AREA, and BCD-POP, Loh and Harmon identified three 'core areas' of global biocultural diversity that include countries of various sizes and populations:

- The Amazon Basin, consisting of Brazil, Colombia and Peru, which ranked highly in BCD-RICH; Ecuador, which ranked highly in BCD-AREA; and French Guiana, Suriname and Guyana, which ranked highly in BCD-POP.
- Central Africa, consisting of Nigeria, Cameroon and the Democratic Republic of Congo (BCD-RICH), Tanzania (BCD-AREA), and Gabon and Congo (BCD-POP).
- Indomalaysia/Melanesia, consisting of Papua New Guinea and Indonesia (BCD-RICH), Malaysia and Brunei (BCD-AREA), and Solomon Islands (BCD-POP).

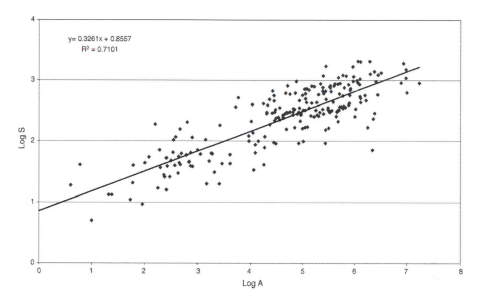

Source: Loh and Harmon, 2005

Figure 3.1 *Species-area relationship for countries of the world (bird and mammal species only)*

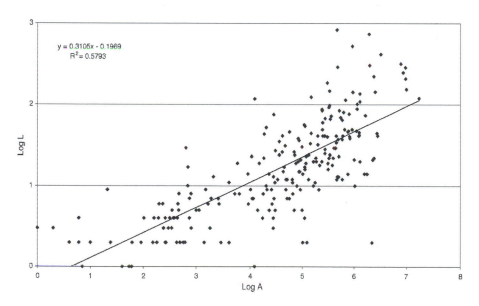

Source: Loh and Harmon, 2005

Figure 3.2 *Language-area relationship for countries of the world*

The world's four most bioculturally diverse countries – Papua New Guinea, Indonesia, Cameroon and Colombia – rank in the top ten for all three components of the index.

The three 'core areas' identified above are in a sense analogous to the results of several schemes that recently have been developed for identifying the world's most important areas for biodiversity conservation and ecoregion protection (Davis et al, 1994; Stattersfield et al, 1998; Myers et al, 2000; Olson et al, 2001).

Loh and Harmon point to a number of limitations of the IBCD and caveats concerning its use, making it clear that this index, like any index, should be used only to measure general conditions and trends and should not be expected to provide an in-depth analysis of within-country variation in biocultural diversity. They also point out that, in its current version, the IBCD only portrays the current state of biocultural diversity, whereas data on trends are as yet missing and are the object of future research (cf. the ILD, outlined above). They conclude that the IBCD data, used in conjunction with careful qualitative analyses, will ultimately provide a more adequate and accurate picture of the global state of biocultural diversity.

A Vitality Index of Traditional Ecological Knowledge (VITEK)   In the last few years, stimulated mainly by the CBD's mandate to develop a series of global indicators for assessing progress towards the reduction of biodiversity loss, a number of proposals have been put forth which seek to identify and define cultural indicators of traditional knowledge, practices and innovations that may be relevant for biodiversity conservation. One proposal is the VITEK assessment. The VITEK is quantitative, comparative, aggregative and directly representative of traditional ecological knowledge (TEK) itself, instead of relying on measures of proxy variables. A precise methodological protocol has already been developed (Zent and Maffi, 2007) and field tests are currently underway.

The VITEK protocol consists of an integrated set of methods:

1   Construction of conceptual categories and items that represent a significant portion of a local body of knowledge.
2   Development of domains that cover a significant portion of TEK, including conceptual knowledge of biodiversity in its different forms (e.g. plants, animals, plant-animal relationships, biological communities, soils, climate, ethnogeography), as well as practical skills associated with the use and management of natural resources. Local domains are established based on local consultation.
3   A list of local domains and categories for each one are then weighted based on local participants' assessment of their cultural importance.
4   Administration of a standardized test designed to rate the TEK aptitude of individuals in the participating local group, with aspects of TEK divided into cosmopolitan (in other words, recognizable across cultures) and localized domains. The aptitude test consists of two equally weighted sections: a conceptual knowledge component and a practical skills component.

Based on this information, the VITEK itself can be calculated. It consists of three related measures: the intergenerational rate of retention (RG), the cumulative rate of retention (RC), and the annual rate of change (CA). All of these rely on very simple statistical calculations and can be done by hand or with a pocket calculator. The first step is to calculate the means of the score results for all age/gender groups included in the aptitude test. The RG indicates the rate of retention between any successive pair of age groups and is calculated as the ratio of the generation mean to that of the generation immediately preceding it. The cumulative rate of retention (RC) essentially reflects the proportion of the baseline aptitude level retained by each succeeding age group. The formula used for RC is adapted from that used to calculate the Living Planet Index (Loh et al, 2005) since they have similar purposes (measuring retention over time based on sample data). The annual rate of change (CA) expresses the average rate and direction of change per year reflected by the target age group. The VITEK measures (RG, RC, CA) can be aggregated according to different scales of inclusiveness, from the local community up to the entire globe, depending on data availability. At the same time, these measures can be disaggregated according to individual TEK domains in order to assess which types of knowledge are more/less susceptible to erosion or change.

Summary and Outlook

The current state of knowledge about diversity While a great deal is known about biological and cultural diversity on a disaggregated basis, we are still a long way from knowing their full extent. Technological advances (such as 'DNA barcode readers') may allow biologists to greatly speed up taxonomic classification of the large number of remaining unidentified species, yet even if this is accomplished there is an immense amount to be learned about the ecological relationships between them. Nor are we anywhere close to cataloguing such basic building blocks of culture as the number of discrete languages, ethnic groups, religions, and so forth, let alone pinpointing trends in their numbers. And again, even if we had such information, it alone would not tell us what we need to know about the dynamics that shape the formation and loss of cultural diversity. Integrated knowledge about global diversity, such as is being pursued by the biocultural diversity approach, is only just now getting underway.

The outlook for diversity The state of the world's biocultural diversity is in crisis. All indications are that the 21st century will witness a large number of species extinctions, the continuing destruction of natural habitat upon which species depend, the degradation or wholesale conversion of ecosystems, further declines in genetic diversity, and the erosion of global cultural diversity. UNEP's Global Environment Outlook offers four scenarios to assess different possible outcomes to 2015 and beyond; under all four, and in every region of the world, the abundance of species is projected to decline significantly (UNEP, 2007). An array of mitigating options have been proposed in global-level initiatives such as the CBD and the MDGs, but no one is expecting losses to be stemmed entirely. That all of this is unfolding in a context of climate change, the ramifications of which we are only beginning to understand, makes the outlook even more precarious.

UNESCO puts the situation with respect to cultural diversity this way:

> *Cultural diversity is in danger. As underlined by the Universal Declaration on Cultural Diversity, adopted by the UNESCO Member States, this threat cannot be reduced to its most widespread and most visible manifestation – namely, the tendency towards the homogenization of cultures, previously believed to be the result of development or 'progress' and now commonly attributed to 'globalization'. The erosion of cultural diversity may in fact assume a variety of forms. Everywhere languages fall into disuse, traditions are forgotten and vulnerable cultures are marginalized or even disappear.* (UNESCO, 2005b)

Languages as a barometer of cultural diversity have received much of the attention so far, but we can expect that as concern for the global state of diversity grows, we will see more sophisticated measures being brought to bear. But under any reasonable scenario we now can foresee, the outlook is the same: a broad decline in cultural diversity. As with biological diversity, initiatives to mitigate this loss are underway, and some, such as the growing capacity of indigenous peoples to take positive steps to resist cultural disruptions, are encouraging (Maffi and Woodley, 2010). But again, no one is expecting losses to be halted anytime soon.

The future of diversity measurement   Most of the world's remaining biological diversity occurs in lands currently occupied by indigenous peoples, and much of the world's cultural diversity resides in indigenous cultures and languages. As described above, indigenous peoples are taking a larger role in international initiatives to inventory and monitor biological and cultural diversity. The methods and models discussed here will be further refined as interest in an integrated approach to diversity conservation continues to grow. As for the future of diversity measurement, however, real advances still remain to be seen in putting the selection and management of indicators in the hands of the people who have the most direct stake in seeing that biological and cultural diversity persist into the future.

References

Bossert, W., D'Ambrosio, C. and La Ferrara, E. (2005) 'A generalized index of ethno-linguistic fractionalization', www.iae.csic.es/konstanz/BossertDAmbrosioLaFerrara.pdf

Brown, J., Mitchell, N. and Beresford, M. (eds) (2005) *The Protected Landscape Approach: Linking Nature, Culture and Community*, IUCN, Gland and Cambridge

Davis, S. D., Heywood, V. H. and Hamilton, A. C. (1994) *Centres of Plant Diversity: A Guide and Strategy for their Conservation*, IUCN, Gland

Florey, M. 2006. 'Assessing linguistic vitality in Central Maluku', Presentation to the 10-ICAL, 18 January

Greenberg, J. H. (1956) 'The measurement of linguistic diversity', *Language*, vol 32, pp109–115, reprinted in A. S. Dil (ed) (1971) *Language, Culture, and Communication: Essays by Joseph H. Greenberg*, Stanford University Press, California, pp68–77

Groombridge, B. and Jenkins, M. D. (2002) *World Atlas of Biodiversity*, Prepared by the UNEP World Conservation Monitoring Centre, University of California Press, Berkeley and Los Angeles

Harmon, D. (1996) 'Losing species, losing languages: Connections between linguistic and biological diversity', *Southwest Journal of Linguistics*, vol 15, no 1/2, pp89–108

Harmon, D. (2002) *In Light of Our Differences: How Diversity in Nature and Culture Makes Us Human*, Smithsonian Institution Press, Washington, DC

Harmon, D. (2007). 'A bridge over the chasm: Finding ways to achieve integrated natural and cultural heritage conservation', *International Journal of Heritage Studies*, vol 13, no 4/5, pp380–392

Harmon, D. and Loh, J. (2004) 'A global index of biocultural diversity', Discussion Paper for the International Congress on Ethnobiology, University of Kent

Harmon, D. and Loh, J. (2010) 'The Index of Linguistic Diversity: A new quantitative measure of trends in the status of the world's languages', *Language Documentation and Conservation,* in press

Heckenberger, M. J., Russell, J. C., Toney, J. R. and Schmidt, M. J. (2007) 'The legacy of cultural landscapes in the Brazilian Amazon: Implications for biodiversity', *Philosophical Transactions of the Royal Society B*, vol 362, pp197–208

Holden, C. J. and Shennan, S. (2005) 'Introduction to Part I: How tree-like is cultural evolution?' in R. Mace, C. J. Holden and S. Shennan (eds) *The Evolution of Cultural Diversity: A Phylogenetic Approach*, UCL Press, London

IUCN (1997) *Indigenous Peoples and Sustainability: Cases and Actions*, IUCN Inter-Commission Task Force on Indigenous Peoples, International Books, Utrecht, The Netherlands

Loh, J., Green, R. E., Ricketts, T., Lamoreaux, J., Jenkins, M., Kapos, V. and Randers, J. (2005) 'The Living Planet Index: Using species population time series to track trends in biodiversity', *Philosophical Transactions of the Royal Society B*, vol 360, pp289–295

Loh, J. and Harmon, D. (2005) 'A global index of biocultural diversity', *Ecological Indicators*, vol 5, pp231–241

Mace, G. M. and Baillie, J. E. M. (2007) 'The 2010 biodiversity indicators: Challenges for science and policy', *Conservation Biology*, vol 21, no 6, pp1406–1415

Mace, R. and Pagel, M. (1995) 'A latitudinal gradient in the density of human languages in North America', *Proceedings of the Royal Society of London B*, vol 261, pp117–121

Maffi, L. (2005a) 'Interim report to the World Bank on socioeconomic, cultural, governance, and human health indicators of the health of the Mesoamerican Reef Region', Unpublished report

Maffi, L. (2005b) 'Linguistic, cultural, and biological diversity', *Annual Review of Anthropology*, vol 29, pp599–617

Maffi, L. and Woodley, E. (2010) *Biocultural Diversity Conservation: A Global Sourcebook*, Earthscan, London

McConvell, P. and Thieberger, N. (2001) *State of Indigenous Languages in Australia – 2001*, Australia State of the Environment Second Technical Paper Series (Natural and Cultural Heritage), Department of the Environment, Canberra

Moore, J. L., Manne, L., Brooks, T., Burgess, N. D., Davies, R., Rahbek, C., Williams, P. and Balmford, A. (2002) 'The distribution of cultural and biological diversity in Africa', *Proceedings of the Royal Society of London B*, vol 269, pp1645–1653

Myers, N., Mittermeier, R., Mittermeier, C. G., de Fonseca, G. A. B. and Kent, J. (2000) 'Biodiversity hotspots for conservation priorities', *Nature*, vol 403, pp853–858

Norris, M. J. (1998) 'Canada's aboriginal languages', *Canadian Social Trends* (Winter), pp8–16

Olson, D. M., Dinerstein, E., Wikramanayake, E. D., Burgess, N. D., Powell, G. V. N., Underwood, E. C., D'amico, J. A., Itoua, I., Strand, H. E., Morrison, J. C., Loucks, C. J., Allnutt, T. F., Ricketts, T. H., Kura, Y., Lamoreux, J. F., Wettengel, W. W., Hedao, P. and Kassem, K. R. (2001) 'Terrestrial ecoregions of the world: A new map of life on Earth', *BioScience*, vol 51, no 11, pp933–938

Oviedo, G., Maffi, L. and Larsen, P. B. (2000) *Indigenous and Traditional Peoples of the World and Ecoregion Conservation: An Integrated Approach to Conserving the World's Biological and Cultural Diversity*, WWF-International and Terralingua, Gland

Pagel, M. and Mace, R. (2004) 'The cultural wealth of nations', *Nature*, vol 428, pp275–278

Rapport, D. J. and Singh, A. (2006) 'An EcoHealth-based framework for State of Environment Reporting', *Ecological Indicators*, vol 6, no 2, pp409–428

Skutnabb-Kangas, T., Maffi, L. and Harmon, D. (2003) *Sharing a World of Difference: The Earth's Linguistic, Cultural, and Biological Diversity*, UNESCO, Paris

State Geological Committee of the Soviet Union (1964) *Atlas Narodov Mira*, Miklukho-Maklai Ethnological Institute at the Department of Geodesy and Cartography of the State Geological Committee of the Soviet Union, Moscow

Stattersfield, A. J., Crosby, M. J., Long, A. J. and Wege, D. C. (1998) *Endemic Bird Areas of the World: Priorities for Biodiversity Conservation*, Birdlife Conservation Series no 7, BirdLife International, Oxford

Stepp, J. R., Cervone, S., Cataneda, H., Lasseter, A., Stocks, G. and Gichon, Y. (2004) 'Development of a GIS for global biocultural diversity', *Policy Matters*, vol 13, pp267–270

Sutherland, W. J. (2003) 'Parallel extinction risk and global distribution of languages and species', *Nature*, vol 423, pp276–279

Taylor, C. L. and Hudson, M. C. (1972) *World Handbook of Political and Social Indicators*, ICPSR, Ann Arbor

UNEP (2007) *Global Environment Outlook: Environment for Development (GEO-4)*, UNEP, Kenya

UNEP/CBD/WG8J/5/1 (2007) 'Indicators for assessing progress towards the 2010 biodiversity target: Status of traditional knowledge, innovations and practices', 5th meeting, Report of the International Experts Seminar on Indicators Relevant for Indigenous Peoples, the Convention on Biological Diversity and the Millennium Development Goals, 5–9 March, Banaue, Philippines

UNESCO (2005a) *Convention on the Protection and Promotion of the Diversity of Cultural Expressions*, http://unesdoc.unesco.org/images/0014/001429/142919e.pdf

UNESCO (2005b) *Measuring Linguistic Diversity on the Internet*, http://unesdoc.unesco.org/images/0014/001421/142186e.pdf

Zent, S. and Maffi, L. (2007) *Final Report on Indicator no 2: Vitality Index for Traditional Ecological Knowledge (VITEK)*, Terralingua Working Document for the 'Global Indicators of the Status and Trends of Linguistic Diversity and Traditional Knowledge' Project

Part II
Landscape and Diversity

4
No Land Apart: Nature, Culture, Landscape

W. M. Adams

The sounds of England, the tinkle of hammer on anvil in the country smithy, the corncrake on a dewey morning, the sound of the scythe against the whetstone, and the sight of a plough team coming over the brow of a hill, the sight that has been in England since England was a land, and may be seen in England long after the Empire has perished and every works in England has ceased to function, for centuries the one eternal sight of England (Baldwin, 1924).[1]

Introduction

In 1924, Stanley Baldwin gave a speech about England, in which he tied his sense of the country to the sounds of the English countryside. England, he said, 'comes to me through my various senses – through the ear, through the eye and through certain imperishable scents' (Baldwin, 1924). To the modern reader, Baldwin's soundscape (like his sentiment) is in every way anachronistic. Corncrake, whetstone and plough team were swept away by tides of agricultural intensification and rural economic change, although the corncrake is being reintroduced in the East Anglian Fenland,[2] having hung on in crofters' fields on Scottish islands, where efforts to conserve it were criticized as English eco-colonialism (Mitchell, 1999).

Yet Baldwin's plundering of Victorian rural imagery still speaks to the power of holistic claims about nature to speak to us. As the radical Welsh novelist and literary critic Raymond Williams (1980) pointed out in *Problems of Materialism and Culture*, ideas from nature are profoundly cultural. People write about 'Nature' as if its meaning was given, but in fact people give it all kinds of meaning – nature 'red in tooth and claw; a ruthlessly competitive struggle for existence; an extraordinary interlocking system of mutual advantage; a paradigm of interdependence and cooperation'. His conclusion is that it can mean, it can be, any and all of these things, because of the ways we load meanings onto it. Nature is never single or simple, a place apart from the human. Ideas of nature are always, as Raymond Williams noted, 'the projected ideas of men'. Nowhere is this more true than in the case of the idea of landscape. All landscapes are the product of human cultures, even those thought of as natural or wild.

Landscape and the Wild

The landscape of the English Lake District is, for a small block of hills, remarkably widely-known for its wildness and beauty. One recent July weekend, I walked with family and friends between the hills called St Sunday Crag and Fairfield. We had started from Patterdale in sunshine, with the long lake of Ullswater shining and blue, but the day had turned, and we were walking in mist, intermittently blasted by a cold wind and raindrops like steel shot. It was, by any standards, a wild scene. The rolling ridge was a barren field of loose rock, and we moved through a grey and shifting world of cloud. Suddenly the view to the south opened, revealing two shining silver ribbons of water, Windermere and, further west, Coniston Water. The sky remained ominously black, but patches of sun and dark shadow flowed over the lines of ridge and hill. West of Coniston Water lay the massed bulk of the Old Man, and beyond that, and to the south, beyond Windermere, lay the sea.

The Lake District owes its fame not only to its physical form, but to its presence in literature, particularly to the Victorian Romantics like Ruskin, and especially the poet William Wordsworth (Bate, 1991). Wordsworth's home at Grasmere – where wild daffodils famously caught his poetic imagination – lay just out of sight below the mountain on which we stood. These mountains still offer the climber 'ridge, and gulf, and distant ocean gleaming like a silver shield' (Wordsworth, 1913). They also still offer a frame that dwarfs human aspiration and hubris: a century after Wordsworth, Norman Nicholson wrote of another mountain, Scafell, looking down at the nuclear plant at Windscale and feeling 'the canker itch between his toes' (Nicholson, 2003).

Writers have inscribed a rich web of formal culture about the Lake District in British, and international, consciousness in the last 200 years. Their ideas and written words have come to take physical form in the landscape as the basis for visitor attractions and tourist enterprises. So, for example, the children's author Beatrix Potter captured Lake District rural life in the characters of Squirrel Nutkin or Jemima Puddleduck, and is captured herself in the fabric of the farmhouse in which she once lived at Near Sawrey, and in the 'World of Beatrix Potter Attraction' and the virtual movie spin-off delights of 'Beatrix Potter's Lake District', exhibitions that celebrate her life, spent weaving wry anthropomorphic fictions about the familiar animal inmates of the Edwardian wood, farm and house.

Nor are the creators of the cultural web that has woven the Lake District into the British imagination innocent of responsibility for the way this representation has been done. Wordsworth's seminal guide book to the Lake District, which famously lamented the 'rash assault' of untutored townsmen on the train to Windermere, was the forerunner of many such profitable compendiums for the tourist, lamenting the spoliation brought by discovery by the hordes from which they themselves profited. In real life, Beatrix Potter was a staunch supporter of the National Trust for Places of Historic Interest and Natural Beauty, established in 1894 in the Lake District by a small group of visionaries. It thrives today – indeed, it has more than three million members and extensive landholdings across England, Wales and Northern Ireland.

From early in the 20th century, it was the need for access to open hills and mountains for the urban working class that drove the long campaign for National Parks (Sheail, 1981). The rounded grassy and heathery hills of the Peak District, between the industrial cities of Yorkshire and Lancashire, were the scene of mass trespass onto private shooting estates in the 1930s, led by the Ramblers Association. After World War II, the Peak District was the first British National Park, declared when legislation was finally passed in 1949, with the Lake District National Park and others soon to follow. Walking in the British hills assumed a social importance out of all proportion to the scale of the mountains, epitomized by the walking guides to the Lake District hand-drawn by Alfred Wainwright, and now themselves part of the ever-growing literary feast associated with the British hills (Wainwright, 1955).

The history of popular and literary engagement with the landscape of the English Lake District is long and the cultural significance of its landscape indisputable. Wordsworth alone supplies the *imprimatur* of high culture, let alone his myriad successors. But, more significantly, the slightest curiosity quickly shows that the Lake District is far from a 'natural' scene. The place names on its maps reflect Celtic and Viking origins (Satterthwaite, Patterdale, Grisedale). The Ordnance Survey maps on which these names are recorded themselves have their origin in the 18th century determination to subdue the Scottish Highlands by creating accurate maps using which the British army could march and drag its cannon. Beneath the green fells lie the barely screened spoil tips and holes from centuries of mining for metals and quarrying for slate, while on the fell sides, vast distances are spanned by now-redundant walls, marking the outfield, and bracken clones spread year by year, recording the changing rural economy and agricultural manpower. Britain lost most of its woodland cover more than a thousand years ago, and the living diversity and beauty of the resulting landscape reflects a deep history of human management (Rackham, 1986; Peterken, 1996).

This beauty of the Lake District's landscape resides not, or not only, in its wildness, but in the juxtaposition of wild and human-made features. The natural beauty for which the National Park was declared is mostly not 'natural' at all, if that is taken to mean that it derives from features and processes that are not directly associated with human society. This landscape is long lived in, used, sometimes abused, and yes, a bit wild.

Wild Landscapes

The Romantics saw in the Lake District, as in other mountain regions such as the Alps, a glimpse of the sublime (McFarlane, 2003). The towering peak, the fearful chasm, the awful clouds were sources of wonder and inspiration. Mountains were places to be visited, and to be climbed, or at least viewed from some lofty vantage point. Such places were re-classified in the Western imagination from being places of fear to places of beauty, or at least of fearful beauty. They were wild, untamed, untrammeled by human constraint and aspiration. This notion excited the Victorian and Edwardian alpinist, and the more timorous elite traveller that thrilled at their tales and drank in the mountain views.

In the US, this sense of the value and importance – and beauty – of wild places was epitomized by the work of the Scottish-born naturalist and essayist John Muir. Muir is best known for his writing and campaigning for the preservation of the Yosemite Valley in California (Muir, 1908). This was first described by European Americans in the 1850s, and the first tourists arrived soon afterwards. A primitive hotel was taking guests at Yosemite Falls by 1857 (Runte, 1990), and a state park was created to protect a small part of the remarkable scenery in 1864. Muir first visited Yosemite in 1868, 19 years after his family emigrated from Scotland. In the ensuing years, he lived and travelled extensively in Yosemite and its mountains, campaigning for its protection as a National Park, a status achieved in 1890. Muir maintained pressure for effective protection, helping found the Sierra Club in 1892 and pushing for restrictions on tourism in the 'wilderness', and on sheep ranching, timber and mineral extraction, and hunting. His most famous campaign was against the construction of a reservoir on the Tuolumne River in the Hetch-Hetchy Valley to supply water to urban San Francisco, a battle of rational conservation in terms of resource use against romantic conservation of the wild beauty of Yosemite. To Muir, 'it appears that the Hetch-Hetchy valley, far from being a plain, common, rock-bound meadow, as many who have not seen it suppose, is a grand landscape garden, one of Nature's rarest and most precious mountain mansions'. Despite his passionate rhetoric, Muir lost his battle against rational conservation (Hays, 1959): the dam was approved in 1913.[3]

Muir ignored the long history of indigenous use of the Yosemite Valley (it was cleared in 1852 by the US Army, who executed five Indians believed to have attacked prospectors, and scattered the remainder (Runte, 1990)). To him, Yosemite, despite the sheep and the tourists, was a sublime wilderness. Muir's vision of Yosemite, and more particularly that of the Sierra Club, helped persuade generations of Americans that wilderness existed, was part of their heritage as colonists and settlers, was threatened by industrial and commercial activities, and deserved protection. That vision was sworn into law in the 1964 Wilderness Act: wilderness became real, at least in law, at that moment.

Wilderness

Of all the ideas that have influenced conservation thinking, this concept of wilderness has been consistently one of the most powerful. The sense of nature's wildness, its independence of human understanding, planning and management, its very complexity that makes it seem unknowable and endlessly novel, are important. More important still is the sense that there are particular places where nature is untrammeled, untrod and free: wilderness. In the closing passage of his essay 'Thinking Like a Mountain' Aldo Leopold (presumably deliberately) misquotes Henry Thoreau's famous dictum about wildness, writing 'in wilderness is the salvation of the world' (Leopold, 1949). Decades later, Edward Wilson observed that wilderness 'settles peace on the soul because it needs no help; it is beyond human contrivance' (Wilson, 1992), capturing the strong American conviction about the proper state of nature.

In Europe, the traditional meanings of wilderness date from the time when people feared nature – feared its teeth and claws, and the blind impartiality of mountain, forest and storm that took and killed people as they wrested a bare living from the face of the Earth. This is the wilderness of European folklore, of dark forests and wolves in woodcutters' cottages, of elves and trolls and highwaymen. The wild was the untamed, lying beyond the tended fields and managed woods, where lawless men roamed and danger lay, when anyone benighted or set upon would not find help (Cronon, 1995; Schama, 1995). Wilderness and wildness were not then virtues, but symbols of barrenness, of lack of harvest, of lack of care. Men like Daniel Defoe, travelling rural Britain in the 18th century, or William Cobbett in the early 19th, spoke of 'wastes', land too wild and rough to be brought under the plough. Land beyond the reach of the hand of agricultural improvement was worthless – worse than that, it was an affront to improvement and should be taken in hand. Wars made wilderness: peace made the fruited plain (Whitney, 1994).

Cronon (1995) argues that the popular idea of wilderness invokes the illusion that 'we can somehow wipe clean the slate of our past and return to a tabula rasa that supposedly existed before we began to leave our marks on the world'. The trouble with wilderness, as Cronon sets it out, is that far from being the 'one place on earth that stands apart from humanity, it is quite profoundly a human creation'. Indeed, Birch (1990) sees wilderness preservation as simply 'another stanza in the same old imperialist song of Western civilization'. He argues that the very act of designating 'wilderness' reserves for wild nature constrains the 'otherness' of wild land. Such 'wilderness' exists at the whim of legislators and government policy and it is 'managed', even if that management is hidden from human visitors. In the US in particular, the concept of wilderness puts nature in a protected area: a place where nature is, and we are not. Yet Cronon points out that when we gaze into wilderness expecting to see pristine un-affected nature, free from all human influence, what we see is a reflection of our ideas, our longings, our own attempt to separate 'natural' from 'human' landscapes.

Despite this unreliable conceptual underpinning, Euro-American ideas of wilderness have been widely disseminated internationally, for example, in Africa and India (Neumann, 1998; Pretty, 2002). Yet they are obviously culturally specific and not universal. A conservation ethic based on the standard Western transcendental and Romantic idea of wilderness is likely to be quite meaningless to people of different tradition and ethnicity (Burnett and wa Katg'ethe, 1994). In Kikuyu thought, for example, wilderness never meant an absence of people. The wild was a place thought of as on the frontier of settlement, a place of some danger but where human activity was intensive. The as-yet unsettled forest was to be approached by people as a group and transformed through a social process of settlement. Closure of the Kikuyu frontier by the Kenyan colonial state, to provide land for white settlers and (ironically) to create reserves for wild nature, disrupted Kikuyu ideas of the wild. This was not a communal process but was arbitrarily imposed. Kikuyu people were denied economic opportunities, and were also denied their understanding of what nature was, as wilderness was converted from social space to 'the domain of beasts, a tourist's pleasuring ground' (Burnett and wa Katg'ethe, 1994).

So, while wild-seeming landscapes have huge cultural resonance in modern Western societies, it is unfortunate that conservationists have tended to imagine that these ideas are universal, and are bound to touch somewhere on indigenous ideas about nature. This is not the case. Ideas about the value of wild landscapes, or nature more broadly, are the products of human culture. They vary, as people vary (Croll and Parkin, 1992; Ingold, 2000).

The Ideology of Landscape

This cultural embeddedness and yet selectivity is just as true of other landscapes. The concept of 'landscape' dates from the dawn of the modern era, and the rise of mercantile capitalism and European empire. The word 'landscape' dates in English from the 16th century, and it is a term taken from Dutch landscape painters to refer to the framing and representation of scenery on canvas. This self-conscious framing of the physical form of the Earth and its surface features as a landscape came to have enormous influence, not only on art, but theatre, law, ideas of nation and nature (Olwig, 2002).

Romanticism about rural life emerged in the UK and US as cities expanded and industrialization took root (Bunce, 1994). Outside the hills and mountains of areas such as the Lake District and the Alps, and the hellish city, nature in Europe was interpreted in pastoral mode. The idea grew (particularly among urban intellectuals) that the agrarian rural life represented an ideal, and that pre-industrial cultures and villages stood as examples of natural communities, and yeoman farmers as 'standard bearers of good Christian, democratic values' (Bunce, 1994). This pastoral ideal framed people within natural landscapes, sometimes maintaining, but more often destroying, its harmonies. George Perkins Marsh noted that 'man is everywhere a disturbing agent. Wherever he plants his foot, the harmonies of nature are turned to discords' (Marsh, 1864).

In the European aristocratic estate of the 17th and 18th centuries, a class insulated by wealth from the rigours of subsistence began deliberately to refashion landscapes as objects of wonder and aesthetic appreciation (Cosgrove and Daniels, 1988). In England, the gentleman's country park was cleared of untidy peasant cottages and landscaped to fit an aesthetic of pastoral beauty by new landscape gardeners (better perhaps termed engineers in the light of their willingness to shift earth, woods and homes) such as Capability Brown and Humphrey Repton (Bunce, 1994). Nature was made genteel and the naked capitalism that created the wealth that sustained it was hidden away.

The whole idea of landscape implies a visual framing, a willful seeing of a seamless and endless context within a specific frame. The classic beauty of a Georgian landscape, of aristocratic pile and landscaped park, trails behind it a comet tail of social and economic debris, and at the least a history of unequal social relations between rich and poor (Monbiot, 2009). The landscape may bear the signature of a draconian restructuring of place and ecology with whole villages moved to improve the vista, or of global trading patterns and commodity chains, often involving the pernicious Atlantic trade in slaves and sugar, or other forms of imperial exploitation. The landscape also bears the shadow of a generation of country workers lost in the Great War, and its countless grim predecessors.

In the 20th century, the pastoral idealization of rural life was a profound influence on conservation thinking, for example, in calls for National Parks in the UK. Here, this view of people and land allowed gross assumptions about the 'wilderness' character of land to be avoided. Thus early ecological studies of British vegetation emphasized the influence of human actions on natural processes (Sheail, 1976, 1987). Arthur Tansley understood very well that vegetation was human-influenced, or 'anthropogenic' (Tansley, 1911, 1939). His *Types of British Vegetation*, and the list of potential nature reserves that he prepared for the Society for the Promotion of Nature Reserves in the First World War, did not distinguish between 'wild' and 'made' landscapes. Conservation was about protecting diversity, and 'unnaturalness' meant the loss of that diversity to plough, pollution or housing estate. Ideas of the pastoral allowed space for human agency.

Elsewhere, the influence of pastoral landscape ideal has been less benign. Today, conservation planners think about rural people in the developing world as engaged in self-sufficient and idealized pre-modern 'communities' (e.g. Agrawal and Gibson, 1999). Any deviation from that pattern, or from scientifically-adjudged 'sustainable' management, implies a fall from grace, a threat to harmonious nature.

Indigenous landscapes

Globally, most 'natural' landscapes are not 'natural' at all, in the sense that they lack the imprint of human hands, any more than the British Lake District. Anderson (2010, this volume) describes the way traditional people can and do create rich, diverse, complex and carefully managed landscapes of food. Fire and livestock have been internal to the ecology of tropical savannas for many centuries, and the extent of farms and dense human settlement on the eve of European contact are only now being revealed by archaeological research (Heckenberger et al, 2008).

reciprocity

Berkes (1999) describes how indigenous communities manage tropical environments: forests, rangelands, islands and coastal lagoons. Without formal science, Berkes argues that diverse communities have learned how nature works, learned to live with it and to persuade it to provide what they need through gathering, hunting, agriculture and other forms of environmental management. Berkes describes the development of the knowledge necessary to do this in terms of contemporary literature on adaptive management and resilience (e.g. Berkes and Folke, 1998; Gunderson and Holling, 2002). The management of land emerges from the cultures and cosmologies of the people who use it. Tyrrell (2010, this volume) records the strong cultural values associated with the procurement and consumption of traditional wild foods such as belugas in the Arctic. Robson and Berkes (2010, this volume) note the long existence of indigenous and community conserved areas and the cultural, spiritual and livelihood values associated with them.

The extent of the cultural association between indigenous peoples and land (and sea, see Tyrell, 2010, this volume) has been widely demonstrated (e.g. Croll and Parkin, 1992; Berkes, 1999; Pretty, 2007). Brody (1987) describes the way that hunters do not recognize the hard boundaries established by Western thought between hunter and hunted, human and animal: humans have souls and spirit powers. Humans and animals are equals. It follows that animals must also have a place in the spirit world. It also follows that animals too must depend on the hunt; they must agree to be killed. From this worldview, a set of principles emerges that guides interactions between hunter and prey that might be described as conservation in the same sense as that developed within the frame of Western thought. Animals, land and nature are not understood by hunters in the same way as they are by scientists, farmers or, for that matter, conservationists (Brody, 1981, 2000).

These kinds of profound cultural links between indigenous cultures and landscape are not confined to hunter-gatherers. Africa's drylands are human-fashioned landscapes, maintained by grazing, browsing and fire. Here, indigenous production systems typically depend on products from live animals (milk or blood) rather than the products of slaughter (meat and hides), and typically involve a mix of species. Herds are managed to exploit spatial and temporal variability in production associated with short rainy seasons and drier and wetter years, adapting herd composition and using movement to maximize survival chances. The Turkana in Kenya, for example, have mixed herds that include camels, which browse resources that are available even in the dry season, and cattle, which are more productive in the wet season but have to move out of the plains into the hills in the dry season (Coughenour et al, 1985; Coppock et al, 1986; McCabe, 2004). Such systems offer a relatively low output compared to modern capitalist systems such as ranching, but they are remarkably robust in terms of providing a predictable, if limited, livelihood.

Much of what has often appeared to development planners to be perversity or conservatism on the part of pastoralists is, in fact, highly adaptive (Behnke et al, 1993; Prior, 1993; Scoones, 1994). In dry rangelands with a strong inter-annual variability, the balance of livestock and range resources changes over time, with drought years first reducing the condition of stock and then (through disease, death and destitution-forced sales) reducing stock numbers. Good rain years then allow pastures to recover, allowing a lagged recovery of herd numbers as pastoralists track environmental conditions. To survive in such environments, herd managers depend on institutions for the exchange and recovery of stock through kinship networks, and also extensive knowledge of the environmental conditions and opportunities in the different areas available to them. In other words, they, like other indigenous groups, need extensive knowledge of the landscapes they fashion, manage, understand and inhabit.

De-Culturing Landscapes

Different cultural ideas of landscapes often clash. Historically, the power to dictate how landscapes were understood and used was dominated, from the 16th century, by successive European imperial powers. Emerging Western ideologies of nature were carried to Europe's newly colonized territories. The doctrine of 'improvement' provided a powerful frame within which the vibrant ecologies of the tropics could be contained (Drayton, 2000). The landscapes of the US, Australia and, in due course, Africa, were seen as under-used and largely not settled. They seemed to lack effective human occupation, because they lacked visible improvement. In legal terms they were defined as *terra nullius*, land without value or property status. In Australia, for example, the importance of aboriginal people's use of fire in creating and maintaining open landscapes was not understood until well into the 20th century, and neither this, nor the deep and complex cultural meanings of land, were accepted until recent decades. The Australian outback was considered to be wilderness in the original European sense, available to be settled and civilized by European farmers and their introduced livestock and technologies (Griffiths and Robin, 1997; Dunlap, 1999; Langton, 2003). Aboriginal people were pushed aside, or actively persecuted, and their impact on the land ignored. Ironically, those landscapes that survived the onslaught of European settlers, and their introduced livestock and wild species, were later re-read according to the new romantic meanings of wilderness, and the rising tide of environmentalist thinking in the later 20th century, as wild places of great natural value.

In North America, European colonists acquired crops and techniques that suited the land (in part from those who lived and worked the lands they occupied), and headed West, into what was for them the unknown. As in South America, indigenous populations shrank before the onslaught of European disease and were fiercely suppressed, relocated from valuable lands into reservations. They were also airbrushed from history, for the story that came to be told about the US was one of a frontier carved from the wilderness.

The idea of empty (indeed emptied) landscapes ripe for colonization by bold individuals became an important element in emergent national identities in colonial settler countries (Dunlap, 1999). The land of the US, Australia and South Africa allowed their settler peoples to see themselves as different from those left behind in Europe precisely because of the distinctiveness and wildness of their landscapes of desert, mountain, bush, savanna and veld (Pyne, 1997). The frontier between wild and settled, sown and developed country was important for what it implied about national character, just as for what it offered in economic opportunity. Unfortunately, such thinking ignored indigenous people, and allowed no recognition that 'wild' landscapes were the fruit of their ideas and labour.

In the plains of East Africa, the first colonial governments encountered landscapes whose creators had recently suffered linked disasters of disease, famine and war (Anderson and Johnson, 1988). Elsewhere, slaving and colonial annexation had disrupted economy, society and environment. Such landscapes (like those in North America) perhaps genuinely seemed empty and wild, in the manner of the garden of an abandoned house. Moreover, to an extent, African populations could be seen to fit such a 'wild' landscape, seen in a sense to be 'natural' themselves, with their primitive use of technology and 'savage' customs (Neumann, 1998). The creation of National Parks like Serengeti, Tanzania, involved the application of an entirely inappropriate Anglo-American nature aesthetic (Neumann, 1995). Africa was imagined as some kind of 'wild' Eden, a paradise to be bounded and protected from those who would defile it; the Maasai pastoralists whose stock grazed the Serengeti plains.

People have lived in and influenced the ecology of almost all habitable regions on Earth. Even in places remote from centres of learning and commerce, people have organized active agricultural, pastoral, manufacturing and trading economies. Colonial authorities, the scientists who advise them, and the soldiers and settlers who fought to survive and recreate the familiar order of home by and large failed to recognize this. They were unobservant and uncaring about the ways people created and used nature. Many conservationists today show a similar blindness to the depth of physical reworking of landscape and the breadth of cultural engagement in nature.

Capitalism's Culture

The instinct to preserve pastoral or wild landscapes, and to exploit their resources, represent opposite faces of the same Western cultures. Both have ridden roughshod over the rights, needs and cultures of indigenous people across the globe (Colchester, 2002; Plumwood, 2003). Of the two, development has led the destructive charge, bringing about a wholesale transformation of land, landscape, ecosystem, society and culture (Adams, 2008).

The scale and seriousness of human impacts on nature became the commonplace of environmentalism in the late 20th century. The statistics are relentless (Adams and Jeanrenaud, 2008, see Table 4.1). The unprecedented scale of human modification of geological and ecological processes is so great that it is now proposed that they be marked by a new geological epoch, the Anthropocene (Crutzen, 2002; Steffen et al, 2007).

The capacity of human society to influence biological and geological processes has grown rapidly over the 20th century, showing particularly sharp change during the 'great acceleration' in industrialization and energy use that

Table 4.1 *Human impacts on the biosphere*

- Three-quarters of the habitable surface of the Earth was disturbed by human activity by the end of the 20th century (Hannah et al, 1994).
- People represent 0.5% of animal biomass on Earth, yet on average human appropriation of net terrestrial primary production (HANPP) is estimated to be 32%. Locally and regionally, impacts are much greater (Rojstaczer et al, 2001; Imhoff et al, 2004).
- Of the nitrogen in the human body, 40–60% is comprised of industrially produced ammonia (Fryzuk, 2004).
- Human activities are now the most significant force in evolution.
- Human activities have increased previous 'background' extinction rates by between 100 and 10,000 times (Wilson, 1992).
- Between 5–20% of the c.14 million plant and animal species on Earth are threatened with extinction.
- Between 1970 and 2003, the Living Planet Index (LPI) fell by around 30%. The terrestrial index (695 species) fell by 31%, the marine index (274 species) by 27% and the freshwater index (344 species) by 29% (Hails et al, 2006).
- In 2005, some 60% (15 out of 24) of ecosystem services evaluated by the Millennium Ecosystem Assessment were being degraded or used unsustainably (Millennium Ecosystem Assessment 2005).
- The population of large predatory fish is now less than 10% of pre-industrial levels. Over-harvesting has devastated both ocean and inshore fisheries (Myers and Worm, 2003; Pauley et al, 2003; Roberts, 2007).
- More than 2 million people globally die prematurely every year due to outdoor and indoor air pollution and respiratory disease
- At the start of the 21st century, we are 'in the midst of one of the great extinction spasms of geological history'.

---

followed World War II (Steffen et al, 2007). Graphs of global population, urban population and consumption (for example, of fertilizer, paper or freshwater), the level of international telecommunications, the number of motor vehicles or the magnitude of international tourism all show steep rises from about 1950. In the last five decades, humans have begun to change the Earth at a rate, on a scale and through a combination of human activities that was fundamentally different from anything that had gone before in human history. Science has barely been able to keep up with our influence. We are changing the Earth more rapidly than we are understanding it (Vitousek et al, 1997).

These now-familiar processes have a long history. Grove describes the development of environmental thought outside the industrializing Europe from the 16th and 17th centuries onwards (e.g. Grove, 1995, 1998; Grove et al, 1998). The pillaging of oceanic islands such as Madeira, Mauritius, and St. Helena by passing ships stimulated an awareness of the ecological price of capitalism. By the early 18th century, surgeons and botanists employed by imperial trading companies (the French *Compagnie des Indes* and the Dutch and English East India Companies) developed and disseminated ideas about desiccation, drought and famine, and environmental limits to human action (Grove, 1995).

Much previous Western environmental thinking was captured by George Perkins Marsh in *Man and Nature* (1864). This reflected a lifetime of observation of forests and rivers of Vermont (fisheries decline, industrial and urban pollution, forest clearance and floods, destruction of forest insects) and his travels in Europe and the Middle East (the clearance of forest, extinction of species, overgrazing, over-ambitious agriculture in the Mediterranean and deforestation, landslides, soil erosion silting of channels in the Alps). Marsh questioned established views about the inexhaustibility of natural resources, for example, of Jefferson or Franklin. He observed government programmes to restore nature (e.g. watershed afforestation in the Vosges and Jura in the 1860s), and he 'urged his countrymen to replace their profligate habits with a programme of public responsibility' (Lowenthal, 1958).

In *Man and Nature*, Marsh was emphatic about the destructive potential of human action: 'a certain measure of transformation of terrestrial surface, or suppression of natural, and stimulation of artificially modified productivity becomes necessary. This measure man has unfortunately exceeded'. This destruction of nature and wholesale transformation of landscape is just as much the product of human culture as the fire-dominated savanna of dryland Africa, or the minutely known landscape of an arctic hunter. Modernity, capitalism and globalized markets create uniform and often dysfunctional landscapes, but they are cultural landscapes nonetheless. In spoil tips, polluted lands, fields without wildlife, cement streets and urban wastelands, capitalism's landscapes are standardized and mass-produced. Their cultural references are those of film and advertising, and they are increasingly imagined and interpreted through popular music, poetry, fiction and above all virtually through the internet and other forms of remote imaging and digital reproduction.

Landscapes, Culture, Choice

Such landscapes are not inevitable. Interest in Marsh's work was rekindled in the mid 20th century, feeding into debates about limits to growth, and eventually about sustainable development (Adams, 2008). In the environmental movement, Western society produced a countervailing ideology to that of unchecked economic growth and material consumption. This movement claimed the various protected landscapes and nature reserves built on ideas of pastoral and wild nature, but added concerns for sustainable energy and cityscapes, for urban farms and parks, clean water and freedom from pollution. The various forms of these landscapes of sustainability remain to be revealed.

Landscapes are manufactured, contested, fashioned and refashioned. To quote Rudyard Kipling (out of context of course): 'Our England is a garden, and such gardens are not made/By singing: "oh how beautiful!" and sitting in the shade' (1911). If all landscapes are made, the critical questions are, what culture – whose culture – is allowed to fashion landscape? How can we make sure that change is somewhat for the better and not the worse? How do we fashion landscapes that are richer, more supportive of human cultural diversity and innovation, more reflective of whatever we believe to be the admirable qualities of the human spirit? These are vital questions. Landscape conservation is a critical challenge, and it is not really about preserving the past, but about the fight for the future.

It is clear that cultural engagement with landscape can be created as well as lost. Cameron (2003) believes that it might be possible to engender love of place as a means of achieving conservation objectives. Mulligan (2001, 2003) argues that experience of a deep relationship with one place can open someone up to a deeper affiliation with all places. Landscapes, and the stories embedded within them, provide an opportunity for conservation in the 21st century to build a new constituency in the public mind.

Thoreau expressed the simplicity of engagement with place and nature in his essay, 'Walking', published in June 1862. Thoreau was a prolific writer, probably best known for *Walden* (1854), the chronicle of his sojourn in the Massachusetts woods. 'Walking' was originally delivered as a lecture in 1851, but was extensively revised before it was published just after his death. Famously, it opens, 'I wish to speak a word for Nature, for absolute freedom and wildness, as contrasted with a freedom and culture merely civil – to regard man as an inhabitant, or a part and parcel of Nature, rather than a member of society'. Thoreau held up the value of wild nature ('in Wildness is the preservation of the world') against the achievements of human civility: 'Now a days, almost all man's improvements, so called, as the building of houses, and the cutting down of the forest, and of all large trees simply deform the landscape, and make it more tame and cheap.' Yet Thoreau's wild nature was not found in remote wilderness, but at hand, within a morning's walk. His essay is a description of the delights of walking, 'sauntering through the woods and over the hills and fields absolutely free of all worldly engagements'. His vision was of nature close to home, lived around and tolerated by human society.

The British naturalist writer Richard Mabey suggested in *The Common Ground* (1980) that conservation 'is concerned ultimately with *relationships*, between man and nature, nature and man'. He emphasizes the seamless integration of nature and culture, for example, in *Flora Britannica* (1997), a meticulous celebration of the cultural engagements with plants, or the auto-biographical *Nature Cure* (2005). Sue Clifford points out that everyday places are as vulnerable as the rare. It is when people lose identification with place that ownership changes hands and a spiral of decline begins. Unless a place has meaning for people, it is unlikely to be well cared for. As Clifford points out, community involvement presupposes the existence of a community: conservationists must re-imagine their thinking and practice to find and support communities that value their local place and local nature (Clifford, 2003).

Nature, culture and landscape are intertwined deep within the DNA of environmentalism. Ultimately, all the choices about society and nature in the 20th century lie within their coils. Nobody has expressed this better than Aldo Leopold, in his book *A Sand County Almanac* (1949). Leopold was trained as a forester (at Yale), and worked in the US Forest Service as a wildlife manager. In the 1920s he was Chair of the American Game Policy Committee, and in 1933 wrote the classic textbook, *Game Management*. However, from 1933, Leopold, taking a teaching post at the University of Wisconsin, began to work a disused bottomland farm along the Wisconsin River, while working on a series of essays on conservation. These were published, as *A Sand County Almanac*, just after his death fighting a bush fire on a neighbour's farm. The book was a runaway success, and has remained continuously in print, becoming a defining text of the environmental movement.

In 'The Round River', Leopold writes with a typically holistic vision, about ecology, the management of soil, the land and the tunnel vision with which it is so often managed. The ecologist, he says 'lives alone in a world of wounds'. For Leopold, conservation emerges from the right management of nature, an understanding founded on a strong sense of place, meaning both the human place in the ecosystem and the individual's place in known and valued landscapes. Leopold's environmentalism grows from specific landscapes, of mountain, marsh and farm. It remains to be seen if these views can shape the fight for the future of both nature and cultures.

Notes

1   I am grateful to Mark Avery for drawing my attention to this quote (www.rspb. org.uk/community/blogs/markavery/default.aspx); http://whatenglandmeanstome. co.uk/?page_id=121.
2   www.rspb.org.uk/supporting/campaigns/nenewashes/corncrakes.asp (accessed 25 August 2009).
3   The Hetch Hetchy reservoir now has no economic function and has outlived its design life: there are proposals to remove it and restore the valley.

References

Adams, W. M. (2008) *Green Development: Environment and Sustainability in a Developing World*, Routledge, London (3rd edition)

Adams, W. M. and Jeanrenaud, S. J. (2008) *Transition to Sustainability: Towards a Humane and Diverse World*, IUCN, Gland

Agrawal, A. and Gibson, C. (1999) 'Enchantment and disenchantment: The role of community in natural resource management', *World Development*, vol 27, pp629–49

Anderson, D. M. and Johnson, D. H. (1988) 'Ecology and society in northeast African history', in D. Johnson and D. M. Anderson (eds) *The Ecology of Survival: Case Studies from Northeast African History*, Lester Crook, pp1–26

Baldwin, S. (1924) 'Speech on England (May 1924)', *On England, and Other Addresses* (1938), pp6–7, http://en.wikiquote.org/wiki/Stanley_Baldwin

Bate, J. (1991) *Romantic Ecology: Wordsworth and the Environmental Tradition*, Routledge, London

Behnke, R. H., Scoones, I. and Kerven, C. (1993) *Range Ecology at Disequilibrium: New Models of Natural Variability and Pastoral Adaptation in African Savannas*, Overseas Development Institute, London

Berkes, F. (1999) *Sacred Ecology: Traditional Ecological Knowledge and Resource Management*, Taylor and Francis, London

Berkes, F. and Folke, C. (eds) (1998) *Linking Social and Ecological Systems: Management Practices and Social Mechanisms for Building Resilience*, Cambridge University Press, Cambridge

Birch, T. H. (1990) 'The incarceration of wildness: Wilderness areas as prisons', *Environmental Ethics*, vol 12, no 1, pp3–26

Brody, H. (1981) *Maps and Dreams: Indians and the British Columbia Frontier*, Faber & Faber, London

Brody, H. (1987) *Living Arctic: Hunters of the Canadian North*, Faber & Faber, London, pp71–85

Brody, H. (2000) *The Other Side of Eden: Hunter Gatherers, Farmers and the Shaping of the World*, Faber & Faber, London

Bunce, M. (1994) *The Countryside Ideal: Anglo-American Images of Landscape*, Routledge, London

Burnett, G. W. and wa Katg'ethe, K. (1994) 'Wilderness and the Bantu mind', *Environmental Ethics*, vol 16, pp145–160

Cameron, J. (2003) 'Responding to place in a post-colonial era', in W. M. Adams and M. Mulligan (eds) *Decolonizing Nature: Strategies for Conservation in a Post-Colonial Era*, Earthscan, London, pp172–196

Clifford, S. (2003) 'Attachment of the ordinary: Valuing local distinctiveness', *Ecos: A Review of Conservation*, vol 24, no 1, pp17–20

Colchester, M. (2002) *Salvaging Nature: Indigenous Peoples, Protected Areas and Biodiversity Conservation*, World Rainforest Movement, Montevideo

Coppock, D. L., Ellis, J. E. and Swift, D. M. (1986) 'Livestock feeding ecology and resource utilisation in a nomadic pastoral ecosystem', *Journal of Applied Ecology*, vol 23, pp573–585

Cosgrove, D. and Daniels, S. (1988) *The Iconography of Landscape*, Cambridge University Press, Cambridge

Coughenour, M. B., Ellis, J. E., Swift, D. M., Coppock, D. L., Galvin, K., McCabe, J. T. and Hart, T. C. (1985) 'Energy extraction and use in a nomadic pastoral ecosystem', *Science*, vol 230, pp619–625

Croll, E. and Parkin, D. (1992) 'Anthropology, the environment and development', in E. Croll and D. Parkin (eds) *Bush Base: Forest Farm; Culture, Environment and Development*, Routledge, London, pp3–10

Cronon, W. (1995) 'The trouble with wilderness, or, getting back to the wrong nature', in W. Cronon (ed) *Uncommon Ground: Toward Reinventing Nature*, W. W. Norton and Co., New York, pp69–90

Crutzen, P. J. (2002) 'Geology of mankind: The anthropocene', *Nature*, vol 415, p23

Drayton, R. (2000) *Nature's Government: Science, Imperial Britain and the 'Improvement' of the World*, Yale University Press, New Haven

Dunlap, T. R. (1999) *Nature and the English Diaspora: Environment and History in the United States, Canada, Australia and New Zealand*, Cambridge University Press, Cambridge

Fryzuk, M. D. (2004) 'Ammonia transformed', *Nature*, vol 427, pp498–499

Griffiths, T. and Robin, L. (1997) (eds) *Ecology and Empire: Environmental History of Settler Societies*, Keele University Press, Keele

Grove, R. H. (1995) *Green Imperialism: Colonial Expansion, Tropical Island Edens and the Origins of Environmentalism, 1600–1800*, Cambridge University Press, Cambridge

Grove, R. H. (1998) *Ecology, Climate Change and Empire: The Indian Legacy in Global Environmental History 1400–1940*, Oxford University Press, New Delhi

Grove, R. H., Damodaran, V. and Sangwan, S. (eds) (1998) *Nature and the Orient: The Environmental History of South and South East Asia*, Oxford University Press, Delhi

Gunderson, L. and Holling, C.S. (eds) (2002) *Panarchy: Understanding Transformations in Human and Natural Systems*, Island Press, Washington, DC

Hannah, L., Lohse, D., Hutchinson, C., Carr, J. L. and Lankerani, A. (1994) 'A preliminary inventory of human disturbance of world ecosystems', *Ambio*, vol 23, no 4–5,
pp246–250

Hails, C., Loh, J. and Goldfinger, S. (eds) (2006) *Living Planet Report 2006*, WWF
International, Zoological Society of London and Global Footprint Network, Gland,
Switzerland

Hays, S. P. (1959) *Conservation and the Gospel of Efficiency: The Progressive Conservation Movement 1890–1920*, Harvard University Press, Cambridge, Massachusetts

Heckenberger, M. J., Russell, J. C., Fausto, C., Toney, J. R., Schmidt, M. J., Pereira, E.,
Franchetto, B. and Kuikuro, A. (2008) 'Pre-Columbian urbanism, anthropogenic
landscapes, and the future of the Amazon', *Science*, vol 321, pp1214–1217

Imhoff, M. L., Bounoua, L., Ricketts, T., Loucks, C., Harriss, R. and Lawrence, W. T.
(2004) 'Global patterns in human consumption of net primary production', *Nature*,
vol 429, pp870–873

Ingold, T. (2000) *The Perception of the Environment: Essays on Livelihood, Dwelling
and Skill*, Routledge, London

Kipling, R. (1911) 'The glory of the garden', in C. R. L. Fletcher and R. Kipling (eds)
A History of England, Clarendon Press, Oxford

Langton, M. (2003) 'The "wild", the market and the native: Indigenous people face
new forms of global colonization', in W. M. Adams and M. Mulligan (eds) *Decolonizing Nature: Strategies for Conservation in a Post-Colonial Era*, Earthscan,
London, pp79–107

Leopold, A. (1949) *A Sand County Almanac and Sketches Here And There*, Oxford
University Press, London and New York

Lowenthal, D. (1958) *George Perkins Marsh: Versatile Vermonter*, Columbia University Press, New York

Mabey, R. (1980) *The Common Ground. A Place for Nature in Britain's Future?*,
Hutchinson, London

Mabey, R. (1997) *Flora Britannica: The Definitive New Guide to Britain's Wild
Flowers, Plants and Trees*, Sinclair-Stevenson, London

Mabey, R. (2005) *Nature Cure*, Chatto and Windus, London

Marsh, G. P. (1864) *Man and Nature; or, Physical Geography as Modified by Human
Action*, Scribners, New York, and Sampson Low, London (reprinted Harvard
University Press, 1965)

McCabe, J. T. (2004) *Cattle Bring Us to Our Enemies: Turkana Ecology, Politics, and
Raiding in a Disequilibrium System*, University of Michigan Press, Ann Arbor

McFarlane, R. (2003) *Mountains of the Minds: A History of a Fascination*, Granta
Books, London

Millennium Ecosystem Assessment (2005) *Ecosystems and Human Wellbeing*, Island
Press, Washington, DC

Mitchell, I. (1999) *Isles of the West: A Hebridean Voyage*, Canongate, Edinburgh

Monbiot, G. (2009) 'Why you'll never find execution or eviction on a National Trust
teatowel', *Guardian*, 24 February

Muir, J. (1908) 'The Hetch Hetchy Valley', *Sierra Club Bulletin*, 6 (4).

Mulligan, M. (2001) 'Re-enchanting conservation work: Reflections on the Australian
experience', *Environmental Values*, vol 10, pp19–33

Mulligan, M. (2003) 'Feet to the ground in storied landscapes: disrupting the colonial
legacy with a poetic politics', in W. M. Adams and M. Mulligan (eds) *Decolonizing
Nature: Strategies for Conservation in a Post-Colonial Era*, Earthscan, London,
pp268–289

Myers, R. A. and Worm, B. (2003) 'Rapid worldwide depletion of predatory fish communities', *Nature*, vol 423, pp280–283

Neumann, R. P. (1995) 'Ways of seeing Africa: Colonial recasting of African society and landscape in Serengeti National Park', *Ecumene*, vol 2, pp149–169

Neumann, R. P. (1998) *Imposing Wilderness: Struggles over Livelihood and Nature Preservation in Africa*, University of California Press, Berkeley

Nicholson, N. (2003) 'Windscale', *Selected Poems 1940–1982*, Faber & Faber, London, p43

Olwig, K. R. (2002) *Landscape, Nature and the Body Politic: From Britain's Renaissance to America's New World*, University of Wisconsin Press, Madison

Palumbi, S. R. (2001) 'Humans as the world's greatest evolutionary force', *Science* vol 293, pp1786–90

Pauley, D., Alder, J., Bennett, E., Christensen, V., Tyedmers, P. and Watson, R. (2003) 'The future for fisheries', *Science* vol 302, pp1359–61

Peterken, G. F. (1996) *Natural Woodland: Ecology and Conservation in Northern Temperate Regions*, Cambridge University Press, Cambridge

Plumwood, V. (2003) Decolonizing relationships with nature', in W. M. Adams and M. Mulligan (eds) *Decolonizing Nature: Strategies for Conservation in a Post-Colonial Era*, Earthscan, London, pp51–78

Pretty, J. (2002) *Agri-Culture: Reconnecting People, Land and Nature*, Earthscan, London

Pretty, J. (2007) *The Earth Only Endures: On Reconnecting with Nature and Our Place in it*, Earthscan, London

Prior, J. (1993) *Pastoral Development Planning*, Oxfam, Oxford

Pyne, S. J. (1997) 'Frontiers of fire', in T. Griffiths and L. Robin (eds) *Ecology and Empire: Environmental History of Settler Societies*, Keele University Press, Keele, pp19–34

Rackham, O. (1986) *The History of the Countryside: The Classic History of Britain's Landscape, Flora and Fauna*, Dent, London

Roberts, C. M. (2007) *The Unnatural History of the Sea*, Island Press, Washington

Rojstaczer, S., Sterling, S. M. and Moore, N. J. (2001) 'Human appropriation of the products of photosynthesis', *Bioscience*, vol 294, pp2549–2552

Runte, A. (1990) *Yosemite: The Embattled Wilderness*, University of Nebraska Press, Lincoln

Schama, S. (1995) *Landscape and Memory*, HarperCollins, London

Scoones, I. (1994) *Living with Uncertainty: New Directions in Pastoral Development in Africa*, IT Publications, London

Sheail, J. (1976) *Nature in Trust: The History of Nature Conservation in Great Britain*, Blackie, Glasgow

Sheail, J. (1981) *Rural Conservation in Inter-War Britain*, Oxford University Press, Oxford

Sheail, J. (1987) *Seventy-five Years of Ecology: the British Ecological Society*, Blackwell Scientific, Oxford

Steffen, W. Crutzen, P. J. and McNeill, J. R. (2007) 'The anthropocene: Are humans now overwhelming the great forces of nature?' *Ambio*, vol 36, no 8, pp614–621

Tansley, A. G. (1911) *Types of British Vegetation*, Cambridge University Press, Cambridge

Tansley, A. G. (1939) *The British Islands and their Vegetation*, Cambridge University Press, Cambridge

Thoreau, H. D. (1862) 'Walking', *The Atlantic Monthly*, vol 9, no 56, pp657–674

UNEP (1995) *Global Biodiversity Assessment*, UNEP, Kenya

UNEP (2007) *Global Environment Outlook: Environment for Development (GEO-4)*, UNEP, Kenya

Vitousek, P. M., Mooney, H. A., Lubchenco, J. and Melillo, J. M. (1997) 'Human domination of Earth's ecosystems', *Science*, vol 277, pp494–499

Wainwright, A. (1955) *Pictorial Guides to the Lakeland Fells Eastern Fells: 1*, Reprinted in 2005, Frances Lincoln, London

Whitney, G. G. (1994) *From Coastal Wilderness to Fruited Plain: A History of Environmental Change in Temperate North America from 1500 to the Present*, Cambridge University Press, Cambridge

Williams, R. (1980) 'Ideas of Nature', in *Problems of Materialism and Culture*, Verso, London, pp67–85

Wilson, E. (1992) *The Diversity of Life*, Harvard University Press, Cambridge, Massachusetts

Wordsworth, W. (1913) 'To — on her first ascent to the summit of Helvellyn', *Selected Poems of William Wordsworth*, Oxford University Press, Oxford, pp189–190

5
From Colonial Encounter to Decolonizing Encounters. Culture and Nature Seen from the Andean Cosmovision of Ever: The Nurturance of Life as Whole

Tirso Gonzales and Maria Gonzalez

Introduction

From the perspective of the Andean cosmovision of ever,[1] culture and nature are not separate. At the core of this cosmovision is the nurturing of life as a whole. Such nurturing takes place within the local *pacha* (meaning the living, natural collectivity of all beings) and comprises the *runas* (humans), *sallqa* (nature) and *apus/huacas* (deities). Learning how to nurture and letting oneself be nurtured are primordial principles and practices in the Andes. Nurturance is carried out through the treatment of all entities as equivalent beings, with respect, empathy, reciprocity and joy. All living beings are considered equivalent persons that complement one another through acts of mutual nurture, manifested in rituals and daily dialogue. Through formal and informal dialogue, Andean indigenous[2] peasants have developed sophisticated responses to the variety of beings inhabiting a particular agricultural place or *chacra*,[3] the small plot of land at the centre of everyday practices and rituality. *Chacra* represents not only the place but also the relations sustaining all equivalent persons (such as the seed, the llama, the rain, the rock) in mutual relations of harmony that procure life continually. The *chacra* is harboured within *pacha*, the landscape that Andean indigenous peoples have become intimate with, the landscape that they have come to know in all its expressions over time. Through ritual, the Andean worldview purports to sustain the creation and recreation of diversity in all of its expressions and practices. The contemporary concept of sustainability is intrinsic to this millenary worldview; the Andean cosmovision is devoted to the procurement of balance and harmony among all living beings demonstrated both in daily and ritual practices (IUCN, 1997; Posey, 1999). This unique approach to life was rarely understood by the colonizer mentality and its dominant Euro[4]-American centred view of the world that has dominated for the last 500 years, thus marginalizing and threatening the Andean way of life (Grillo, 1998b).

Indigenous Andean Peasant Communities in Regional and Historical Context

The spaces designated as the Americas (North America, Latin America, South America and Central America) are historically recent creations within a Western colonial imagery. The precolonial foundations of today's Latin American land-cultures can still be found within the indigenous groups that persist, namely in their ever persistent relationship to their appropriated and exploited homelands. The sustainability blueprint for today's Latin America is present, alive and regenerating within the various strongholds of indigenous communities' land-cultures, languages and cosmovisions. This is rarely visible, however, to the population at large, nor to the developmentalist intellectual elites. Consequently, the past and present colonization process in Latin America has had a significant and deleterious impact upon the lives, cultures, lands, territories and nature of this region.

Latin America's total population is some 580 million people, who live today mostly in urban areas, detached and alienated from nature or their land-cultures (IAASTD, 2009a,b), with little distinct sense of place (Gonzales, 2008). According to some estimates, 40 million people can be formally identified as indigenous peoples (Deruyttere, 1997). But the total area controlled by indigenous peoples throughout the Americas has shrunk significantly (Toledo et al, 2001) due to economic globalization, nation-state building and internal colonialism. We have thus seen, particularly from the 1940s onwards, the outgrowth of space-based, non-sustainable monocultures of the mind, land and spirit. This erosive colonial process has challenged sustainability rooted in indigenous places. However, while the non-sustainable monocultures purveyed by dominant cultures continue to expand across South America, Andean *ayllu* (cultural places) continue to be nurtured through the spiritual values of indigenous communities. *Ayllu* is a Quechua and Aymara word that implies all living beings are harboured in a place where the natural collectivity of visible and non-visible living beings (people, llamas, rocks, mountains, rivers and so on) is nurtured by *pachamama* (Earth mother). In this way, *ayllu* is the regional land based order around which indigenous communities base their ethnic organization and, according to the Andean cosmovision, *ayllu* is also the seed of all life (Choque and Mamani, 2001).

For Latin America, and particularly for the Andean-Amazonian region, indigenous places or *ayllu* are considered to be the core primary and secondary

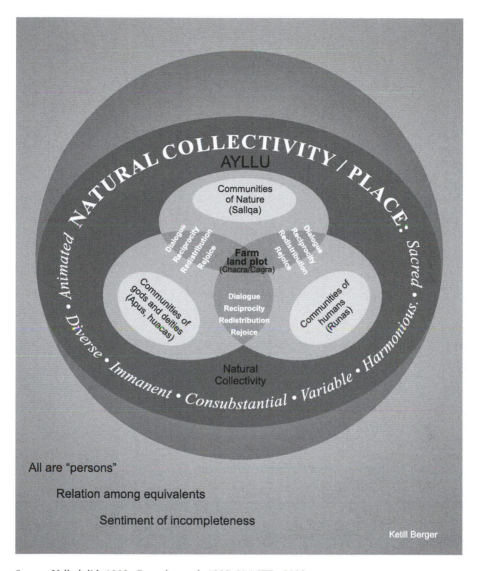

Source: Valladolid, 1992; Gonzales et al, 1999; IAASTD, 2009a

Figure 5.1 *The Andean cosmovision*

centres of the origin of biodiversity (Fowler and Mooney, 1990; Diversity, 1991; Greenpeace, 1999). But these centres have been reduced and marginalized by the colonial, mechanistic space-based worldview embedded within dominant industrial cultures. Today, indigenous peoples' territories in the Latin American region are populated by more than 400 ethnic groups and more than 800 cultures (Toledo, 2007), each with its own distinct language, social organization and cosmovision, as well as distinct forms of economic organization and ways of production adapted to their local ecosystems (Deruyttere, 1997). The inextricable link between indigenous peoples' cultural diversity and biological diversity is demonstrated through the frequent classification of indigenous peoples' lands as 'gene-rich' protected areas. It has been estimated that indigenous peoples live in 80 per cent of ecologically protected areas in Latin America (Oviedo, 1999; Colchester, 2003).

Latin America has been susceptible to bioprospecting since the 'discovery' of the Americas through the Columbian exchange. However, several recent historical events and social movements articulate the shifting relations of power back from the republican nation state to indigenous peoples (Quijano, 2000). For example, the right to land promulgated in state legislation in Peru in 1969 empowers a legal entity, the *Comunidad Campesina* (CC) or 'peasant community' (based on the Agrarian Reform Law D.L. 17716), to make a legal claim to land through the state's juridical apparatus. This continues to take place well after the developmentalist land reform of the 1970s inaugurated this legal instantiation through which this indigenous right could be claimed. The number of indigenous communities or collectivities organized in accord with traditional indigenous knowledge and institutions has grown over recent decades (Robles, 2007; Gutierrez, 2007). In Peru, as of July 2001, there were 5,827 CCs and, of these, 4,224 had their property land titles registered at the Public Registrar and claimed ownership of land, covering a land area of more than 18 million hectares. Today, however, CCs own only 10 per cent of the total agricultural land (Maffi and Woodley, 2010), despite *chacras*, *ayllu* and peasant communities making up 90 per cent of the agricultural and pastoral units in the Peruvian Andean territory.

The Bolivian social movement which grew in momentum, particularly in the 1980s, around *la reconstitución del ayllu* (the reconstitution of the *ayllu*) continues today. It takes the form of an active coalition of traditional Andean indigenous leaders, termed *malkukuna*, that represent the interests of the multiple *ayllus* and the *markas* – the unit of governance representing multiple *ayllus* as one indigenous governmental unit before the republican Bolivian state, *Consejo de Ayllus y Markas del Collasuyu* (CONAMAQ).

Several examples of Andean traditional life persist today, and integrate what industrial epistemologies divide into the discrete concepts of 'economy', 'environment', 'society', and 'politics'. Through this integration, nature and culture are inextricably related through the continuing rituals and daily practices of dialogue, mutual nurture and regeneration (Posey,1999; Grillo, 1991, 1998a,b). This integrative worldview occurs within the majority of Andean indigenous peoples in a ritual cycle of cultivating the land, *pachamama*, and continues to extend into urban sectors by the indigenous ethnic diaspora.

The Dominant Western Mechanistic Worldview and the Agrocentric Andean Worldview of Ever

The dominant Western mechanistic worldview that spread throughout the world from the 1700s to the present day has been widely scrutinized

(Merchant, 1992; Noorgard, 1994). It is the blueprint that Western(ized) societies have used to build their scientific institutions and to construct capitalist development. Such a worldview has made central the dichotomy and premises of the subject-object relationship. It views nature as inert, whether animate or inanimate, as malleable for profit and as an endless source of resources. It relies heavily on reductionist science by detaching the material world from the non-material world. It views the past as primitive, traditional and backward. The concept of sustainability is not part of the thread of such a worldview, as its basic impulse is the relationship of dominance implicit in the subject-object split. Its key characteristics are in stark contrast with those of the Andean worldview of ever.

The language and practice of sustainability has increasingly become part of the development paradigm, however, the latter is still heavily attached to a military industrial complex, seeking to secure survival through the domination of nature and the continuous removal of indigenous peoples from their territories (Radcliffe, 2007; World Bank, 2008). To a significant extent, sustainability has been foreign to the dominant, colonial, Euro-American worldview which has spanned the last 500 years. We suggest that this is propelled by a broken relationship between culture and nature, one that has become toxic and violently dominant and destructive of global ecological systems previously sustained by the rule of equilibrium and harmony, a rule outside of human dominion. The environmental crisis we currently face suggests that the past and current Euro-American worldview, and way of being, urgently needs to be revisited.[5]

One of the primary distinctions between the dominant modernist worldview and the Andean cosmovision of ever is to be found in the difference between languages and their loss of meaning in translation. We put forth that Andean literacy and understanding is necessary in order to truly distinguish the specificity of an expression from and within its language, and especially from a particular language that stems from a unique worldview. We suggest that for the Andean region, throughout history as well as in contemporary times, indigenous cultures have sustained a relationship with nature which is inextricably interwoven with the land itself. This biocultural practice is intrinsic to a particular way of knowing, being and relating to the world which centres on the land.

The region's *ayllu* or *chacra*-based cultural practices could be said to centre on an Andean order, where the Quechua language develops through intimate conversation and practices among *sallqa*, *runa* and *apu* collectivities. This Andean biocultural order is shared amongst Quechua, Aymara and many other indigenous languages. It marks a pathway which traverses the coastal region, the Andean highland region and the Amazonian region. The Quechua order implies that every living being lies along this transversal pathway, cyclically regenerating all life, in response to all beings.

The Quechua/Aymara peasant ('chacarero', peasant of the chacra) perceives their work to be cyclical and in a state of permanent motion. Furthermore, it is perceived as trans-territorial and trans-regional. The Quechua and Aymara person walks with kawsay mama (the living mother, the living seed)

Table 5.1 *Two contemporary worldviews*

Dominant western mechanistic worldview	Andean indigenous place-based worldview
1 Western epistemology, ontology, cosmovision. Western way of knowing, of being and being related to the world	1 Andean epistemology, ontology, cosmovision
2 Grounded in the Judeo-Christian and Cartesian cosmovision	2 Grounded in indigenous, pre-colonial cosmovision
3 Man dissociates from nature (subject-object)	3 No subject-object separation. Human beings are part of life as whole. Mutually reliant
4 Anthropocentric vision of the world: man is the centre of the world	4 World is *pacha*, which is made by *ayllu* – natural collectivity of equivalent beings (*runas*/humans, *sallqa*/nature, *apus*/*huacas*/deities). All are relatives
5 Mechanistic worldview: world viewed as a machine	5 World is viewed as an organism/animal world
6 Life moves around men's material needs	6 Natural collectivity interacts through loving, filial, and respectful dialogue/conversation to procure balance/harmony and nurture life as a whole
7 Egocentric ethic: what is best for the individual is best for society as a whole	7 Agrocentric. Life is nurtured in reciprocal work: *ayni* and *minka*, joyfully
8 Based on Western mechanistic science and capitalism. Lab based	8 *Chacra* (small plot of land: 1-2ha) is the center of rituality and festivity. Mimics nature
9 Earth is dead and inert, malleable from outside, and exploitable for profits	9 *Pacha mama*/Earth mother and its *ayllu* members are alive. It is consubstantial, immanent, harmonious, dynamic, variable. All are 'persons'
10 Innovation protected by Individual Property Rights	10 Communal Rights. Customary Law (Rule of balance and harmony for the regeneration of life as whole: well being of all, *allin kawsay*)
11 Linear vision of history (past-present-future)	11 Circular vision of history: past-present-future are together
12 Knowing is scientific: specialized/fragmented/reductionist	12 Knowing is cosmovisional: in *chacra*-making, holistic
13 Production: homogenizing/standardizing	13 Regeneration: local *pacha*/place-based biodiversity oriented
14 Often non-sustainable	14 Often sustainable

Note: Adapted by T. Gonzales.
Sources: Grillo, 1990, 1998a,b; Merchant, 1992; Valladolid, 1998, 2001

along its multiple paths, through which diversity is cultivated as a spiritual practice of biocultural sustainability (Choque and Mamani, 2001; Valladolid, 2001, 2005). Viewed from the West, this world order is a paradigmatic shift: a cycle of life that is conversant with the vitality of all beings, in an elliptical transversal motion tilted in accord to the movements of all living beings, runa, sallqa, and apus in pacha, in a specific place and time. The Andean landscape is protected by the Apu, or mountain deity, responsible for overseeing a specific pacha, which is in turn harbored within pachamama, mother nature.

The Andean landscape is thus governed by the engagement between the *runa* and the *haqi/jaque* (community), in a practice dedicated to conscientiously procuring harmony among all living beings, for the welfare of all. Whereas the ontology of Western thought would halve the world between being and beings, the Andean world order renders the world as a whole. All this is manifest in the vitality of all beings, one distinct from the other, yet in integrative union where all are alive and are equally important for life to be regenerated. The equal importance of all beings to the cycle of regeneration of life as a whole is expressed through the daily devotion to conversation.

Language is infused with a worldview but, by the same token, language plays a significant part in expressing and shaping our view of the world. Just as there is not one worldview, nor one way of undertaking agriculture, words such as seed, culture, kinship, health, illness, nurturing, nature, biodiversity, resource management, Earth and living world do not find equivalent meanings in every language or cosmovision. The multiple words for seed to be found in the Quechua or Aymara languages, for instance, find meaning in each moment the seed is encountered, in relation to all the communities the seed engages. If transcribed to a written expression, the word used for seed in a specific instance would have to describe this specificity of relations in a particular place and time. Culture, in the Andes, therefore, has a wider definition than commonly used:

> *The particularity of the Andean world is that this capacity to nurture or cultivate (which we would define here as culture) cannot be exclusively attributed to the human community, but is rather the attribute of all the beings that dwell in the chacra and thereby in nature as a whole. This means that we are in a world where everyone is cultivated, nurtured, and every living being has culture. Culture could thus be defined here as the commitment on the part of all living beings to lovingly perpetuating life as a whole* (translation M. Gonzalez) (Rengifo, 1991a).

This worldview is intrinsic to the belief systems of indigenous land-cultures cultivated for millennia and is fundamentally different from the dominant Euro-American worldview. In efforts to protect Andean knowledge systems and their related land-based practices, Andean *saberes*, ways of knowing, have been researched and recorded over the last two decades, and land-cultural practices are slowly being reinvigorated as part of the ongoing learning partnership between *Proyecto Andino de Tecnologias Campesinas* (Project of Andean Peasant Technologies, PRATEC), *Nucelos de Afirmación Cultural Andina* (Nuclei of Andean Cultural Affirmation, NACA), and indigenous Andean peasants. Some examples of the *saberes* recorded by PRATEC have recently been shared with Agroecologia Universidad Cochabamba (AGRUCO)[6] in Bolivia (AGRUCO, 2008; BioAndes, 2008a,b).

Western Scientific Knowledge

The dichotomy of subject–object, culture–nature and man–nature is central in the dominant scientific discourse to obtain knowledge. For the individual to

BOX 5.1 IMPORTANCE OF CONTEMPORARY LATIN AMERICAN AND INDIGENOUS LATIN AMERICAN CULTURAL AND AGRO-BIOLOGICAL DIVERSITY IN THE REGION

Mesoamerica and the central Andes hosts two out of eight centres of origin of agriculture in the world

Population
- Total in 2005: 569 million (77.6% urban; 22.4% rural)
- Total indigenous people*:
  ✓ 48 million; 700 ethnic groups; 800 cultural groups
  ✓ Central Andes (Peru, Bolivia, Ecuador): more than 15 million
- Most peasants are indigenous peoples**

Surface Area
- Total: 2018 million ha
- Total arable land: 576 million ha
  ✓ 28.5% of the region's total arable land
  ✓ 30% of Earth's total arable land

Units of Peasant Production, UPP (end of 1980s)
- Total: 16 million UPP
- Surface under control: 60.5 million ha (34.5% of total cultivated land)
  Peasant population: 75 million people or two-thirds of Latin America's total rural population
- Average size UPP: 1.8ha peasant plot (minifundia)
- Agricultural contribution to general food supply
  ✓ In the 1980s approximately 41% of the agricultural output for domestic consumption
  ✓ Responsible for producing, at the regional level, 51% of maize, 77% of beans, 61% of potatoes
    ➢ **Peru** Minifundia in 1994: 84% from a total of 1.7 million agricultural units (AU). Peasant Communities own only 10% of the agricultural land
    ➢ **Ecuador** Minifundia: 91% from a total of 843,000 AU. Peasant sector occupies more than 50% of the area devoted to food crops (i.e. maize, beans, wheat and okra)
    ➢ **Mexico:** Peasants occupy at least 70% of the area assigned to maize and 60% of the area under beans

Notes:
* The authors use the UN Declaration on the Rights of Indigenous Peoples and ILO Convention 169 definitions on Indigenous Peoples
** Campesino/peasant concept originates in Europe and stresses class over ethnicity. It conflicts with indigenous people(s) concepts. Development and Marxist theories neglect ethnicity/identity/culture. Official national and international statistics (FAO, IDB, World Bank) illustrate this. The concept favoured the old, unproductive 'peasantization-depeasantization' debate.

Adapted by T. Gonzales.
Sources: Diversity, 1991; Deruyttere, 1997; Posey, 1999; IDB, 2004; Hall and Patrinos, 2005; Estrada, 2006; Toledo, 2007; Altieri and Koohafkan, 2008; IAASTD, 2009b; Maffi and Woodley, 2010

Table 5.2 *Comparison of conocimiento/knowledge and saber*

Conocimiento/knowledge	Saber (sapere)
Universal	Local/contextual
Cerebral	Sensuous/nurturing
Impersonal/disembodied	Embodied in the *ayllu* (Andean extended family including natural entities and deities)
Analytic	Holistic
Unique access as the result of the application of a specific method	Multiple ways of access: watching and doing; through dreams; in master–apprentice relationship
Distance	Immediacy
Technical gaze	Involvement
Hierarchical opposition of subject/object	Equivalence
Theoretical	Lived

Source: Ishizawa, 2006

know the object under study, they have to detach from it, order it, and ultimately dominate it (Rengifo, 1991a, b, 1998a, b; Merchant, 1992; Noorgard, 1994). In this process of knowing, the scientist tends to delimit what their senses perceive, thereby apprehending what they want to know. The scientist does not deal with the whole, but with an abstraction of it. Knowledge and truth are pursued in order to manipulate, transform, dominate and perfect reality, whereby the mind and reason impose order upon the chaotic outside world. The modern scientific approach seeks to generate a universal knowledge in that truth is the inviolable and universal outcome. However, the sum of the fragments of reality often fails to make the whole.

In this approach, the scientist is the only one who knows and the object is inert; reality has to be measurable, quantifiable and tangible. Spirituality or metaphysical issues are rarely part of the scientific realm (Merchant, 1992; Rengifo, 1998a). In this cosmovision, Rengifo (1998a) quoting Pannikar (1992, p20) notes 'modern man fears that reality is his enemy. He trusts only in his power, his intelligence, in what he can control'. Rengifo (1998a) notes that 'from these perspectives we can advance very little in our understanding of learning in cultural contexts different from the Western one' (Rengifo, 1998a). Ermine (1995) describes the Western search for truth and knowledge as a journey into 'outer space', where man is separated from nature. The anthropologist Narby (2005) notes what his incursion into the ritualized practices of the Peruvian Amazon revealed to him:

> *Western science has some difficulty with the possibility of both non-human intelligence and the subjective acquisition of objective knowledge... By digging into history, mythology, indigenous knowledge, and science, I had found clues pointing to intelligence in nature. This seemed like a new way of looking at living beings.*

I had grown up in the suburbs and received a materialist and rationalist education – a worldview that denies intention in nature and considers living beings as 'automatons' and 'machines'. But now, there was increasing evidence that this is wrong, and that nature teems with intelligence.

Peasant *Saberes* and Experimentation

In the Andean-Amazonian indigenous world, learning and knowing, as well as their underlying purposes, are fundamentally different than in the industrial world. Such activities are not circumscribed to a research laboratory and its methods and theories (Grillo 1990; Rengifo, 1991a, b, 1998a, b, 2004; Narby, 2004, 2006, 2007; Ishizawa, 2006). With regard to language in Andean culture, and in particular, the relationship between Andean cosmovision and language, Grillo (1991) concurs that 'the Andean cosmovision confers, transmits its way of being to Andean language' (translation M. Gonzalez). In a world that is alive such as the Andean world is, everything is alive, including language: 'We are presented with a language that is alive in a world that is alive. The word, the phrase, is alive' (translation M. Gonzalez) (Grillo, 1991). Language does not belong exclusively to the human community. Under the Andean cosmovision, everything speaks; everyone speaks.

To the Andean peasant, learning is a result of the process of cultivating and letting oneself be cultivated by a world that is alive. You dialogue with and nurture disease (ABA, 2008), water, animals, pestilence, colonization and so on. The Andean peasant's learning is not the result of a separation between subject and object. In the Andean world, *los saberes* (knowing) is a result of the here and now, of living in conversation with and between everyone and everything. Conversing and cultivating are not the exclusive privilege of the human collectivity. Andean indigenous culture is one of nurturance through a flowing and continual conversation among the three collectivities that comprise the local *pacha*: *runas*, *sallqa* and *apus* (see Figure 5.1). These three collectivities cultivate *ayllu* (natural collectivity). Living in *ayllu* is living among relatives, as everyone is a daughter or son of *pachamama*. Kinship goes beyond bloodlines. In the Andean world everything is alive: mountains, clouds, rivers, wind and hail. Conversation takes place among equivalent beings, each with its own culture.

Humans are not outside of or above nature. The *chacra*, cultivated land, is the centre of rituality where all the members of the natural collectivity (the *ayllu*) interact. During the agricultural year, *wata* in Quechua, life-learning or *saberes* emerge in line with the signs and seasons of the agrofestive ritual year as it evolves, and *chacareros* respond to all the groups of living beings who communicate through these signs. The agro-festive ritual year is ordered in accord with how distinct *pacha* in the past have understood and interpreted the signs that tell the *chacarero* when to plant, when to harvest, and so on. *Chacareros* celebrate these milestones, and rituals are the result of a constant tuning into the expansion and contraction of the cosmos in its entirety, phases of the moon and the sun, climate, soil, rain and insects. This Andean knowing is not a rational outcome resulting from the separation of subject-object or culture-nature, nor is there a separation of the subject from emotion. Rengifo points out:

> [Andean indigenous] *Knowing is not a rational act in which one proposes goals, such as is the case for someone who goes to a school to learn a technique. Tuning in to something commits* [the Andean peasant] *in a sensorial, affective, and emotional way. It is her/his senses which are at play when she/he cultivates, when she/ he lives* (translation M.Gonzalez) (Rengifo, 1998b).

Grillo (1991) and Rengifo (1998b) clarify this concept further:

> *When an Andean peasant pronounces a word, the word does not allude to a universal, a symbol, a concept, but rather to a concrete thing which makes itself present when it is pronounced. The word mentions the attributes of that to whom one is referring; it is not an image, nor a representation, but the thing itself. The* [Andean] *word names the particular without there being a hiatus between the word and that which is named* (Grillo, 1991).

> *In* [Andean] *living, an abstract thought does not issue forth to order a represented reality. What is there, simply is. For the Andean peasant a rock that is present in a ritual mesa is not the representation of an Apu,* [Andean deity/mountain protector] *and neither is the illa of a llama the metonymic representation of the llama, in the way that an amulet might be, but rather, it is the llama itself* (translation M. Gonzalez) (Rengifo, 1998b).

Knowing is dependent on what takes place in each *chacra*, where specific *saberes* are given. It is the result of the unique cultivation occurring in that place, the response to local *pacha*'s conversation and the conscientious, and the continual process of procuring harmony among all beings which is the Andean peasant's task. Cultivating does not guarantee a result. Cultivating implies *prueba*, trial and sustained conversation. Trial involves 'the process of "accustoming" the members of one collectivity, within the heart of members of her/his own collectivity, or another collectivity'. Conversation would be 'the mutual relationships which are established among collectivities in order to encourage and accompany one another in the re-creation of life. The trial is an expression of this conversation' (translation M. Gonzalez) (Rengifo, 1991). Thus a trial, in the case of the incorporation of a new seed (that is, the incorporation of this new person/being) may imply a period of three or five agricultural years between the seed becoming accustomed to the family plot and the *chacras*. This is a time to become acquainted through constant dialogue and courtship (Gonzales et al, 1998, 1999; Gonzales, 2000) with all the members who make up the *chacra*: the water, the soil, the *runa*, the sun, the moon, the climate. After this period, the seed will decide whether to stay, having been well received, or whether to leave, having been badly received (Gonzales et al, 1998, 1999; Gonzales, 2000).

Given that Andean indigenous peasant culture is centred in the *chacra*, it is necessary to invigorate the *chacra* from within the Andean vision of the world (Grillo, 1990), which is part of an urgent process of cultural affirmation and the regeneration of the Andes as a whole.[7] As Valladolid (2008) notes, the Andean peasants who work their *chacras* 'do not need to re-indigenize themselves. We, the agricultural technicians who come from these rural areas and have gone off to university – we are the ones who need to re-indigenize ourselves'. Cultural affirmation is 'the nurturing of harmony most adequate to the plenitude of the living world that we are' (Grillo, 1998b, p139), where the person who decides to enter this process does not establish the agenda, nor the focus. Instead, the role of the person and the skills that are required are fundamentally different from those of professional technicians. One of the central objectives is to 'recover the knowledge and wisdom of nurturance still present in the memory of the elders of the community' (Ishizawa, 2006). One tool that has helped to evoke dialogue between peasants is the use of *cartillas* (booklets) of Andean indigenous peasant *saberes*. The accompaniment and conversation that procured the booklets also led to the production of Agrofestive Ritual Calendars, constructed by locality, month, signs, crop, festivity, ritual, positioning of the moon, the sun, the constellations, and the Pleiades (PRATEC, 2006). In this rich diversity of agri-cultural, ritualized and festive practices, we see the sophistication and complexity of the Andean agro-centric life.

The *cartillas de saberes*[8] (booklets recording Andean knowing) became the conduits for a revalorization long overdue in indigenous Andean communities. They were a response to the Andean *ayllu* members' need to dialogue with the Western world. The booklets do not replace the local *saberes*. Instead, their central goal is to 'stimulate reflection, to recreate the practice agreed upon, and to remember concomitant practices' among community members (Gomel, 1998). However, the knowledge contained within *cartillas* is not part of a Western scientific research process; it does not follow the standard scientific protocol or the subject-object detachment associated with modern science.

A number of fields, disciplines and sub-disciplines are now moving out of the fragmented, mechanistic Western worldview towards ecocentric, holocentric worldviews and paradigms (Pimbert, 1994b; Santamaria, 2003; Pretty et al, 2008; IAASTD, 2009a, c). These will begin to break down the barriers imposed as part of the dominant Euro-American centrism, widespread throughout colonial and neocolonial times. The fields that emerge in response to the possibility of sustainability are accepting of difference (Pimbert, 1994a), intrinsically pluricultural, and pluri cosmovisional. More importantly, they preserve a place for all of us, in one world.

Today, more than 90 per cent of the world's total population is non-indigenous. The culture-nature dichotomy for these people has led to a placelessness (Escobar, 2001) requiring reindigenization, remembering and reconnecting, for the health of us all (Gonzales, 2008; Nelson, 2008; Valladolid, 2008). In other

Table 5.3 *Ways Andean and Western society relate to nature*

Andean cosmovision (Comprises living visible and invisible natural collectivity – 3 communities (humans, nature, deities))	Dominant western mechanistic cosmovision
Prueba(s)/trial(s)	*Modern experimentation*
• Process of getting 'accustomed to one another' by members of one community, in the heart of members of their own community, or of a different community • Is an expression of 'dialogue' • Does not imply dominion of a collectivity over another • It is not a technique, part of an essay, nor a tentative examination, nor a rationale to demonstrate something • Process through which a member of a collectivity passes to another one • Have different durations associated with the agricultural practice of the peasant • Holistic: relates all aspects of life (agricultural, ritual, social, natural). The way of being of each participant is respected • *Chacra:* scenario of *pruebas* and an expression of *pruebas* made by the human community in its dialogue with nature. *Chacra* mimics nature and establishes intimate relations among the living beings 'mirroring' what happens in the natural collectivity • Dialogue: all three living communities 'speak' • Mutual relationships established among the collectivities with the goal of encouraging and accompanying each other to re-create life • Dialogue and reciprocity (*ayni, minka*) with nature • Involves 'speaking', contemplating, listening, nurturing, whistling, dancing, gesticulating • There is not an appropriation criteria of nature, but an alliance of reciprocities • Win-win reciprocal game: nature 'wins' → biomass increases. Human community 'wins' → favours life among its members • Nurture and let yourself be nurtured • Feeling of incompleteness is precondition for the manifestation and establishment of an environment of trust and search for completeness • Learning: no subject-object separation • Symbiotic, filial, empathetic conversation	• Mechanism to have access to the 'facts' of nature • World is viewed as a machine • Nature is not viewed as a living world with which we have to live • Nature is viewed as an external object, inert and manipulable for profits • Experimentation has surpassed the lab/greenhouse environment. The world has become the experimental field • Man is above nature • Nature is considered dead or, in the best case scenario, is made up by biotic and abiotic components • Nature is an endless source of resources, and is to serve man's needs • Homogenizing: monocultures of the agricultural fields, the mind, life • Based on dominant Euro-American centric view of life, culture and agriculture • Emphasizes science based approach to agriculture, production and productivity • The agricultural technician is the specialist on agriculture • This approach has overemphasized the non-sustainable conventional productivist system of agricultural production to the detriment of the agroecological and indigenous peasant systems • This system is supported by a well organized and funded transnational network of institutions from the international to the local • Progressive approaches are still generated from outside the community and by external actors. Local ways may be considered, but only partially and out of a frame of cultural affirmation and decolonization

Note: Adapted by T. Gonzales.
Sources: Rengifo, 1991b; Merchant, 1992; IAASTD, 2009c

words, 'becoming native to this place' (your locality, your country) (Jackson, 1996) should be prioritized on the sustainability agenda. What is more, it is vital in order to end the unsettling placelessness which permeates the Western world (Berry, 1996). For dominant Western societies and industrial agriculture, the challenge is to learn from holocentric and ecocentric organic farming, permaculture and agroecology. All these approaches are steps in the direction of suturing the Western rift between human culture and nature. However, opening up to intra- and intercultural dialogue among partners, allowing each to speak on an equal footing, with a view towards the regeneration of the planet as a whole, remains a major challenge.

Notes

1   The term 'the Andean cosmovision of ever' expresses the constancy and permanence of the regenerative cycle sustaining all indigenous Andean life, through language, nurturance, conversation, daily and ritual practice, centred on *pacha*, the Quechua word for place, nurturing regenerative world. This term has been developed through the work of Proyecto Andino de Tecnologías Campesinas (PRATEC) and its associated Nuclei of Cultural Affirmation (NACA) throughout the Peruvian Andes during the last 20 years.

2   This chapter uses the UN Declaration on the Rights of Indigenous Peoples and ILO Convention 169 definitions on Indigenous Peoples. *Campesino*/peasant concept originates in Europe and stresses class over ethnicity. It conflicts with indigenous people(s) concepts. Development and Marxist theories neglect ethnicity/identity/culture. The concept favoured the old, unproductive 'peasantization-depeasantization' debate.

3   'In a broad sense *chacra* is all that is nurtured, thus the peasants say that the llama is their *chacra* that walks and whereof wool is harvested. We ourselves are the *chacra* of the *wakas* or deities that care for, teach, and accompany us' (Rengifo, 2005).

4   Eurocentrism 'is the imaginative and institutional context that informs contemporary scholarship, opinion, and law. As a theory it postulates the superiority of Europeans over non-Europeans. It is built on a set of assumptions and beliefs that educated and usually unprejudiced Europeans and North Americans habitually accept as true, as supported by "the facts", or as "reality". 'A central concept behind Eurocentrism is the idea of diffusionism. Diffusionism is based on two assumptions; (i) most human communities are uninventive, and (ii) a few human communities (or places or cultures) are inventive and are thus the permanent centres of cultural change or "progress"' (Battiste and Youngblood Henderson, 2000). On this concept also see Quijano (2000, 2005) and Lander (2000).

5   Former US Vice-President and 2007 Nobel Peace Prize recipient Al Gore, in the film *An Inconvenient Truth* missed the opportunity of further educating the lay public by not going beyond the symptoms of climate change. He never suggested, much less pointed out directly, that the drivers of climate change may very well be found in the template/worldview/blueprint underlying capitalist development.

6   In Bolivia, Agroecologia Universidad Cochabamba, AGRUCO, (Agroecology University of Cochabamba) is also gathering Andean *saberes*, within an approach different to PRATEC's but with a similar methodological tool. AGRUCO, is a centre of the Universidad Mayor de San Simón, Cochabamba, that generates and disseminates concepts, methodologies, techniques and strategies for sustainable

agro-ecological development throughout Bolivian and Latin American universities; it also implements, development programmes within municipalities and rural-based organizations. AGRUCO promotes agro-ecology and sustainable endogenous development based on local knowledge through revalued participatory research and the wisdom of indigenous peoples, providing training, support and advice to rural communities and municipalities. AGRUCO COMPAS (Comparing and Supporting Endogenous Development) is Latin American coordinator and regional programme coordinator of BioAndes. BioAndes is a regional programme of the Swiss Agency for Development and Cooperation (SDC). It is executed by a consortium of three local institutions, under the lead of AGRUCO. AGRUCO is part of the COMPAS Platform Latinoamerica: AGRUCO-Bolivia, Kume Felen-Chile, Jatun Yachay Wasi-Ecuador, Oxlajuj Ajpop-Guatemala, Surcos Comunitarios-Colombia, CEPROSI-Peru and Los Pasos del Jaguar-El Salvador (BioAndes, 2008a,b, www.agruco.org).

7 The quantitative and qualitative importance of this sector within the Latin American region is recognized through serious reflections stemming from processes of accompaniment and strengthening of the *chacra* in order to affirm the indigenous culture and all of life in the Andes (PRATEC, 1998, 2001; Ishizawa, 2006) as well as by scientific studies from various fields and disciplines – agroecology, ethnoecology, ethnobiology, the biology of conservation (Netting, 1974; Altieri, 1995; IAASTD, 2009c) and by important international conservationist organizations (IUCN, 1997; Oviedo, 1999).

8 PRATEC has gathered 3500 booklets (*cartillas de sabidurías campesinas*) recovered directly from the agriculturalists, women and men of diverse localities in the Peruvian, Ecuadorian, Bolivian and Argentine Andes, throughout many years of work and reflection, firstly, from the students of the *Curso de Formación y Agricultura Campesina Andino-Amazónica*, and secondly from the participants and graduates of the certificate and masters programmes developed by PRATEC. The *cartillas* reflect the nurturing/cultivating vision of the Andean person, as much as their valuing and caring for nature (Rengifo, 2008; PRATEC, 2008; Ishizawa and Rengifo, 2009).

References

ABA (Asociación Bartolomé Aripaylla) (2008) *Los Animos de la Enfermedad, Plantas Medicinales, manos y Sitios Sanadoras*, prepared by Magdalena Machaca, Asociación Bartolome Aripaylla, ABA-Ayacucho, Perú

AGRUCO (Agroecología Universidad Cochabamba) (2008) *Experiencias Colectivas de las Comunidades. Revalorizacion de la Sabiduria de los Pueblos Indigenas Originarios de los Andes*, Universidad Mayor de San Simon, Agencia Suiza para el Desarrollo y la Cooperacion, COMPAS, Bolivia

Altieri, M. (1995) *Agroecology: The Science of Sustainable Agriculture*, Westview Press, Boulder, CO

Altieri, M. A. and Koohafkan, P. (2008) *Enduring Farms: Climate Change, Smallholders and Traditional Farming Communities*, Environment and Development Series 6, Third World Network, Geneva

Battiste, M. and Youngblood Henderson, J. (2000) 'Introduction' in *Protecting Indigenous Knowledge and Heritage: A Global Challenge*, Purich's Aboriginal Issues Series. Purich Publishing Ltd. Saskatoon, Saskatchewan, Canada

Berry, W. (1996) *The Unsettling of America: Culture and Agriculture*, University of California Press, Berkeley

BioAndes (2008a) 'Experiencias colectivas de las comunidades: Revalorizacion de la Sabiduria de los Pueblos indigenas originarios de los Andes', *Biodiversidad y Cultura en los Andes, Bolivia, Peru*, p36, www.bioandes.org

BioAndes (2008b) 'Los Andes: La montana que se ilumina y los caminos bioculturales', *Biodiversidad y Cultura en los Andes*, Bolivia, Peru, pp4–7, www.bioandes.org

Choque, M. E. and Mamani, C. (2001) 'Reconstitución del Ayllu y derechos de los pueblos indigenas: El movimiento indio en los andes de Bolivia', *Journal of Latin American Anthropology*, vol 6, no 1, pp202–224

Colchester, M. (2003) 'Indigenous peoples and protected areas: Rights, principles and practice', *Nomadic Peoples*, vol 7, no 1, pp33–51

Deruyttere, A. (1997) *Indigenous Peoples and Sustainable Development: The Role of the Inter-American Development Bank*, Report IND97-101, Washington, DC

Diversity (1991) 'Latin American centers of diversity', *Diversity*, vol 7, no 1–2

Ermine, W. (1995) 'Aboriginal epistemology', in M. Batiste and J. Barman (eds) *First Nations Education in Canada: The Circle Unfolds*, University of British Columbia Press, Vancouver

Escobar, A. (2001) 'Culture sits in places: Reflections on globalism and subaltern strategies of localization', *Political Geography*, vol 20, pp139–174

Estrada, D. (2006) 'Latin America: Family farms – durable but fragile', 4 October, http://ipsnews.net/news.asp?idnews=34996, accessed 31 March 2009

Fowler, C. and Mooney, P. (1990) *Shattering: Food Politics and the Loss of Genetic Diversity*, The University of Arizona Press, Tucson

Gomel, Z. (1998) 'Contemos escribiendo lo que sabemos', in G. Rengifo (ed) *La Regeneración de Saberes en los Andes*, Proyecto Andino de Tecnologías Campesinas, PRATEC, Perú

Gonzales, T. (1996) 'Political ecology of peasantry, the seed, and NGOs in Latin America: A study of Mexico and Peru, 1940–1996', PhD thesis, University of Wisconsin, Madison

Gonzales, T. (2000) 'The cultures of the seed in the Peruvian Andes', in S. B. Brush (ed) *Genes in the Field: On-Farm Conservation of Crop Diversity*, IDRC/IPGRI/Lewis Publishers, Ottawa

Gonzales, T. (2008) 'Renativization in North and South America', in M. Nelson (ed) *Original Instructions: Indigenous Teachings for a Sustainable Future*, Bear and Company, Vermont

Gonzales, T., Chambi, N. and Machaca, M. (1998) 'Nurturing the seed in the Peruvian Andes', *Seedling*, vol 15, no 2

Gonzales, T., Chambi, N. and Machaca, M. (1999) 'Agricultures and cosmovision in contemporary Andes', in D. Posey (ed) *Cultural and Spiritual Values of Biodiversity*, Intermediate Technology Publications, UNEP, Nairobi

Greenpeace (1999) *Centres of Diversity: Global Heritage of Crop Varieties Threatened by Genetic Pollution*, Greenpeace International, Berlin

Grillo, E. (1990) 'Cultura y Agricultura Andina', in AGRUCO and PRATEC (eds) *Agroecología y Saber Andino*, PRATEC, Lima

Grillo, E. (1991) '*El Lenguaje en las Culturas Andina y Occidental Moderna*', in PRATEC (ed) *Cultura Andina Agrocéntrica*, Proyecto Andino de Tecnologías Campesinas, PRATEC, Lima

Grillo, E. (1998a) 'Cultural affirmation: Digestion of imperialism in the Andes', in F. Apffel-Marglin with PRATEC (eds) *The Spirit of Regeneration: Andean Culture Confronting Western Notions of Development*, Zed Books, New York

Grillo, E. (1998b) 'Development or cultural affirmation in the Andes?', in F. Apffel-Marglin with PRATEC (eds) *The Spirit of Regeneration: Andean Culture Confronting Western Notions of Development,* Zed Books, New York

Gutiérrez Leguía, B. (2007) 'La formalización de la propiedad rural en el Perú: Período 1996–2006, lecciones aprendidas', www.catastro.meh.es/esp/publicaciones/ct/ct60/60_5.pdf

Hall, G. and Patrinos, H. (2005) *Pueblos Indígenas, Pobreza y Desarrollo Humano en América Latina: 1994–2004,* Banco Mundial, Washington

IAASTD (2009a) *Agriculture at a Crossroads: International Assessment of Agricultural Knowledge, Science and Technology for Development,* Synthesis of the Global and Sub-Global IAASTD Reports, Island Press, Washington, DC

IAASTD (2009b) *Agriculture at a Crossroads: Summary for Decision Makers,* International Assessment of Agricultural Science and Technology for Development LAC SDM Latin America and the Caribbean, Island Press, Washington, DC

IAASTD (2009c) *Agriculture at a Crossroads: Volume III,* International Assessment of Agricultural Science and Technology for Development Latin America and the Caribbean, Island Press, Washington, DC

IDB (2004) *Operational Policy on Indigenous Peoples,* Report GN-2296, www.iadb.org/sds/doc/IND-GN2296aE.pdf

Ishizawa, J. (2006) *What Next? From Andean Cultural Affirmation to Andean Affirmation of Cultural Diversity-Learning with the Communities in the Central Andes,* Draft Thematic Paper, What Next Forum, www.dhf.uu.se/whatnext/papers_public/Ishizawa-Draft-01Sep2006.pdf

Ishizawa, J. and Rengifo, G. (2009) 'Biodiversity regeneration and intercultural knowledge transmission in the Peruvian Andes', in P. Bates, M. Chiba, S. Kube and D. Nakashima (eds) (2009) *Learning and Knowing in Indigenous Societies Today,* UNESCO, Paris, http://unesdoc.unesco.org/images/0018/001807/180754e.p

IUCN (1997) *Indigenous Peoples and Sustainability: Cases and Actions,* IUCN Inter-Commission Task Force on Indigenous Peoples, Netherlands

Jackson, W. (1996) *Becoming Native to this Place,* Counterpoint, Washington, DC

Lander, E. (2000) 'Eurocentrism and colonialism in Latin American social thought', *Nepantla: Views from South,* vol 1, no 3, pp519–532

Maffi, L. and Woodley, E. (2010) *Biocultural Diversity Conservation: A Global Sourcebook,* Earthscan, London

Merchant, C. (1992) *Radical Ecology: The Search for a Livable World,* Routledge, New York, and Chapman & Hall, London

Narby, J. (2004) 'Shamans and scientists', in J. Narby and F. Huxley (eds) *Shamans Through Time: 500 Years on the Path to Knowledge.* New York: J.P. Tarcher/Penguin

Narby, J. (2006) *Intelligence in Nature: An Inquiry into Knowledge,* Penguin, New York

Narby, J. (2007) 'An anthropologist explores biomolecular mysticism in the Peruvian Amazon', *Shaman's Drum,* no 74, pp43–46

Nelson, M. (2008) 'Original instructions: Indigenous teachings for a sustainable future', in M. Nelson (ed) *Inner Traditions,* Bear & Company, Vermont

Netting, R. M. (1974) 'Agrarian ecology', *Annual Review of Anthropology,* vol 3, pp21–56

Noorgard, R. (1994) *Development Betrayed: The End of Progress and a Coevolutionary Revisioning of the Future,* Routledge, London and New York

Oviedo, G. (1999) '*La perspectiva del WWF sobre la conservación con los Pueblos Indígenas*', WWF-Fondo Mundial para la Conservación, Documento de Trabajo, WWF International, Switzerland

Pimbert, M. (1994a) 'Editorial', *Etnoecológica*, vol 2, no 3, pp3–5

Pimbert, M. (1994b) 'The need for another research paradigm', *Seedling*, vol 11, pp20-32

Posey, D. A. (1999) 'Introduction: Culture and nature – the inextricable link', in D. Posey (ed) *Cultural and Spiritual Values of Biodiversity*, UNEP's Global Biodiversity Assessment Volume, Cambridge University Press, Cambridge

PRATEC (1998) *La Regeneración de Saberes en los Andes*, Proyecto Andino de Tecnologias Campesinas, PRATEC, Perú

PRATEC (2001) *Comunidad y Biodiversidad. El Ayllu y su organicidad en la crianza de la diversidad en la chacra*, PRATEC, Perú

PRATEC (2006) *Calendario Agrofestivo en Comunidades y Escuela*, PRATEC, Perú

PRATEC (2008) *Proceedings Primer Encuentro Nacional sobre Cambio Climatico, Soberania Alimentaria, y Conservacion In situ de Plantas Nativas y sus Parientes Silvestres*, 14–18 December, Huamanga, Ayacucho, Perú

Pretty, J., Adams, W., Berkes, F., Ferreira de Athayde, S., Dudley, N., Hunn, E., Maffi, L., Milton, K., Rapport, D., Robbins, P., et al (2008) 'How do biodiversity and culture intersect?' Plenary paper for 'Sustaining cultural and biological diversity in a rapidly changing world: Lessons for global policy', 2–5 April, AMNH, IUCN and Terralingua, New York

Quijano, A. (2000) 'Coloniality of power, eurocentrism and Latin America', *Nepantla: Views from the South*, vol 1, no 3, pp533–580

Quijano, A. (2005) 'El "Movimiento indígena" y las cuestiones pendientes en América Latina', Universidad Nacional Mayor de San Marcos, http://sisbib.unmsm.edu.pe/BibVirtualData/publicaciones/san_marcos/n24_2006/a01.pdf, www.revistapolis.cl/polis%20final/10/quija.htm

Radcliffe, S. A. (2007) 'Latin American indigenous geographies of fear: Living in the shadow of racism, lack of development, and anti-terror measures', *Annals of the Association of American Geographers*, vol 97, no 2, pp385–397

Rengifo, G. (1991a) 'El Saber en la cultura andina y en occidente moderno', in PRATEC (ed) *Cultura Andina Agrocéntrica*, Proyecto Andino de Tecnologías Campesinas, PRATEC, Perú

Rengifo, G. (1991b) 'Prueba y diálogo en la cultura andina: Experimentación y extensión en occidente moderno', in PRATEC (ed) *Cultura Andina Agrocéntrica*, Proyecto Andino de Tecnologías Campesinas, PRATEC, Perú

Rengifo, G. (1998a) 'Education in the modern west and in the Andean Culture', in F. Apffel-Marglin with PRATEC (eds) *The Spirit of Regeneration: Andean Culture Confronting Western Notions of Development*, Zed Books, New York

Rengifo, G. (1998b) 'Hacemos Así, Así: Aprendizaje o empatía en los Andes', in PRATEC (ed) *La Regeneración de Saberes en los Andes*, Proyecto Andino de Tecnologías Campesinas en los Andes, PRATEC, Perú

Rengifo, G. (2004) 'Saber local y conservacion in situ de plantas cultivadas y sus parientes silvestres', Serie Kaway Mama, Madre Semilla, Proyecto in Situ, PRATEC, Perú

Rengifo, G. (2005) 'The educational culture of the Andean-Amazonian community', *INTERculture*, vol 148, pp1–36

Rengifo, G. (2008) 'Notas sobre el camino de la revalorización de lo saber-hacer Andino-Amazónico en PRATEC', Presented at Primer Encuentro Nacional sobre Cambio Climático, Soberanía Alimentaria, y Conservación *In situ* de Plantas Nativas y sus Parientes Silvestres, 14–18 December, PRATEC, Huamanga, Ayacucho, Perú

Robles, R. (2007) 'Legislacion peruana sobre comunidades campesinas', http://sisbib.unmsm.edu.pe/BibVirtual/libros/2007/legis_per/contenido.htm

Santamaria, J. (2003) 'Institutional innovation for sustainable agriculture and rural resources management: Changing the rules of the game', PhD thesis, Wageningen University, Netherlands

Toledo, V. M. (2007) 'Indigenous people and biodiversity', in S. Levin (ed) *Encyclopedia of Biodiversity*, Academic Press, Elsevier, Missouri

Toledo, V. M., Alarcón-Chaires, P., Moguel, P., Olivo, M., Cabrera, A., Leyequien, E. and Rodriguez-Aldabe, A. (2001) 'El atlas etnoecológico de México y Centroamérica: Fundamentos, métodos y resultados', *Etnoecológica*, vol 6, no 8, pp7–34

Valladolid, J. (1992) 'Las plantas en la cultura Andina y en occidente moderno', Serie Documentos de Estudios N-23, Marzo, Proyecto Andino de Tecnologías Campesinas, Lima, Peru

Valladolid, J. (1998) 'Andean peasant agriculture: Nurturing a diversity of life in the Chacra', in F. Apffel-Marglin with PRATEC (eds) *The Spirit of Regeneration: Andean Culture Confronting Western Notions of Development*, Zed Books, New York

Valladolid, J. (2001) 'Andean cosmovision and the nurturing of biodiversity in the peasant Chacra', in J. Grim (ed) *Indigenous Traditions and Ecology: The Interbeing of Cosmology and Community*, Harvard University Press, Cambridge, Massachusetts

Valladolid, J. (2005) 'Kawsay Mama: Madre semilla', Serie Kawsay Mama, Proyecto in Situ, PRATEC, Peru

Valladolid, J. (2008) 'Re-indigenization defined', speakers G. Cajete, J. Mohawk and J. Valladolid, in M. K. Nelson (ed) *Original Instructions: Indigenous Teachings for a Sustainable Future*, Bear and Company, Vermont

World Bank (2008) *The Role of Indigenous Peoples in Biodiversity Conservation: The Natural but Often Forgotten Partners*, The World Bank, Washington, DC

6

The Dual Erosion of Biological and Cultural Diversity: Implications for the Health of Ecocultural Systems

David Rapport and Luisa Maffi

Introduction

There is strong evidence of a global decrease in both biological and cultural diversity. The Living Planet Index (LPI) estimates that biological diversity has declined by more than 30 per cent between 1973 and 2006 (WWF, 2006). During this time, cultural diversity has also been declining, as measured in terms of loss of global linguistic diversity. A newly developed Index of Linguistic Diversity (Harmon and Loh, 2010, in press), which reports on trends in linguistic diversity as measured by changes in the number of mother-tongue speakers of a globally representative sample of languages, indicates that global linguistic diversity has declined by 20 per cent since 1970. In both magnitude and trajectory, this trend tracks closely the trend in loss of biodiversity as measured by the LPI over the same period (see also Harmon and Loh, 2004; Loh and Harmon, 2005). Studies have suggested that the decline in biodiversity and cultural diversity are not simply happening in parallel, but rather represent a 'converging extinction crisis' of the diversity of life in both nature and culture (Harmon, 2002).

Further, biological and cultural diversity (including linguistic diversity) should be understood as interrelated and interdependent manifestations of the diversity of life, and considered as a single 'biocultural' whole, emerging from long-standing interactions between humans and the environment (Maffi, 2001, 2005, 2007). At the global scale, notable patterns of spatial overlap of these diversities have been observed, pointing to a global 'geography of biocultural diversity' (Oviedo et al, 2000; Skutnabb-Kangas et al, 2003; Stepp *et al.*, 2004, 2008). It has been proposed that both ecological and socio-cultural factors may account for the occurrence of these patterns, and that the permanence, or loss, of biocultural diversity is affected by similar causes, again of both an ecological and a socio-cultural nature (Maffi, 2001; Harmon, 2002). While such global-scale correlations have been identified, the complex dynamics of human-environment interactions and their changes over time are best detected at regional or local scales. At such scales it is possible to gain a more detailed understanding of the underlying factors inducing change in social-ecological systems or, as we refer to them here, 'ecocultural systems'.

Ecocultural systems are complex dynamic systems of interaction between humans and the environment. In this chapter, we argue that the maintenance of biocultural diversity is essential to the health of ecocultural systems – allowing for co-evolution of humans and the environment without compromising either critical ecosystem processes or the vitality of cultures. We then examine several case studies in which the health of ecocultural systems and the state of biocultural diversity have been compromised by anthropogenic stress, leading to what may be termed an 'ecocultural distress syndrome'. We explore the factors responsible for the loss of ecocultural health and identify some of the mechanisms by which these factors, singly and collectively, have led to the degradation of ecocultural systems. We conclude with identifying potential strategies to prevent further degradation in ecocultural health and loss of biocultural diversity at regional scales, and by implication, at the global level.

Ecocultural Health and Its Relationship to Biocultural Diversity

The concept of health has been applied in turn to individuals, populations, ecological systems and social systems. Health as a property of individuals has, of course, been well understood for millennia. It was a major leap forward when, a little more than half a century ago, the concept of health began to be used as descriptive of the state of populations, both in veterinary medicine and in public health. Expanding the realm of application of this concept also changed the metrics. Individual health is determined by physiological and behavioural characteristics. Population health is measured in terms of the capacity of the group to maintain itself over time (most often reflected in the size of the population, its reproductive potential, the burden of pathogens found within the population, and so forth). More recently, a further expansion in the use of the health concept has seen its application to ecological and social systems. While this has sometimes been regarded as metaphoric, here we argue that health is, in fact, a general systems property, defined as the dynamic potential of complex systems to maintain their full function.

In the early 1940s, the famed American naturalist Aldo Leopold wrote an essay on 'land sickness' (Leopold, 1941), in which he catalogued common signs of degradation in the rural countryside of his adopted state, Wisconsin. These included, among others, the loss of biodiversity, diminished crop yields, a spread of invasive species and non-native species, and an increase in pathogens. In his crowning work, which was not published until half a century after his death, Leopold penned a series of pioneering essays on 'land health' (Leopold, 1999).

Independently, closely related concepts of ecosystem health and ecosystem pathology were developed in the late 1970s and throughout the 1980s. This body of work was stimulated by the growing interest of national and international statistical agencies for measures that could be useful in statistical reports on the state of the environment at regional and national levels (Rapport and Friend, 1979; Bird and Rapport, 1986). Comparative studies of the behaviour of ecosystems under anthropogenic stress led to the identification of common signs of pathology found in heavily impacted systems, collectively termed 'ecosystem distress syndrome' (Rapport et al, 1985). Field studies of ecosystem pathology (Rapport, 1983, 1989a; Hildén and Rapport, 1993) led progressively to related questions of what constitutes ecosystem health (Rapport et al, 1979, 1995, 1998a, b, 2001, 2003, 2009; Rapport, 1989b, 2007a, b; Mageau et al, 1995; Rapport and Whitford, 1999).

The further step toward conceptualization of ecocultural health (and ecocultural pathology) came from the intersection of the fields of ecosystem health and biocultural diversity. Defining ecocultural health as a dynamic, co-evolutionary interaction between humans and the environment, that does not compromise either critical ecosystem processes or the vitality of cultures, reflects the convergence of these two fields (Rapport and Maffi, 2010, in press). The ecocultural health approach brings a holistic perspective to the key question raised by Aldo Leopold (1949): how can we humanly occupy the Earth without rendering its life systems dysfunctional?

Health in an ecocultural context can be understood in terms of the same fundamental properties that were earlier identified as characterizing healthy ecosystems: namely, organization (structure), vitality (function) and resilience (Rapport and Maffi, 2010, in press). A healthy ecocultural system is one that maintains the following:

- Organization: the system's socio-cultural, economic and governance institutions and practices are aligned with the maintenance of its biotic and socio-cultural composition, interactions and integration.
- Vitality: the system is able to maintain its full biological, socio-cultural, economic and governance functions and to reproduce itself, so that its adaptive potential is undiminished.
- Resilience: the system retains its biological, socio-cultural, economic and governance coping mechanisms, so that it has the capacity to withstand and rebound from disturbances such as drought, floods, fires, epidemics, socio-economic or cultural pressures, and conflict.

Ecocultural health can thus be seen as a multifaceted state, encompassing ecological, socio-cultural, economic and governance dimensions. In addition, there is growing recognition that healthy ecocultural systems also foster the maintenance of population health in humans and other biota, as they tend to keep pathogens in check and to stave off other health risks.

From a systems perspective, diversity provides redundancy, thus contributing to a more robust structure and function and to greater resilience (Levin, 1998). In this sense, biocultural diversity is essential for the robustness of ecocultural systems and thus for ecocultural health. At regional and global scales, biocultural diversity is a key indicator of ecocultural health. At local scales, ecocultural health can be assessed in terms of the capacity to maintain the diversity of biota found in a given landscape and of the full complement of cultural values, beliefs, knowledge, practices and languages held by the local culture or cultures. In compromised ecocultural systems, one can expect that significant aspects of both culture and the diversity of biota will be diminished.

Impacts of Anthropogenic Stress on Eco-cultural Health: Some Examples

Over past decades, a growing body of research has shown that ecosystem health has been compromised (Regier and Hartman, 1973; Rapport and Whitford, 1999; Rapport et al, 2003; Bails et al, 2005). At a global scale, indicators include declines in biodiversity, loss of marine fisheries and decreases in critical habitat such as wetlands and forest cover (MEA, 2005; UNEP, 2007). At a regional scale, a number of assessments of the health of large-scale ecosystems have been carried out, such as the Laurentian Great Lakes (US/Canada), the Baltic Sea (Northern Europe), Moreton Bay (Australia), Chesapeake Bay (US), the desert grasslands of New Mexico and Arizona (US), the grasslands of Inner Mongolia (China), among many others (Liu et al, 2003; Rapport and Whitford, 1999; Rapport et al, 2003). The loss of biocultural diversity has been documented both globally and regionally (Maffi, 2001; Harmon and Loh, 2004, 2010, in press; Loh and Harmon, 2005), as well as in many ethnographic and ethnobiological studies that have focused on the breakdown of cultural relationships with local ecosystems and biodiversity. A key aspect of this breakdown is the loss of intergenerational transmission of traditional ecological knowledge (TEK) in indigenous and rural communities – a phenomenon that is beginning to be systematically documented and quantified (Zent, 2008). The documentation of changes in *ecocultural health*, on the other hand, is in its infancy.

By examining three case histories from an ecocultural health perspective, we show how these systems are being negatively transformed by ecological, socio-economic and political forces. The case studies we have chosen range from quintessential 'cultural landscapes' (the *satoyama* rural countryside of Japan) to landscapes characterized by long-standing traditions of human adaptation to environments (the grasslands of Inner Mongolia, China and the Mesopotamian marshlands). In all three cases, the data reveal that there has been a marked deterioration in ecocultural health.

Satoyama: The rural countryside of Japan   The rural landscape of Japan has been utilized intensively for at least two millennia (Seguchi et al, 2008). Forests were harvested for building materials and as a source of fuel from at least 300 BC. By the end of the Nara era (800 AD), secondary forests surrounding human settlements, which comprised mainly oak and pine, were culturally-modified and maintained through intensive harvesting. Over the centuries, and well before the Edo period (1603–1867), what many Japanese people regard as their traditional countryside was established: a human-created and maintained landscape comprising a mosaic of paddy fields, managed woodlands and rural settlements (Yokohari and Kurita, 2003). Such landscapes are found throughout Japan, in valleys and the foothills of mountains. These are known as *satoyama,* which refers to 'mountainous landscapes close to rural villages' or 'agricultural woodlands' (Takeuchi, 2003a). These traditionally represented a model of integration of nature and culture. Of course, these landscapes were not always maintained in a sustainable state. During some periods and in some regions, over-harvesting of woodlands has occurred, resulting in shrubby or denuded landscapes. Over time, *satoyama* has undergone many dynamic changes in response to human demands and utilization.

While *satoyama* landscapes are diverse (dependent on local ecology, topology and climate, as well as human use), they are generally characterized by natural forests in the upper reaches of the mountains or hillsides, managed coppiced woodlands in the hills closer to the villages, and terraced paddy fields. Villages are situated linearly at the base of the hills, and both flooded and dry fields are found in the valleys, as well as grasslands and shrublands in some areas. These landscapes were created and maintained to provide diverse sources of food – from the cultivation of rice and vegetables to foods harvested from the forest. Coppiced woodlands provided important sources of fuel (wood and charcoal), as well as food (e.g. wild berries, mushrooms). They also had a critical role in mitigating the potentially damaging impacts of the torrents of runoff that poured down from steep mountain slopes during the intense monsoon season. Runoff lost erosive power as it passed through the managed woodlands and then through terraced paddy fields, gardens, villages and lower paddies and fields, before flowing to streams and rivers.

Satoyama was a fully integrated system of food production, fuel production, timber supply and environmental protection against the potential flooding and other damage brought about by the effects of a monsoon climate in steep mountain terrain. *Satoyama* also provided a high degree of biodiversity, as flooded rice paddies were habitat for fish, amphibians and an array of wading bird species. Leaves from trees and other organic materials from the forests provided nutrients for rice paddies, while coppiced woodlands provided sustainable supplies of wood for charcoal making and for direct use for heating. Thus, *satoyama* fostered sustainable use of resources, high biodiversity and provided for all of the basic needs of the surrounding rural communities.

Post-World War II, the vitality of the *satoyama* and the cultural practices that sustained it precipitously declined, and this trend continues to the present time. Early on the countryside suffered from large-scale depopulation as farmers were drawn to the cities in search of better pay and less arduous jobs. As those who remained on the land began to age, they became less able to carry out the labour-intensive activities required to maintain paddy fields and woodlands. Further, the children of ageing farmers generally showed less interest in continuing this increasingly socially-stigmatized work. There were also changes in the economic landscape that hastened the demise of the *satoyama*. With the availability of cheaper sources of fuel and power after the war, there was a drop in demand for fuel from the coppiced woodlands. Charcoal and firewood production peaked in 1955 and fell thereafter. Production levels were negligible after 1980 (Tsunekawa, 2003). There was also a decline in the demand for domestic lumber as cheaper sources of imports from southeast Asia became available. Other important economic considerations were the dietary transition from rice to bread and a surplus of rice production (motivated by agricultural protectionist policy to keep the price of rice high), coupled with a rice-acreage-reduction policy, all of which have contributed to the abandonment of paddy fields in *satoyama*. Price competition with Chinese rice is another factor, particularly in preventing significant exports of Japanese rice to China and to a lesser extent competition in domestic markets with Chinese rice. These stresses have been compounded by heavy pressures on *satoyama* lands from urban, industrial and infrastructure development sectors.

These changes have resulted in a considerable loss of vitality and health of these agro-ecosystems. With the abandonment of cultural practices, woodlands became thickets, in which invasive species such as Chinese bamboo now dominate. Where the terrain was suitable, small paddy fields have been amalgamated into large industrial fields and are now drained seasonally for harvest by heavy machinery. This has reduced the habitat available to amphibians and fish, and with the loss of these species came the loss of avian diversity. Heavy use of inorganic fertilizers has resulted in nutrient loading into lakes and streams. Furthermore, as elder *satoyama* farmers age and pass away without having transmitted their knowledge and skills to younger generations, the cultural vitality of the *satoyama* is diminished and a fundamental management resource is being lost.

Today, there is growing recognition nationally that the diminishing vitality of rural areas is causing an ecological and cultural loss to Japan. Moreover, it is posing a threat to the well-being of the rural population and is a likely source of future economic problems. Several initiatives are now underway to assess the conservation and restoration potential of *satoyama*, such as the Sub-Global Assessment of the *Satoyama* (Morimoto et al, 2009). While these efforts are geared to meeting a limited range of specific objectives, such as restoration of local woodlands, re-establishment of local grasslands and shrublands, or restoring specific rice paddies – what is missing is an overall strategy with compatible legal frameworks, governance mechanisms and socio-economic incentives to effectively bring back ecocultural health to the *satoyama*. Furthermore, these efforts are often overtaken by the impacts of rapid industrial and urban growth which continue to erode the land base for *satoyama*. As urban areas expand, fragmentation of *satoyama* landscapes continues. Landscape ecologists in Japan argue that 'lost nature must be regenerated and the city must be restored to the condition of a healthy ecosystem with a link to the preservation of *satoyama* landscapes' (Takeuchi, 2003b). This will require major innovations in economic, social, administrative, governance and legal frameworks and institutions. And even with such changes, restoring ecocultural health of rural areas of Japan is likely to remain an ongoing challenge, given the shift in cultural values that has come about with urbanization and the breakdown of intergenerational transmission of knowledge and practices related to *satoyama* management.

The Mesopotamian Marshlands   The Mesopotamian Marshlands are located in southern Iraq and southwestern Iran, within the Tigris-Euphrates Basin, one of the great cradles of early civilization. The marshlands were once the most extensive in western Eurasia (Takeuchi, 2006) and among the world's largest wetlands. At their peak, the wetlands covered nearly 15,000km^2 (Richardson et al, 2005). Fed by snowmelt from the mountains of Turkey through the Tigris and Euphrates rivers, a complex network of marshes and ephemeral and permanent lakes were formed around the confluence of these rivers.

Ecologically, the marshlands comprised a vast and complex mosaic of habitats that supported high biological diversity. Vegetation of the permanent marshes was dominated by common reed. Reed mace was found in the ephemeral areas, and temporarily inundated mudflats provided habitat for low sedges and bullrush. Deeper permanent lakes supported submerged vegetation such as hornwort, eel grass and pondweed (UNEP, 2001). The marshlands were habitat for millions of birds, and a flyway for millions more travelling between river basins in western Siberia and Africa (Richardson et al, 2005). A variety of mammals inhabited the marshes, including the honey badger, striped hyena, jungle cat, goitered gazelle, Indian crested porcupine, grey wolf, long-fingered bat and smooth-coated otter. Furthermore, the southern portion of the marshlands served as a major spawning ground and nursery for marine life (Richardson *et al.*, 2005; Takeuchi, 2006).

The cultural history of the marshlands similarly stretches back thousands of years. Mounds that can be seen today rising above the marsh waters are thought to be the sites of ancient cities (UNEP, 2001). For more than six millennia, the marshes were home to an indigenous people, the Ma'adan or 'Marsh Arabs', heirs of the Sumerian and Babylonian civilizations. The Ma'adan evolved a unique lifestyle, establishing settlements within the marshes on platforms and fortifying them with reeds and mud, or living on the margins of the marshland. Their dwellings constructed entirely of reeds are depicted in Sumerian clay tablets dating back at least 5000 years. Their guesthouses, with cathedral-like arches made from reeds, appear to be a cultural legacy of ancient times (UNEP, 2001).

The Ma'adan were mainly fishers, hunters and gatherers, and moved about the marshes in dugout canoes. They also maintained water buffalo that fed on young reed shoots. The buffalo were a source of milk, butter and yoghurt as well as a source of energy in the form of dung fuel and manure. Thus the resources that the Ma'adan depended upon for survival derived directly from the marshes: materials for their reed dwellings, fuel for their fires, abundant fresh water, and fish, wildlife and berries for food, in addition to the cultiva-tion of rice and millet. So successful were they that by the late 1970s their population was estimated to be between 350,000 and 500,000 (UNEP, 2001).

Maintenance of this complex ecocultural system depended upon the massive through-flows of water along the Tigris and Euphrates rivers, as well as substantial seasonal fluctuations in the water levels of the marshes. But the conditions for the flourishing of the marshes and their inhabitants came to an abrupt end in the 1970s. The construction of dozens of large dams and diver-sions in Turkey, as well as significant hydrological projects in Syria and Iran, diverted waters from the Tigris and Euphrates rivers away from the marsh-lands. By 2001, there were 32 major dams on the Euphrates and Tigris, with 8 more under construction and at least 13 more planned (UNEP, 2001). Inflow was reduced to less than one third of historic volumes, and the sharp seasonal fluctuations in water inflows were diminished. Collectively, these projects resulted in the desiccation of much of the marshlands.

Comparisons of satellite imagery from 1973 and 2000 indicate that 85 per cent of the nearly 9000km^2 of permanent marsh had vanished over this period (Richardson et al, 2005). Remaining marshes became highly fragmented and isolated. Furthermore, in the absence of the moderating influence of marshland reeds on the regional climate (which served to break the dry hot summer winds that funnel over the marshlands), surrounding areas have become desertified and polluted from toxic chemicals in the soils that are dispersed by hot summer winds.

To this already devastating outcome was added the politically-motivated drainage of the marshes of southern Iraq by the Saddam regime (Glanz, 2005). In early 1993, Saddam Hussein, convinced that the southern marshlands served as a refuge for deserters from the army as well as harbouring the largely Shi'ite Marsh Arabs, ordered the construction of the 'Glory River' that drained the Central Marshes by cutting off inflows of water from the Tigris (Richardson et al, 2005). By 2003, most of the marshland lay barren and the population had fled (Takeuchi, 2006). After partial re-flooding of the marsh in 2004, some 40,000 Marsh Arabs returned to take up life in what was left of the marshes, while others had become 'Field Arabs' on the periphery of the former marshes. Owing to the pollution, lack of potable water, loss of fisheries and wildlife, and a harsher regional climate, living conditions remain extremely precarious.

Immediately after the fall of Saddam, ambitious international efforts were launched to restore the marshlands. One of the key projects, Eden Again, had the support of the World Bank as well as American, Canadian, British, Italian and Iraqi agencies and foundations. Through this consortium, coop-erative agreements were signed with Turkey, Syria, Iran and Iraq to facilitate the partial re-flooding of the marshes. According to the estimate of Azzam Alwash, director of the Eden Again project at the Iraq Foundation, some 20–40 per cent of the original marshes had been partially re-flooded by 2005 (*The Economist*, 2005), and in some areas reed beds, some species of fish, and water buffalo have returned.

However, re-flooding has been haphazard, and restoration results have been mixed. In some areas, marsh functions have been partially restored; in others, there has been no significant rehabilitation (Takeuchi, 2006). Considerable uncertainties remain. For example, it is not known whether present inflows will be sufficient to sustain restoration in drought years. Also unknown is the fate of many threatened species. In partially re-flooded areas, key amphibian, fish, bird and mammal species are either not present or are present only in very low numbers. Moreover, the construction of dams, dykes and diversions in Turkey, Iran and Syria continues to reduce inflows to the marshes and threatens to cause further desiccation to major marshland areas (Richardson et al, 2005). Ecologically, the accumulation of salt in many areas is likely to limit significant re-colonization of the original biota, even after re-flooding. Even in those regions in which reed beds have returned, the reeds appear to be far from healthy, at perhaps only 50 per cent of their former density (Glanz, 2005). Culturally, the prospects are dim. While one of the early goals of Eden Again was the establishment of green villages in which the Ma'adan could resume their traditional ways of life, this goal now seems unrealistic. The polluted and saline waters are not conducive to re-establishing any large-scale settlements. Furthermore, most of the former Marsh Arabs have been forced by circumstances to adopt a different, if more precarious, way of life on land. This diaspora makes it hard to imagine that the unique cultural aspects of this ecocultural system can be re-established any more readily than the ecological aspects can.

The Inner Mongolian grasslands   Some of the world's most extensive grasslands are found in Inner Mongolia Autonomous Region, China. There are five major grasslands within Inner Mongolia. We focus primarily on one of these, the Horqin Sandy Lands in north-eastern Inner Mongolia. This grassland has been periodically occupied by Mongolian nomadic herders for many thousands of years. They developed their nomadic lifestyle to adapt to the ecology of the grasslands, moving their herds (sheep, goats, horses and camels) as seasonal grasses became available, and thus maintaining low grazing pressure.

With the advent of agriculture in the region some 5000 years ago, the harmony between nomadic culture and grassland ecology began to break down. Initially, the grasslands were shared between farmers/pastoralists and nomads. However, over time the grassland ecology was destabilized, with shifting cultivation becoming predominant (whereby an area of grassland was ploughed and farmed for a brief period), along with over-harvesting of woodlands for firewood and overgrazing by domestic animals. Presently, the Horqin Sandy Lands are into a third cycle of desertification. In addition to the pressure on the landscape from farming, nomadic life itself is coming under increasing restriction by government regulations limiting the nomads' traditional movement and thus adding to grazing pressure in some areas. As of 2006, for Inner Mongolia as a whole, an estimated 90 per cent of grasslands were degraded to some degree – more than twice the amount of a decade ago – while productivity of disturbed grasslands is only 50 per cent of the productivity of the undisturbed steppe (Jiang et al, 2006). The present cycle can be traced back to early in the Qing Dynasty (1644–1911), when traditional light grazing by Mongolian nomadic herders gradually gave way over the centuries to an influx of farmers.

Since the 1950s, more than 2.5 million hectares of grasslands in the Horqin Sandy Lands have been converted to farmland, and farming activities have expanded to the northern frontier areas. In recent times, farms have been abandoned owing to nutrient decline, while seasonal wind storms blow off remaining soils and expose underlying sand, leading to dune mobilization. By the late 1950s, 28 per cent of the land in the Horqin steppe had become desert; by the middle of the 1970s, this had increased to 53 per cent, and by the end of the 1980s it stood at 78 per cent (Liu et al, 2003). Today, blowing and drifting sands threaten remaining farmlands, and roads are often buried by the wind-borne sand particles. This ongoing and intensifying desertification of the Horqin Sandy Lands appears far more extensive than the two previous historical periods of desertification. In other sandy lands, such as the Hunshandak Sandy Land, half of the grassland has become shifting sand dunes, covering an area 20 times that of 50 years ago (Liu et al, 2003).

Over the past five decades, there has also been a rapid increase in population in the Horqin Sandy Lands; from 1 million in 1947 to 3.5 million in 1996. During this same period, stock intensity increased to 4.6 sheep per hectare by 2000, nearly five times the intensity that prevailed in 1949 and well above proposed sustainable levels of 0.9–1.5 per hectare. The livestock population in Inner Mongolia was 9.2 million in 1947, but now surpasses 62 million. At the same time, the amount of total usable pasture declined from 88 to 63 million hectares (Chuluun and Ojima, 2002). In other grasslands of Inner Mongolia, pressures also arise from the breeding of goats for the cashmere wool industry, logging, mining, road building, the creation of instant cities, and industrial development.

Within Inner Mongolia, desertification has resulted in reduced populations of megafauna, including Mongolian gazelle, roe deer, foxes and wolves. Desertification has also threatened the livelihoods of both the farmers/pastoralists and the Mongolian nomadic peoples. As croplands are abandoned and dunes mobilized, people are being forced to exploit marginal lands for growing crops (such as along the margins of river beds), even if the prospects for sustaining food production under these conditions are dim. As the sands advance, towns have been overtaken and abandoned, while the political and environmental conditions for continuing nomadic ways of life have rapidly declined.

Various efforts are being made to halt desertification in the sandy lands, including the growing of crops that tolerate extreme arid sandy conditions, planting of trees as shelter breaks to stop the wind, and restrictions on grazing. Many of the trees have died – and in any case they are proving ineffective in stopping the movement of mobile dunes. Yuanchun Shi, former President of China Agricultural University, suggests another solution, namely that 'overcultivation, grazing and logging must be restrained in the source area of desertification. Grazing should be forbidden in those seriously degenerated and desertified grasslands... The purpose of stopping cultivation and grazing is to activate the ecosystem's self-repair function, so that vegetation cover and soil conditions can be improved on a large scale' (Shi, 2002). In a few exceptional cases, the exclusion of grazing has resulted in grassland regeneration; however, this outcome appears to be the exception rather than the rule (Jiang et al, 2006). Where soil conditions are less favourable, the exclusion of animals has not resulted in significant recovery.

The grasslands of many regions of Inner Mongolia are likely to be well beyond the point of 'self-repair' – unless we are considering time scales from centuries to millennia – even more so with advancing climate change. However, this should not preclude undertaking positive steps to reduce sources of anthropogenic stress (intensive farming, overgrazing and so forth) and limit further damage. Encouraging revegetation with shrubs may prove more effective in holding back the sands than planting trees. Shrubs, however, bring problems of their own, as they are efficient at corralling limited soil nutrients, and thus depleting nutrients in shrub-free areas. This exposes inter-shrub surfaces to wind and rain erosion (Rapport and Whitford, 1999). In summary, the process of transformation of the grasslands of Inner Mongolia to mobile dunes is ongoing, and with this comes the loss of both the grasslands and the unique culture of Mongolian nomadic herders.

The Collapse of Ecocultural Systems and the Ecocultural Distress Syndrome

These case studies are representative of a phenomenon that is occurring in many parts of the world: the collapse of contemporary ecocultural systems, driven by a complex of ecological, socio-economic, cultural and political forces. In each of the examples, different combinations of factors were at work, but ultimately the results were the same: deterioration of the biophysical landscape with loss of biodiversity, accompanied by loss of cultural vitality in the human component of the landscape, including loss of traditional knowledge and practices that maintained the landscape and sustained its inhabitants over generations.

In the *satoyama*, the deterioration of ecocultural health was triggered by changing economic conditions, making urban jobs more attractive and lessening the necessity to maintain traditional paddy fields and woodlands. Subsequently, large-scale out-migration of farmers, combined with an ageing farmer population, led to the near complete loss of a way of life and the cultural landscape upon which it is based. In Mesopotamia, the abandonment of the landscape by its inhabitants followed the erosion of the landscape, caused by economic and political forces over which the Ma'adan had no control. The massive deterioration of the landscape also made return to former ways of life for the Ma'adan impossible or at least highly unlikely, even after the political conditions changed and restoration efforts were made. In the grasslands of Inner Mongolia, degradation of the landscape and loss of co-adapted ways of life were also related to demographic change; but rather than as an exodus, the latter change came as a massive influx of farmers and their animals, combined with economic and political factors (including the curtailment of the freedom of movement of Mongolian nomads). All these factors led to inappropriate land use (farming, overgrazing, and urban and industrial development) and ultimately to the collapse of the ecocultural system. In all three examples, there was a clear interplay between the loss of ecosystem health and the demise of cultural traditions that had facilitated landscape maintenance. Millennia of ecoculturally sustainable relationships between people and their environments have been lost in only a few decades.

These cases bear the signs of what we call an 'ecocultural distress syndrome'. This syndrome is characterized by a loss of organization (indicated by factors such as decreases in biodiversity, erosion of cultural knowledge and practices, and decline of local languages), loss of vitality (manifested in aspects such as diminished ecological regeneration potential and breakdown of intergenerational transmission of cultural knowledge and practices), and loss of resilience, in other words, increased vulnerability to catastrophic failure (signaled by the degradation of the biological robustness of the landscape and by a weakening of the social, cultural and economic coping capacity of the people involved). While healthy ecocultural systems are able to rebound from perturbations such as drought, floods, fires, epidemics, socio-economic or cultural pressures, and conflict, ailing ecocultural systems are likely to be pushed over a tipping point beyond which recovery in ecological time is no longer possible.

Assessing Change in Biocultural Diversity and Ecocultural Health at Regional Scales

Many activities are taking place around the world in efforts to restore damaged ecocultural systems (Rapport et al, 2003; Carlson and Maffi, 2004; Maffi and Woodley, 2010). Yet, the overall trends at global and sub-global levels remain alarmingly negative (MEA, 2005; UNEP, 2007). There are at least three reasons for this: (i) the results of ecological restoration activities have been disappointing on the whole and often fail to fulfill their promise (Roberts et al, 2009). Likewise, once human cultures are on the brink, their vitality can rarely, if at all, be brought back by cultural salvage operations; (ii) economic growth, development and consumption with attendant destructive impacts on both nature and culture far out-pace ecological restoration and cultural revitalization activities; (iii) the limited self-interest displayed in the behaviours of government, industry and consumers, in pursuit of narrowly defined goals, contributes interactively to a collective momentum that is destructive of ecocultural health.

Clearly, preventive action is needed to avoid further degradation and potential collapse of ecocultural systems. In order to be forewarned when ecocultural systems are at risk, it is essential to devise appropriate and effective indicators to continually assess and monitor the ecocultural health of large-scale ecosystems and landscapes. Indicators of ecocultural health enable early detection of pathology in ecocultural systems and are meant to trigger corrective actions. There is already a history of the use of indicators to assess the ecological health of large-scale ecosystems (Rapport and Regier, 1980; Rapport, 1992; Rapport and Whitford, 1999; Rapport et al, 2003), as well as a long history of their application in state of environment reporting (Bird and Rapport, 1986). There is less of a history of indicator development to assess cultural health. However, over the last few years, various indicators have been developed that make a contribution in this direction. These include indicators of key aspects of cultural diversity, such as the state and trends of languages and TEK (Zent, 2008; Harmon and Loh, 2010, in press). Efforts have also been made to combine measures of cultural diversity and biodiversity into a single index (Harmon and Loh, 2004; Harmon et al, 2010, this volume). From these various approaches, integrative indicators of ecocultural health are beginning to emerge.

A recent study on the health of the Mesoamerican Coral Reef extended previous work on ecological indicators of reef health to include socio-economic, governance, cultural, and public health dimensions and indicators (World Bank, 2006). In this report, various policy questions were raised, and specific indicators proposed to address these. For example, among socio-economic indicators, measuring the percentage conversion of natural habitat as a consequence of coastal development might inform a policy question on the sustainability of local economic development. For governance indicators, a policy question requiring evidence of good governance at the sub-national level might be addressed by looking at indicators of changes in land tenure. For cultural losses, indicators might include the rate of language loss, the proportion of young people speaking native languages, and the availability of bilingual education. This exercise could provide a starting point for a concerted effort to devise indicators of ecocultural health at regional scales and for promoting their use as key policy and implementation tools. Yet, while more integrative indicators of ecocultural health have the potential to play a valuable role in assessment and monitoring, even the most informative set of indicators would be insufficient to restore ecocultural health in the absence of the political will to act. What is needed, beyond the science-based information that select ecocultural indicators can provide, is a concerted global strategy to protect and restore the organization, vitality and resilience of ecocultural systems.

Global Strategies to Prevent Erosion of Biocultural Diversity and Ecocultural Health

At both global and national levels, a large number of initiatives now underway have the restoration of ecocultural health as their primary goal. International conventions have also been established with the purpose of halting declines in biodiversity, ecosystem health and cultural diversity: the CBD, the Convention on Wetlands, the Convention to Combat Desertification, the Kyoto Protocol on Climate Change, the Declaration on the Rights of Indigenous Peoples, and the Convention on Cultural Diversity. Why is it then that the erosion of biocultural diversity and ecocultural health continues unabated? And what more could be done to stop this erosion?

One key stumbling block is that many international agreements, as well as national and state legislation, fail to have sufficient powers to ensure that targets are being met. For example, the 2010 targets set up by the CBD to combat the loss of biodiversity are nowhere near being met; nor are the undertakings to meet the Kyoto Accord for reducing CO_2 emissions by 2012, even though these were binding to signatory countries. Other international instruments, such as the Declaration on the Rights of Indigenous Peoples, are not binding and thus can only exercise moral persuasion. At the same time, international economic and trade activities put massive pressures on ecocultural systems worldwide. In general, activities undertaken at the national or state level to advance economic and/or political objectives almost invariably result in compromising ecocultural health. There are few, if any, effective sanctions against such practices.

Clearly, the international community and the world's governments must address this fundamental failing if we are to stem the pandemic of ecocultural collapse. Effective action requires audits of ecocultural health coupled with mechanisms for corrective interventions. The key to achieving successful mitigation will be to establish solid indicators of ecocultural health, institute rigorous and independent periodic monitoring of conditions and trends in regional and global ecocultural health, set binding timelines and targets for achieving marked improvements in the state of ecocultural health, undertake periodic assessments to ensure that targets are being met or, if they are falling short, ensure that more stringent steps are taken to achieve them. Ecocultural health should become one of the overarching goals of global society and inform a new incarnation of global initiatives.

Acknowledgements

We thank Dr Toshiya Okuro, Dr Dehui Zeng and Ashbindu Singh for their review of and thoughtful comments on an earlier draft of this chapter.

References

Bails, J., Beeton, A., Bulkley, J., DePhilip, M., Gannon, J., Murray, M., Regier, H. and Scavia, D. (2005) 'Prescription for Great Lakes ecosystem protection and restoration: Avoiding the tipping point of irreversible changes', www.miseagrant.umich.edu/downloads/habitat/PrescriptionforGreatLakes.pdf

Bird, P. M. and Rapport, D. J. (1986) *State of the Environment Report for Canada*, Canadian Government Publishing Centre, Ottawa, 264 pp

Carlson, T. and Maffi, L. (2004) *Ethnobotany and Conservation of Biocultural Diversity*, Botanical Garden Press, New York

Chuluun, T. and Ojima, D. (2002) 'Land use change and carbon cycle in arid and semi-arid lands of East and Central Asia', *Science in China* (series C), vol 45, supplement October, pp48–54

Costanza, R., Norton, G. and Haskell, B. (eds) (1992) *Ecosystem Health: New Goals for Environmental Management*, Island Press, Washington, DC

Glanz, J. (2005) 'For Iraq's great marshes, a hesitant comeback', *New York Times*, 8 March

Harmon, D. (2002) *In Light of Our Differences: How Diversity in Nature and Culture Makes Us Human*, Smithsonian Institution Press, Washington, DC

Harmon, D. and Loh, J. (2004) 'The IBCD: A measure of the world's biocultural diversity', *Policy Matters*, vol 13, pp271–280

Harmon, D. and Loh, J. (2010) 'The Index of Linguistic Diversity: A new quantitative measure of trends in the status of the world's languages', *Language Documentation & Conservation*, in press

Hildén, M. and Rapport, D. J. (1993) 'Four centuries of cumulative cultural impact on a Finnish river and its estuary: An ecosystem health approach', *Journal of Aquatic Ecosystem Health*, vol 2, pp261–275

Jiang, G., Han, X. and Wu, J. (2006) 'Restoration and management of the Inner Mongolia grassland requires a sustainable strategy', *Ambio*, vol 35, no 5, pp269–70

Leopold, A. (1941) 'Wilderness as a land laboratory', *Living Wilderness*, vol 6, no 3

Leopold, A. (1949) *A Sand County Almanac and Sketches Here and There*, Oxford University Press, London and New York

Leopold, A. (1999) *For the Health of the Land: Previously Unpublished Essays and Other Writings Edited by J. Baird Callicott and Eric T. Freyfogle*, Island Press, Washington, DC

Levin, S. A. (1998) 'Ecosystems and the biosphere as complex adaptive systems', *Ecosystems*, vol 1, no 5, pp431–436

Liu, M., Jiang, G., Li, Y., Shun-Li, Y., Niu, S., Peng, Y., Jiang, C., Leiming, G. and Li, G. (2003) 'The control of land degradation in Inner Monoglia: A case study in Hunshandak Sandland', UNESCO, www.unesco.org/mab/doc/drylands/ChinaRep.pdf

Loh, J. and Harmon, D. (2005) 'A global index of biocultural diversity', *Ecological Indicators*, vol 5, pp231–241

Maffi, L. (ed) (2001) *On Biocultural Diversity: Linking Language, Knowledge, and the Environment*, Smithsonian Institution Press, Washington, DC

Maffi, L. (2005) 'Linguistic, cultural, and biological diversity', *Annual Review of Anthropology*, vol 29, pp599–617

Maffi, L. (2007) 'Biocultural diversity and sustainability', in J. Pretty, A. Ball, T. Benton, J. Guivant, D. Lee, D. Orr, M. Pfeffer and H. Ward (eds) *Sage Handbook on Environment and Society*, Sage Publications, Los Angeles, London, New Delhi and Singapore, pp267–277

Maffi, L. and Woodley, E. (2010) *Biocultural Diversity Conservation: A Global Sourcebook*, Earthscan, London

Mageau, M. T., Costanza, R. and Ulanowicz, R. E. (1995) 'The development and initial testing of a quantitative assessment of ecosystem health', *Ecosystem Health*, vol 1, pp201–213

MEA (2005) *Ecosystems and Human Well-being: Current State and Trends*, Millennium Ecosystem Assessment, Island Press, Washington, DC

Morimoto, J., Kondo, T. and Miyauchi, T. (2009) 'Satoyama-satoumi sub-global assessment in Japan and involvement of the Hokkaido Cluster', *Landscape Ecology*, vol 5, pp91–96

Oviedo, G., Maffi, L. and Larsen, P. B. (2000) *Indigenous and Traditional Peoples of the World and Ecoregion Conservation: An Integrated Approach to Conserving the World's Biological and Cultural Diversity*, and accompanying map *Indigenous and Traditional Peoples and the Global 200 Ecoregions*, WWF-International and Terralingua, Gland

Rapport, D. J. (1983) 'The stress-response environmental statistical system and its applicability to the Laurentian Lower Great lakes', *Statistical Journal of the United Nations ECE*, vol 1, pp377–405

Rapport, D. J. (1989a) 'Symptoms of pathology in the Gulf of Bothnia (Baltic Sea): Ecosystem response to stress from human activity', *Biological Journal of the Linnean Society*, vol 37, pp33–49

Rapport, D. J. (1989b) 'What constitutes ecosystem health?' *Perspectives in Biology and Medicine,* vol 33, pp120–132

Rapport, D. J. (2007a) 'Healthy ecosystems: An evolving paradigm', in J. Pretty, A. S. Ball, T. Benton, J. S. Guivant, D. R. Lee, D. Orr, M. J. Pfeffer and H. Ward (eds) *Sage Handbook of Environment and Society,* SAGE Publications, London, pp431–441

Rapport, D. J. (2007b) 'Sustainability science: An EcoHealth approach', *Sustainability Science,* vol 2, pp177–184

Rapport, D. J. and Friend, A. M. (1979) *Towards a Comprehensive Framework for Environmental Statistics: A Stress-Response Approach,* Statistics Canada, Ottawa, p87

Rapport, D. J., Thorpe, C. and Regier, H. A. (1979) 'Ecosystem medicine', *Bulletin of the Ecological Society of America,* vol 60, pp180-182

Rapport, D. J. and Regier, H. A. (1980) 'An ecological approach to environmental information', *Ambio,* vol 9, pp22-27

Rapport, D. J., Regier, H. A. and Hutchinson, T. C. (1985) 'Ecosystem behavior under stress', *American Naturalist,* vol 125, no 5, pp617–640

Rapport, D. J., Gaudet, C. and Calow, P. (eds) (1995) *Evaluating and Monitoring the Health of Large-Scale Ecosystems,* Springer Verlag, Heidelberg

Rapport, D. J., Costanza, R., Epstein, P., Gaudet, C. and Levins, R. (eds) (1998a) *Ecosystem Health,* Blackwell Science, Oxford

Rapport, D. J., Costanza, R. and McMichael, A. J. (1998b) 'Assessing ecosystem health', *Trends in Ecology and Evolution,* vol 13, no 10, pp397–401

Rapport, D. J. and Whitford, W. G. (1999) 'How ecosystems respond to stress: Common properties of arid and aquatic ecosystems', *Bioscience,* vol 49, no 3, pp193–203

Rapport, D. J., Fyfe, W. S., Costanza, R., Spiegel, J., Yassie, A., Bohm, G. M., Patil, G. P., Lannigan, R., Anjema, C. M. Whitford, W. G. and Horwitz, P. (2001) 'Ecosystem health: Definitions, assessment and case studies', in M. Tolba (ed) *Our Fragile World: Challenges and Opportunities for Sustainable Development,* EOLSS, Oxford, pp21–42

Rapport, D. J, Lasley, W., Rolston, D. E., Nielsen, N. O., Qualset, C. O. and Damania, A.B. (eds) (2003) *Managing for Healthy Ecosystems,* Lewis Publishers, Boca Raton, London, New York and Washington, DC

Rapport, D. J., Daszak, P., Froment, A. Guegan, J. F., Lafferty, K. D., Larigauderie, A., Mazumder, A. and Winding, A. (2009) 'The impact of anthropogenic stress at global and regional scales on biodiversity and human health', in O. E. Sala, L. A. Meyerson and C. Parmesan (eds) *Biodiversity Change and Human Health,* Island Press, Washington, DC, pp 41–60

Rapport, D. J. and Maffi, L. (2010) 'Eco-cultural health, global health, and sustainability', *Ecological Research,* in press

Regier, H. A. and Hartman, W. L. (1973) 'Lake Erie's fish community: 150 years of cultural stresses', *Science,* vol 180, pp1248–1255

Richardson, C. J., Reiss, P., Hussain, N. A., Alwash, A. J. and Poo, D. J. (2005) 'The restoration potential of the Mesopotamian Marshes of Iraq', *Science,* vol 307, pp1307–1311

Roberts, L., Stone, R. and Sugden, A. (eds) (2009) 'Restoration ecology' (Special Issue), *Science,* 325, pp555–576

Seguchi, R., Brown, R. D. and Takeuchi, K. (2008) 'Land use change from traditional Japan', in S. K. Hong, N. Nakagoshi, B. J. Fu and Y. Morimoto (eds) *Landscape Ecological Applications in Man-Influenced Areas: Linking Man and Nature Systems,* Springer, The Netherlands, pp113–128

Shi, Y. C. (2002) 'Reflections on twenty years: Desertification-control in China', *Science and Technology Daily*, 25 February

Skutnabb-Kangas, T., Maffi, L. and Harmon, D. (2003) *Sharing a World of Difference: The Earth's Linguistic, Cultural, and Biological Diversity*, and companion poster *The World's Biocultural Diversity: People, Languages, and Ecosystems*, UNESCO, Paris

Stepp, J. R., Cervone, S., Castaneda, H., Lasseter, A., Stocks, G. and Gichon, Y. (2004) 'Development of a GIS for global biocultural diversity', *Policy Matters*, vol 13, pp267–270

Stepp, J. R., Castaneda, H., Reilly-Brown, J. and Russell, C. (2008) 'Set of 13 maps of global and regional biocultural diversity', Biocultural Diversity Mapping Project, University of Florida, Gainesville

Takeuchi, K. (2003a) 'Satoyama landscapes as managed nature', in K. Takeuchi, R. E. Brown, I. Washitani, A. Tsunekawa and M. Yokohari (eds) *Satoyama: The Traditional Rural Landscape of Japan*, Springer, Berlin, Heidelberg, New York, pp9–16

Takeuchi, K. (2003b) 'National land planning of Satoyama landscapes', in K. Takeuchi, R. E. Brown, I. Washitani, A. Tsunekawa and M. Yokohari (eds) *Satoyama: The Traditional Rural Landscape of Japan*, Springer, Berlin, Heidelberg, New York, pp200–208

Takeuchi, K. (2006) 'Report of the Special Committee on Environmental Cooperation for Iraq', March 2006, www.env.go.jp/earth/report/h18-02/e002.pdf

The Economist (2005) 'One third of paradise: The marshes of southern Iraq', 26 February, pp 77–78

Tsunekawa, A. (2003) 'Strategic management of Satoyama landscapes' in K. Takeuchi, R. E. Brown, I. Washitani, A. Tsunekawa and M. Yokohari (eds) *Satoyama: The Traditional Rural Landscape of Japan*, Springer, Berlin, Heidelberg, New York, pp179–191

UNEP (2001) 'The Mesopotamian marshlands: Demise of an ecosystem', Early Warning and Assessment Technical Report, UNEP/DEWA/TR.01-3 Rev 1, Division of Early Warning and Assessment, UNEP, Nairobi, Kenya

UNEP (2007) *Global Environment Outlook 4*, UNEP, Nairobi

World Bank (2006) *Measuring Coral Reef Ecosystem Health: Integrating Societal Dimensions*, Report No 36623, World Bank, Washington, DC

WWF (2006) *Living Planet Report*, WWF, Gland

Yokohari, M. and Kurita, H. (2003) 'Mechanisms of Staoyama landscape transformation' in K. Takeuchi, R. E. Brown, I. Washitani, A. Tsunekawa and M. Yokohari (eds) *Satoyama: The Traditional Rural Landscape of Japan*, Springer, Berlin, Heidelberg, New York, pp71–79

Zent, S. (2008) 'Methodology for developing a vitality index of traditional environmental knowledge (VITEK)', Technical Report to The Christensen Fund, Terralingua, Salt Spring Island, British Columbia

Part III
Hunting

7
Biodiversity and Cultural Diversity: The Interdependent and the Indistinguishable

Martina Tyrrell

Nunavut, July 2006 *'My old man carved this,' the Inuit hunter said as he handed me a piece of rock that had begun to take on the character of a seal. 'He was sitting here, waiting for a seal to come by. Probably thinking about seals. Thinking, and waiting, and carving this stone.'*

Nunavik, May 2006 *'They used to wait for* [beluga] *whales at the Point,' the hunter's wife told me. 'Now everyone runs down* [to the beach] *with their guns and trucks, and they start shooting from every direction. They stand around, and they're not supposed to. You're supposed to lay low and wait for the whales to come in.'*

Introduction

The current global emphasis on biodiversity conservation is arguably more a process of human, rather than ecosystem, management. Conservation laws and policies are enacted to curtail or alter human activity for the benefit of ecosystems by restricting development through the creation of national parks, wilderness areas and green belts, for example, and by placing prohibitions on human actions that may prove harmful to plants and animals. Many of us take this mode of management for granted. We see restrictions and prohibitions on certain forms of agriculture, industry or construction as the necessary price to pay for the survival of species at risk. We walk on paths, purchase environmentally-sensitive foods, and return captured fish to rivers because we believe in the greater good of the environment. Those whose livelihoods are directly affected by conservation law, such as farmers and fishers, often have to embrace enforced changes to their practices – some willingly, others grudgingly – even as they face economic hardship and upheavals as they adjust to these changes.

Across the industrialized world, and increasingly elsewhere too, the management of human behaviour for the sake of biodiversity conservation has become a norm. Our culture has been transformed. We think, talk and act in ways unheard of a generation ago, adapting our lives and livelihoods for the sake of biodiversity conservation. This enormous transformation of our culture is generally not considered a diminishment. We have come to accept that such changes to cultural practice are necessary for the maintenance of biodiversity and thus for the enhancement of human experience through our interactions and engagement with a species-rich environment.

In this chapter, however, I argue that we must not overlook the important link between cultural diversity and biodiversity. Across the world there are people and communities who have for generations inhabited and depended upon the same ecosystems as many of the world's species currently at risk. There are farmers and fishers, hunter-gatherers and horticulturalists, birdwatchers and beekeepers with intimate knowledge of animals and plants, and their interconnecting webs of relatedness. Their knowledge of ecosystems is crucial to scientific understanding of how ecosystems work and often alludes to how best to establish practices that ensure continued health of those ecosystems.

For the vast majority of the world's peoples, such environmental knowledge is embedded in non-Western worldviews and interpretations of the role of humans within the environment (Nelson, 1986; Bird-David, 1990; Ingold, 2000). The concept of human stewardship over the environment is alien to many peoples whose cultures commonly dictate that the Earth, in fact, holds stewardship over humanity (Bird-David, 1990; Johannes et al, 2000). For many indigenous and tribal peoples in particular, humans and animals are held to be equal in their habitation of the Earth and, for them, such concepts as 'biodiversity management' seem disrespectful to free-living and sentient creatures that can never and should never be managed.

All too often conservationists have sought to impose scientifically-informed biodiversity management practices on peoples who have for generations lived intimately with their environments. Conservationists have done so without adequate regard for the detailed knowledge possessed by individuals and communities who rely on these environments for their livelihoods. And they have done so with scant regard for the role of culture in the maintenance of biodiversity and environmental health. So much, therefore, is lost in the attempt to turn local people into environmental stewards and conservationists. The culturally-informed relationships between humans and the environment that is at the heart of customary conservation practices can so easily be diminished in the attempt to make people think about ecosystems in different ways. Thus, the outcome of imposing culturally inappropriate management policy has often been detrimental to both biodiversity and cultural diversity.

Taking two examples from the Canadian Arctic, I explore how the relationship between Inuit and certain marine mammals has been altered by the imposition of wildlife management laws that have transformed the role of these animals in Inuit culture. This has diminished both Inuit culture and these at risk species. As these case studies demonstrate, biodiversity and cultural diversity are indistinguishable and ultimately the vitality and richness of one depends on that of the other.

Inuit and Animals: Neighbours in a Shared Environment

Approximately 150,000 Inuit[1] reside in small communities scattered across the Arctic, from eastern Siberia, through Alaska and Canada, to eastern Green-land. The ancestors of contemporary Inuit inhabited the Arctic for at least 4000 years (Hoffecker, 2005), hunting both marine and terrestrial mammals. In the circumpolar north, the relationship between Inuit and animals has long been one of co-habitation and mutual respect, where encounters occur only through both parties' willing participation in the hunt. The animals of the Arctic (such as seals, whales, bears, caribou, musk oxen and fish) are believed to only give themselves to those Inuit who have proved, by their actions, to be generous humans (Bodenhorn, 1990; Turner, 1990). In the past, strict taboos surrounded the hunting of animals in Inuit society; women behaved in prescribed ways while their husbands hunted and men thought and acted respectfully towards animals before, during and after the hunt. A hunter did not hunt beyond his needs.[2] Prohibitions on the preparation and consumption of certain foods or the transformation of animal skins into clothing at certain times and places were also observed. Rituals were performed on the animal to appease its soul and to assure the return of that soul into the body of another animal in the future. At the core of this relationship between humans and animals was the sharing of the harvest of meat and pelts once it was brought back to the community. While 21st century Inuit no longer adhere to many of these taboos and prescriptions, sharing continues to be at the heart of Inuit life, and central to the continuing relationship between Inuit and animals. Animals continue to be perceived as willing participants in the hunt and most Inuit continue to think about, talk about and interact with animals in ways that are deemed respectful to these co-inhabitants of the Arctic environment.

The recent history of Inuit has been marked by repeated waves of contact with Europeans; from Basque whalers in the early 1500s (Barkham, 1980) to the arrival of other European and Euro-North American traders, whalers and missionaries from the 17th century onwards. Each arrival brought changes to Inuit culture, with the introduction of guns, iron and steel, timber, and European technologies and foodstuffs. These tools and technologies transformed the lives of Inuit, with both positive and negative outcomes. Contact with whalers and traders brought changes to nomadic practices, as families incorporated the fur trade or seasonal work on whaling ships into their migrations across their lands. Mutual dependencies developed, whereby traders came to rely on Inuit for the provision of pelts, from animals such as the Arctic fox and wolf that had rarely figured in Inuit hunting previously, and Inuit relied on the newcomers for consumables such as tea, flour, sugar and tobacco.

land rich → economically poor
resource

Until the 1950s, and in some cases even later, some Inuit still lived nomadically, although many aspects of this life had been transformed by the incursion of traders, missionaries and law enforcement officers (Brody, 1975; Tester and Kulchyski, 1994; Stevenson, 1996). Throughout the 20th century, governance from southern Canada, the US, Denmark and the Soviet Union fast encroached on nomadic life. Inuit were increasingly forced to live in permanent settlements, many of which were established around old trading posts or whaling stations. The formal education of Inuit children became compulsory and children were removed from their families to attend boarding schools. A growing dependency on economic markets contributed to the abandonment of nomadic life. A shift towards consumable foods and modern hunting equipment, such as rifles and bullets, required Inuit to supplement traditional livelihoods with wage labour or apply for government welfare. Neither choice was compatible with life lived exclusively on the land (Williamson, 1974; Damas, 2002).

In the Canadian Arctic, the focus of this chapter, the impetus for the sedentarization of Inuit onto permanent settlements in the 20th century also came from concerns raised by the federal Canadian Wildlife Service (CWS). It speculated that continued hunting and gathering would rapidly lead to the demise of Arctic species such as the caribou, musk ox and polar bear (Kulchyski and Tester, 2007; Sandlos, 2007, 2009). The CWS, alongside other government agencies, became active in promoting the transformation of Inuit culture. These agencies saw their role as assisting Inuit to become 'modern Canadians' through access to formal education and employment, for instance, as miners, nurses, mechanics and clerical staff. Contemporary Inuit reflect on this as a dark period in the history of their colonization, and one that reverberates through to the present day, as Inuit continue to struggle to come to terms with their (often) forced resettlement (Tester and Kulchyski, 1994), the loss of cultural knowledge, spiritual beliefs and practices (Laugrand, 2002), and the denial of a long-held right to participate in the hunt (Kulchyski and Tester, 2007).

The Arctic is a very different place today. Since the late 1970s, Inuit have acquired various levels of self-governance across their homelands, such as Greenlandic Home Rule in 1979, the creation of the Inuvialuit Settlement Region within Canada's Northwest Territories in 1984, the foundation of Nunavut Territory in the Canadian eastern Arctic in 1999, and the signing of land claims agreements in Nunatsiavut, northern Labrador, in 2005 and Nunavik, northern Quebec, in 2007. The creation of these self-governing homelands has led to far greater levels of autonomy for Inuit, with the indigenous governments of these regions striving to create public institutions that are organized around the principles of Inuit culture. Despite this, Inuit governance is heavily structured on Western models, with bureaucracy, budgets and transparency concerns, not dissimilar to other democratic governments across the world.

Within the tiny settlements sparsely scattered across these homelands, Inuit seek to get on with their lives. All settlements today have schools, health centres, at least one (often cooperative) store, churches and many of the other features of contemporary industrialized life, albeit on a small scale. Elected mayors and village or hamlet councils run the day to day affairs of these communities, with other elected institutions responsible for supporting hunting and trapping activities, and other aspects of Inuit life. Interlinked at various levels of local and regional government are other institutions, from federal or regional departments of environment, wildlife, economic development and education, many of which have offices situated in these small settlements. Thus an essentially post-modern picture emerges across the Arctic, where the lives of individual Inuit are influenced by and linked with a much broader world beyond the boundaries of the local community, where decisions made not only in regional capitals such as Iqaluit and Nuuk, but in Ottawa, Washington or Copenhagen can and do impact on Inuit life. Families are thus attempting to live culturally and socially meaningful lives while at the same time engaging with the market economy, and in the face of globalizing processes that neither understand Inuit life nor are willing to accommodate alternative non-Western approaches to the conservation of animals and ecosystems.

Traditionally, a variety of terrestrial and marine mammals, birds and fish were harvested for human and dog food, as well as for clothing and shelter. With the shift to settlements and the introduction of mechanized forms of transport in the latter half of the 20th century, sled dog teams have dramatically decreased in number.[3] The availability of store-bought synthetic materials has resulted in a decline in dependency on fur-bearing animals such as seals, caribou and polar bears. Many hunters, however, still prefer a combination of synthetic and natural garments; pelts are still used as a comfortable and warm lining for sleds, and are used on overnight excursions as either a warm base underneath synthetic sleeping bags or as sleeping bags themselves. Despite the radical changes to Inuit life over the past century, reliance on Arctic animals remains strong. And in the modern Arctic, craftsmen and women transform caribou antlers, and walrus and narwhal tusk, into carvings and jewellery that are sold door to door or in southern Canadian or US stores specializing in 'native art'. Seal and caribou skins are also used by women to make wall-hangings depicting scenes of Inuit life and these are also sold for income.

However, the biggest single use of harvested animals among Inuit continues to be food. The list of animals that comprise the Inuit diet includes four species of seal, walrus, narwhal, beluga whale, bowhead whale, polar bear, caribou, musk ox, Dall sheep, various species of goose and duck, Arctic char, trout, grayling, and other fresh and salt-water fish. Furthermore, across the Arctic, Inuit gather a wide variety of goose, duck and other birds' eggs, and collect mussels and clams from the sea shore.[4] All of these meats and fish can be consumed fresh (raw or cooked) or can be preserved by drying, freezing or fermenting for use at a later date. Due to the vast geographic area they inhabit, not all of these species are hunted by every Inuit community but most have access to some combination of seal, whale, caribou, birds and fish.

The economic and nutritional value of these foods is substantial. Kuhnlein et al (2003) found that the 450 residents of Qikiqtarjuaq on Baffin Island consume an average of 140kg of locally harvested foods each day, or a total annual consumption of more than 50,000kg. The daily diet of men in this community comprises 38 per cent country food, and, in all, local foods provide

12–40 per cent of dietary energy for Inuit of both sexes (Kuhnlein et al, 2001). These foods account for a large proportion of Inuit nutrient intake, with marine mammals in particular being a rich source of protein, energy and fat, as well as vitamins A and C, omega-3 fatty acids, zinc, iron and many other nutrients (Downie and Fenge, 2003; Northern Contaminants Programme, 2003).

Beyond economics and nutrition, strong cultural values are attached to the procurement and consumption of traditional wild foods (Borré, 1991; Wein et al, 1996; Pars et al, 2001; Jolles, 2002; Searles, 2002). Inuit often say that *qablunaat*[5] food is less nutritious, and express a physiological and psychological need for traditional foods. Today, there are ever more store-bought choices available to Inuit consumers but, for many, identity remains entwined with the procurement and consumption of wild foods. The consumption of these foods is a political, social and individual expression of the abiding relationship that exists between animals and people in their shared environment.

The knowledge and skill required to successfully hunt this array of animals throughout the seasons is great. Experienced hunters have a deep understanding of animal behaviour and physiology, and of the relationship between these and differing environmental phenomena. They combine this with knowledge of the dynamic marine and terrestrial environment, as treacherous open water, sea ice, river or tundra conditions continuously change throughout the year, requiring Inuit to have a range of travelling and hunting skills. As a result, Inuit knowledge of Arctic biodiversity is vast, with understandings of the ways different species interact and influence each other, and of the role of changing environmental phenomena on animal behaviour, being key to safe and successful hunting practice. Once harvested, Inuit further display their knowledge and skill as animals are transformed into food and clothing, with, for example, animal health (the presence of parasites or infections) being determined by the condition of fur, meat and internal organs.

Anthropological literature from the Arctic has many references to the social and metaphysical roles of animals (Guemple, 1986; Fienup-Riordan, 1988; Bodenhorn, 1990; Turner, 1990; Nuttall, 1997; Lowenstein, 1999; Brody, 2000; Pelly, 2001). Bodenhorn (1990), for example, writes that human behaviour among Alaskan Iñupiat draws or repels whales. Whales are aware of the generosity of hunters' wives and will only give themselves to those hunters whose wives are socially responsible. Similarly, Turner (1990) observes that the initiative lies with the animal being hunted, as it decides whether or not to give itself to the hunter. The good behaviour and manners of humans, she writes, both towards each other and towards the animals they hunt, are essential to safe and successful hunting.

On the west coast of Hudson Bay and in northern Quebec, public displays of pride following the success of a hunt are discouraged, as are public displays of excitement and anticipation prior to a hunt. Inuit beliefs require that the successful hunter is humble in his success and does not brag about the size or number of his harvest, or of his own bravery or prowess during the hunt.[6] Rather, he is expected to share the harvest with his immediate and extended family, and with those others who have contributed to the success of the hunt.

The hunter should spend time quietly thinking about animals. Many hunters say that while at sea or on the tundra they focus their thoughts on the animals they hope to encounter. For some, whittling or carving[7] provides the opportunity to quietly reflect on the animal as it emerges out of wood or stone, to think about its shape and its movements, and to prepare physically and psychologically for the hunt. This too is a humble act, as the animal emerges from the material in the hands of the carver, just as it emerges from the sea or appears on the land to give itself to the hunter.

Thus the relationship between Inuit and animals is not one of resource user and resource. The act of hunting is to engage in a reciprocal dance. The reliance by Inuit on the participation of animals is expressed in respectful human behaviour – sharing the harvest, not harvesting more animals than one requires, and accepting the bounty offered when an animal presents itself. Yet this relationship, based on a unique worldview, the richness of Arctic biodiversity and Inuit knowledge and expertise, is increasingly under threat from wildlife conservation policies and laws. These regulatory tools make little effort to understand Inuit culture or the role that human-animal interactions play in the continued health of Arctic biodiversity.

Wildlife Conservation in the Canadian Arctic

Since the 1920s, wildlife scientists employed by the Canadian federal government (such as the CWS) have sought to monitor, protect and conserve many of the species upon which Inuit rely for subsistence (Tester and Kulchyski, 1994; Kulchyski and Tester, 2007; Sandlos, 2007). Among these are the caribou, musk ox, polar bear, narwhal, beluga whale, bowhead whale and some species of goose. The perceived decrease in species population numbers has been linked to everything from commercial whaling by British and American whalers to predation by other species, and from hydroelectric dam-building projects to climate change (Stirling et al, 1999; Doidge et al, 2002; Hammill et al, 2004). But at the heart of all of these concerns has been an emphasis on Inuit hunting practices, and Inuit reliance on these animals for food and/or clothing (Hammill, 2001; Bourdages et al, 2002). From the 1920s through to the present day, it is the management of Inuit hunting practices and, by extension, Inuit culture that are viewed as priorities in the race to save Arctic species.[8]

The problematization of Inuit hunting and, by extension, Inuit culture has been a recurrent feature of wildlife conservation in the Canadian Arctic (Usher, 2004; Kulchyski and Tester, 2007). Sandlos (2007) frames this reaction within broader colonial attitudes to hunting. In Europe in the early years of the 20th century, hunting had become a pastime of the rich, who were instrumental in the development of large game parks in Africa and south Asia. These big game hunters were often contemptuous of local hunting practices – objecting both to the utilitarian nature of the hunt and to specific hunting methods, which they viewed as violating the sportsman's code (for example, only using one bullet per animal or carefully stalking individual animals compared to indigenous practices of taking large numbers of animals at places such as river crossings). Evidence

misperception

of this ethnocentrism is to be found in many of the wildlife documents from this time[9] and such attitudes towards Inuit hunting practices continue to dominate federal management of Arctic wildlife into the 21st century. The sight of large numbers of caribou killed at river crossings, of carcasses left to rot on the land (which biologists interpreted as 'wanton slaughter' but which were actually carcasses left to freeze or ferment for later use) and of Inuit consumption practices, were all signs to wildlife biologists that uncultured and uneducated Inuit were at fault for the perceived demise of Arctic species.[10]

The social histories of wildlife management in the first half of the 20th century reveal how little has changed since the early days of Arctic wildlife biology. The fact remains that 21st century biologists continue to misinterpret and problematize Inuit culture and hunting practice in a similar manner. Contemporary Inuit argue that the research methods employed by today's wildlife biologists are flawed and biased (Tyrrell, 2008), failing repeatedly to make use of expert Inuit knowledge concerning animal behaviour or the most appropriate times and places to undertake population surveys, regarding local expertise simply as 'anecdotal evidence'. Wildlife biologists employed by the Canadian government have consistently failed to understand the relationship between Inuit and Arctic animals. Today, a simplistic relationship between resource-user and diminishing resource is still assumed. Individual Inuit are often seen to take far more animals than they require, as wildlife authorities fail to understand the importance and prevalence of sharing beyond the confines of the household. Thus, for the past half century or more, the Canadian government has taken various steps to divert Inuit away from the hunting of Arctic animals and towards the market economy, where they can purchase low quality and overpriced alternatives to their traditional foods.

The perceived simplicity of the resource-user/resource relationship is mirrored in the assumptions made by many biologists and policy-makers regarding the protection of Arctic species. One positivistic assumption is that conservation can take place through systematic study accompanied by the placing of restrictions on harvesting or so-called 'adaptive management' (COSEWIC, 2004; DFO, 2005a). By recording the numbers of animals that are present at given times and places, and comparing those numbers to other times and places, the health of a species population can be gauged. Thus, where the hunting of species is seen as a threat to biodiversity, the simple solution is to limit or ban hunting practice of a particular species in that area. And so, in some instances in the Canadian Arctic, moratoria have been placed on the hunting of certain sub-populations, while in others, the hunting of sub-populations has been curtailed through the imposition of hunting quotas, open and closed hunting seasons, and detailed guidelines on the organization of the hunt and the hunting techniques which may be employed. Thus the long-enduring relationship between Inuit and animals has been transformed to the detriment of both Inuit culture and Arctic biodiversity. By reducing Arctic animals to mere numbers, hunting has been transformed from an act embedded in all aspects of Inuit culture, to one that is now perceived as an Inuit 'right', thus severing the reciprocal human-animal relationship which lies at the heart of

Inuit culture. What follows are two case studies from the Canadian Arctic, which explore what can happen to biodiversity and cultural diversity when strict wildlife management laws are imposed.

The Nunavik Beluga Whale Hunt

Some 15,000 Inuit live in 14 small coastal villages in Nunavik, an autonomous region in the northern reaches of the province of Quebec. Subsistence marine mammal hunting predominates in these communities, and the beluga whale[11] is among the most important species. Villages along the northern coast of Nunavik take advantage of the twice yearly mass migrations of belugas through Hudson Strait, while smaller summer populations of belugas along the shores of the east coast of Hudson Bay and in Ungava Bay are hunted on a more small-scale and sporadic basis.

Belugas have long played an important role in the lives of Nunavimmiut,[12] with the hunting complex[13] – preparation of hunting clothes, tools and equipment, the hunt itself, and the butchering, processing and consumption of the harvest – being inclusive of all villagers, irrespective of age or gender. Older Inuit refer to belugas as 'neighbours' with whom relationships of respect must be maintained, similar to the reciprocal respect that should be maintained with human neighbours (Tyrrell, 2008). Until the mid-1990s, extended families travelled varying distances to communal hunting camps, spent time living off the land in tents or purpose-built cabins, and engaged in a range of social hunting/gathering activities that coincided with the beluga hunt. The hunt could last for weeks, with discerning hunters taking their time to harvest the best whales based on their knowledge of beluga physiology and behaviour. This was a time to teach children the skills of the beluga hunting complex, based on active engagement in a range of activities from hunting to flensing. Villagers along the Hudson Strait coast often harvested upwards of 35 belugas in a year (Lesage and Doidge, 2005), with occasional bumper years of 100 animals or more, while east coast Hudson Bay and Ungava Bay hunters generally harvested fewer than ten belugas per year. It was not uncommon in some years for the latter to harvest none at all.

Since 1996, the Inuit hunting of belugas in the waters around Nunavik has been actively managed by the Canadian Department of Fisheries and Oceans (DFO). DFO biologists estimate the east coast Hudson Bay population of belugas to be 3100 (DFO, 2005b), while the population in Ungava Bay is estimated to number no more than 200 (Hammill et al, 2004). Hudson Strait is classed as a migration route for various beluga sub-populations and therefore no statistics have been recorded for belugas in this area. However, twice each year, both the small east Hudson Bay beluga population, and the massive west Hudson Bay population (the largest sub-population in the world, estimated at 57,000 animals; Richard, 2005) migrate through Hudson Strait, swimming to summer calving grounds in spring and to winter feeding grounds in autumn. Villagers along the Hudson Strait coast of Nunavik have traditionally taken advantage of these mass migrations, harvesting and processing belugas for use throughout the year.

In the mid-1980s, the DFO became concerned about the small numbers of belugas in the east coast Hudson Bay and Ungava Bay populations. Intensive commercial whaling by American and British whalers from the 1750s to the early 1900s is considered to be the primary reason why numbers are now so low (Hammill et al, 2004). The DFO argues that the failure of these populations to recover from commercial whaling is compounded by more recent mass hydro-electric dam-building projects by Hydro-Québec, which have radically altered the estuarine habitats that belugas rely on to birth and nurse their young each summer, as well as predation by killer whales and competition from the Greenlandic halibut fishery at the beluga wintering grounds. Furthermore, the DFO argues that a growing Inuit population in Nunavik, with improved hunting technologies, is also limiting the ability of beluga populations to recover and return to pre-commercial whaling numbers. However, of all of these perceived threats, it is only Inuit hunting practices that have been actively managed.[14]

In 1996, the DFO implemented its first five-year beluga management plan for Nunavik (DFO, 2005a).[15] Harvesting was limited to 240 belugas per year for the entire region. However, between 1996 and 2001, an average of 282 animals was harvested per year, 42 more than the quota allowed. In 2001, a second management plan was introduced. The annual quota was cut to 125 belugas, but in the first year alone Inuit reported harvesting 395 whales, and it was suspected that many more went unreported. In 2002, quotas were further slashed in reaction to the non-compliance of Inuit hunters the previous year. Villages that previously had a quota of 35 were reduced to 15, and those with quotas of 25 were reduced to 10 (Kishigami, 2005). The 2005 management plan proved to be even more detailed, outlining not only how many belugas Inuit were permitted to harvest, but also the methods by which they were to be harvested and the manner in which hunters should behave and organize their hunts.

From the point of view of wildlife managers, the 2005 season proved to be an unmitigated disaster. Hunters from one village harvested more belugas in one day than their village was allowed for the entire year (Tyrrell, 2007), leading to conflicts with other villages over access rights to resources, when the hunting season across the region was brought to an abrupt close. The 2006–2008 beluga management plan further reduced the quota to nine belugas per village in some cases, to reflect the 'over-hunting' that had taken place in previous years. Hunters were warned that harsh measures would be taken against those who failed to comply with these regulations. The management plan stated that hunters were 'absolutely forbidden to hunt' in eastern Hudson Bay and Ungava Bay (George, 2006). Ignoring the regulations once again, hunters in various communities took more than their allowed harvest. In August 2006, the hunting season was declared closed by the DFO, effectively denying belugas to those villages whose hunt had not yet even begun. As a result of the continued disregard for the quota by some Nunavimmiut, in 2007 the DFO warned that, during the 2008 hunting season, it would send enforcement officers into villages to 'make sure the quotas are respected' (Siku News, 2007). By mid-2008, seven hunters from across the region were under investigation for poaching, with each hunter facing a maximum fine of

C$250,000 or five years' imprisonment. It seems the greater the level of regulation imposed by the DFO, the more defiant Nunavik hunters have become.

The management of the beluga hunt has had far-reaching consequences on Inuit life (Tyrrell, 2008). The majority of Nunavik villages adhere to the quotas each year, and for these the loss of belugas has been acutely felt. In one village, the first beluga landed is equally divided between all 72 households in order to ensure that everyone has at least a token share of beluga each year. Furthermore, the strict management of belugas and the size of the quota means that hunting no longer takes place at the traditional hunting camps, and families are no longer involved in activities that once surrounded the hunt – such as clam-digging, seal-hunting, and the butchering and processing of harvested belugas. There have been cases where villages have inadvertently surpassed their quota as a result of different groups of hunters harvesting belugas in different locations at the same time. In order to avoid this, hunters now gather together in one location, standing virtually shoulder to shoulder with their rifles aimed at the water. With only nine belugas allowed to be harvested in each village, the entire hunt is potentially over in one day. There is no longer a reason for entire families to relocate to beluga hunting camps along the coastline, and no opportunity for young hunters to learn the skills of the hunt. Today, only the most skilled hunters participate, to reduce the chances of returning home empty-handed.

Prior to the introduction of quotas, hunters were discerning in their choices, knowing that if they were unsuccessful one day, they could take their time and try again on a subsequent day. This is on longer the case. Hunters now race down to the beach at the first news of whales and fire indiscriminately, paying little attention to the size, age or sex of the animals they shoot. This change in behaviour arises entirely from the spiral of restrictions imposed by the DFO.

Moratoria have also been placed on beluga hunting in the southern portion of Ungava Bay and along all of the east coast of Hudson Bay. Here, beluga hunting was traditionally more sporadic, but since the imposition of the new management regime, hunters in these villages believe it is their 'right' to have access to belugas, just like their more northerly Nunavik neighbours. As a result, the government of Nunavik has been forced to distribute the quota evenly among all villages. And so, each year, hunters from these Ungava and Hudson Bay villages fly to villages on Hudson Strait (at the expense of their local communities) in order to fulfil their new found 'right' to harvest the same number of belugas as all other Nunavik communities.

The DFO's aim in imposing strict quota limits and managing the hunt has been to give Ungava Bay and east coast Hudson Bay beluga whale populations the chance to recover in number. Yet the means of achieving this may well be counter-productive, potentially altering beyond recognition the relationship between Inuit and belugas in this part of the Canadian Arctic. The anger hunters feel at being told how to manage an activity that has been an integral part of their culture for centuries has manifested itself in a changed and potentially detrimental relationship. This altered relationship is probably the greatest threat to the conservation of beluga whales in this part of the Arctic.

The respect Nunavik Inuit once held towards belugas, and which they continue to exhibit towards seals, walruses, polar bears and other animals, is now almost entirely absent. Some older men refuse to participate in the hunt for this reason, and many young men who have never had the opportunity to learn the skills of the beluga hunt do not want to be involved. Many Inuit, particularly elders, express sadness at the changing relationship between humans and beluga. In the past, no hunter had the 'right' to hunt any animal, and could only be successful if he proved himself to be a generous and good man who shared his harvest. Through the external management of whale hunting, belugas have been transformed from sentient neighbours who engage in the hunt at will, into basic resources that Inuit defend as their right to hunt.

Consequently, care and respect for belugas is fast disappearing, and the discernment and skill that was once held essential to proper hunting is no longer present. As this unique aspect of Nunavimmiut culture quickly perishes, so too does an essential element of beluga conservation. The DFO has missed a valuable opportunity to engage with Inuit hunters, to learn from their skills and knowledge of beluga whales, and to build on Inuit customary hunting laws and hunting practices. Instead, it has transformed a relationship of mutual respect into one based on Western concepts of rights and privileges, in which there is no room for Inuit and beluga whales to be the neighbours they once were.

The Nunavut Polar Bear Hunt

On the other side of Hudson Bay lies the Kivalliq region of Nunavut. Nunavut is a 1.9 million km² self-governing territory within federal Canada, and has a population of approximately 30,000 people (85 per cent of whom are Inuit) divided among 29 towns and villages. Stretching along the west Hudson Bay coast from northern Manitoba and into Kivalliq is a sub-population of approximately 1200 polar bears. The world's largest denning site for pregnant polar bears is found just outside the town of Churchill, northern Manitoba, where 100–150 cubs are born each year (Fikkan et al, 1993). Traditionally, Inuit living to the north of Churchill in Kivalliq rarely encountered polar bears, and bears were only sporadically hunted, usually in self-defence or occasionally in order to trade hides with *qablunaat* at established trading posts. In recent decades, Inuit in this region, and particularly in the hamlet of Arviat, have encountered polar bears much more frequently, in close proximity to and actually within their communities. Elders say they rarely, if ever, saw polar bears in their youth, but today hunters and travellers regularly encounter bears at close quarters while on land and sea, and each autumn bears wander the streets of Arviat causing concern for the safety of the local population of 2000 people (Tyrrell, 2006, 2009a).

For Inuit, polar bears are physically and symbolically the most powerful animals with whom they share the Arctic environment (Saladin D'Anglure, 1990; Tyrrell, 2006) and traditionally Inuit sought to maintain relationships of respect with these almost supernatural animals. In particular, they strove to avoid turning their thoughts to bears or hoping to see a bear. Today, many Inuit are troubled by the intense interest shown in polar bears by tourists, journalists, biologists and social scientists, believing this interest to be disrespectful and potentially dangerous.

Since the early 1960s, polar bears have been subject to management in Nunavut (then part of the Northwest Territories), with the first community quotas introduced in 1963. In the hamlet of Arviat, between 1962 and 1971, the average yearly harvest was six bears (McEachern, 1978). In the mid-1970s, Arviat's first quota was set at four bears per year, and by 1978 this had risen to 15. Prior to this time, the number of harvested bears varied from year to year, but once the quota was introduced, the number of bears harvested almost always reached the quota target. The introduction of quotas saw Arviarmiut[16] begin actively to hunt bears, while prior to this bears were killed only opportunistically or in self-defence.

In the mid-1980s, a flexible quota system was introduced across Arctic Canada in an attempt to better respond to evolving threats to polar bears and their habitat.[17] By the mid-1990s, Arviat's quota had risen to 20 bears per year, and remained static until 2004. Since 2005, the management of polar bears on the west coast of Hudson Bay has altered drastically. The reasons include the changing role of Inuit knowledge in wildlife management, scientific findings (from south of the border with Manitoba) which suggest that this sub-population of polar bears may be in decline, the status worldwide of polar bears as the poster species for climate change, and the decision by the US government to upgrade the status of polar bears under its Endangered Species Act. All of these have had a negative impact on Canada's management of its polar bear populations (Tyrrell, 2009b).

In January 2005, based on a combination of Inuit *qaujimajatuqangit*[18] and the research findings of polar bear biologists of the government of Nunavut, the territory-wide quota was increased by 28 per cent (Wiig, 2005). In Arviat, this signified a quota increase from 20 to 22 bears. Two years later in 2007, the government of Nunavut was forced to bow to international pressure (predominantly from the US Fish and Wildlife Service), leading to a decrease in Arviat's quota from 22 bears in 2006 to 15 bears in 2007. In 2008, the quota for Arviat was further slashed to just three bears. In the space of just over 40 years the relationship between Arviarmiut and polar bears has undergone a dramatic shift. In the mid-1960s, an average of six bears was harvested each year and people rarely encountered bears close to or within the community. With the introduction of quotas, Arviarmiut began actively hunting and, as the quota gradually increased over the decades – from 15 to 20 and finally peaking at 22 in 2005–2006 – so too the number of bears harvested also increased. Of course we cannot know if, in the absence of external polar bear management, Arviarmiut would have gradually increased the number of bears they harvested, as bears became a more regular feature of community life, particularly in the 2000s. But from observing the ways in which Arviarmiut continue to engage with other, non-managed species, it would be accurate to surmise that a respectful relationship would have continued to inform the hunting of polar bears.

Local hunters and trappers organizations (HTOs) across Nunavut claim that there is little or no community consultation, and local knowledge of bear populations and behaviour is overlooked by wildlife managers and policy-makers who are responsible for quota allocation.[19] Each year, once the quota has been decided at the territorial level, each community's locally elected HTO is allocated its share of the quota. Once allocated, it is then up to each HTO to decide how to distribute the quota among local hunters. The method of distri-bution varies between communities and is determined by the size of the quota and the number of hunters over the age of 16 living in each community. Arviat has a relatively large population of approximately 2000 people. In 2007, 1047 Arviarmiut were eligible to hunt polar bears. Arviat's HTO decided many years ago that the fairest way to distribute the quota was through a public lottery. Therefore, on the night of 31 October each year, the names of all Arvi-armiut over the age of 16, male and female, are entered in a draw. Taking 2007 as a typical year, the first five people whose names were drawn were offered the opportunity to sell their hunting tag back to the HTO for C$2500, which the HTO then used to outfit the hunting of polar bears by US trophy hunters. There were then ten quota tags remaining. The next ten people whose names were drawn had 48 hours from midnight to hunt and retrieve a polar bear. A list of alternate names was also drawn and if, at the end of the first 48-hour window, the quota was not filled, then these people had the next 48-hour window to go bear hunting. This continued until the entire quota was filled.

This scheme has led to a shift in peoples' relationships with polar bears. For many older Arviarmiut the loss of respect for polar bears is troubling. Hunting parties, eager to harvest a bear during their brief window of opportu-nity,[20] often race to the locations where there have been recent bear sightings. It is not uncommon to have five or more hunting parties in pursuit of one bear. Elders say that this is disrespectful and, like all other animals, hunters should wait for the bear to come to them, showing that it is a willing participant in the hunt. Unlike in the hunting of other species, hunters are no longer discerning in their polar bear hunting practice and often return home with bears that are barely over the legal age at which they can be hunted. Anecdotal evidence also suggests that each year some bears escape into the water with bullet wounds, and on one recent occasion a hunter shot a bear up to 20 times before it finally died. Such occurrences trouble many Arviarmiut, and suggest that the relation-ship between human and bear is almost beyond repair.

Polar bear hunting has also come to be perceived as a 'right' by Inuit. Prior to the polar bear lottery at the community hall each year, Arviarmiut are offered the opportunity to publically air their views to the HTO board on various wildlife issues. Heated debate over the polar bear hunt can often carry on for hours before the draw takes place, with many people in the crowded hall publicly voicing their opinions. What emerges most strongly each year from these debates is that Arviarmiut now view the bear hunt as a right. Young men protest that the lottery should be skewed so that those who have never hunted a bear will have a greater chance of having their names drawn. Older men protest that it has been many years since they had the opportunity to hunt a bear and the lottery should be skewed to favour them. Others protest that those who have successfully hunted a bear in the past five years should have their names removed from the draw. Men of all ages protest that women should not be included. Polar bears are no longer the spiritually and symbolically powerful animals they once were. Their role in Inuit culture has in many ways been demoted and they are now merely a symbol of Inuit rights.

In 2008, Arviat's polar bear quota was slashed to three. The HTO decided that its best option was to set this quota of three aside, in case they were required for defence kills. Shooting a bear in self-defence is extremely rare, with one such killing occurring perhaps once every three to four years. When a defence kill does occur, that bear is removed from the quota. The HTO reasoned that if no defence kills occurred then later in the year these three tags could be distributed in the usual way. In the first week of November 2008, when Arviat received its quota, an unprecedented four defence kills of polar bears took place.

This altered relationship has a detrimental effect on polar bear populations. As Inuit have come to view the bear hunt as their right, they have become determined to fulfil the quota each year. Rather than hunting up to 15 or 20 bears, local communities harvest exactly the number allocated. The occasional opportunistic or self-defence kill is a thing of the past, and hunters often pursue bears in a frenzied manner, eager to make the most of their small window of opportunity, thus paying little regard to the age or sex of the polar bears they pursue. All of this is counter-productive to the aims of wildlife management. Rather than leading to greater care and smaller numbers of bears being taken, management of the bear hunt has had the opposite effect.

Making Space for Non-Western Perspectives

It could be argued that Inuit culture has changed so dramatically in the past half century, it is inevitable that the relationship between humans and animals will alter too, and that maintaining relationships of respect is no longer a priority for today's more globalized Inuit. While it is true that many Inuit no longer adhere to the prescriptions and taboos that were once integral to hunting, the core values of sharing, avoiding waste and not taking more than one requires continue to be important aspects of Inuit society. Some Inuit do transgress these customary laws but they are quickly brought to task by being publicly chastised and reprimanded by their elders and peers via local CB and FM radio. In general, species that are not subject to wildlife management practices continue to be shown respect. This respect may be informed by traditional beliefs or, more commonly today, by the belief that animals have been given to Inuit by the Christian God and treating the animals with respect is a means of demonstrating respect for God.

What is intriguing about these two case studies is how closely one mirrors the other. The beluga hunt has never been managed on the west coast of Hudson Bay in Arviat, nor has the polar bear hunt been managed in Nunavik, thus providing an opportunity to observe how Inuit communities hunt these species under different regimes. In Arviat, beluga whale hunting is a relaxed affair. The number of belugas harvested has remained static for the past 30–40 years. Hunters take to the sea in their small boats during the two-month beluga migration each summer. Boats are scattered along the coastline, with hunters enjoying the opportunity to get away from the community and relax, often in the company of younger family members (such as sons and brothers). Hunters rarely actively seek out belugas, but rather wait for belugas to come to them before engaging in the hunt. It is rarely, if ever, a frenzied or rushed affair. Sometimes a beluga is hunted by one boat, or if large pods migrate past on a particular day, upwards of 20 small boats might take to the water to maximize the harvest. When belugas are killed close to the community, flensing and butchering becomes a communal affair, with word quickly spreading across the hamlet via CB and FM radio. Men, women and children all descend on the site to participate and share in the celebration of the harvest. An unsuccessful hunt is not viewed as a failure, as hunters also use their time at sea to fish for arctic char, gather mussels along exposed reefs and occasionally harvest a ringed or harp seal. Beluga hunting is a social event, a time for banter and conversation and, ultimately, a time of celebration.

Meanwhile, across Hudson Bay in Nunavik, the management of polar bears is virtually non-existent. Bears[21] in Nunavik are only killed in self-defence. When bears do enter communities, every effort is made to turn them away using bangers and other loud noises, or firing shots in the air. Only as a last resort, or if someone faces imminent danger of a bear attack, is a polar bear killed. Nunavimmiut do not speak of having a 'right' to hunt polar bears, and as one hunter said, there is a general sense that 'if you leave them alone, they'll leave you alone'. Polar bear hunting is, thus, a relatively rare occurrence in Nunavik.

The manner in which Arviarmiut hunt beluga whales and Nunavimmiut hunt polar bears reflects how Inuit from both regions recall hunting prior to the advent of external management and the imposition of hunting quotas. In both examples, the perverse management of Inuit hunting practices has had a detrimental impact on the enculturation and enskilment of young people, and in turn on Inuit perceptions of these animals. The imposition of wildlife regulations has brought about cultural, perceptual and relational changes. And each of these changes has arguably had a negative impact on the animals that conservationists strive so hard to protect. Those who set the rules of management continue to misunderstand and misinterpret the cultural and symbolic role of animals locally, assuming that through the act of hunting, animals are a mere commodity for Inuit.[22] In fact, it would appear that the imposition of management practices actually serves to transform these animals into commodities and no longer sentient co-inhabitants of the Arctic environment. In their transformation into commodities, they have become a medium of expression of the Inuit right to cultural self-expression and self-determination.

For biodiversity conservation to achieve its goals, conservationists must develop an awareness and understanding of non-Western relationships with the environment and its constituent components. It is all too easy to assume that common access to wildlife is a simple case of Hardin's Tragedy of the Commons (1968), and thus that human reliance on the natural environment for livelihoods negatively affects biodiversity in all circumstances. There are many lessons to be learned from the way indigenous peoples think about and relate to the animals and plants with which they share their environment. In the Canadian Arctic, some steps have been taken to incorporate Inuit *qaujimajatuqangit*, and other forms of indigenous knowledge, into wildlife management practices. Most of these can be attributed to the growing levels of self-governance among indigenous groups in Canada, and their growing voice around the wildlife management table.

However, the elements of indigenous knowledge that are incorporated into wildlife management are all too often in the form of quantifiable knowledge (e.g. where, when and how many). What is much more difficult to record and make use of are the social, cultural and semiotic roles that animals, and the environment in general, play in the lives of indigenous peoples. Cultural barriers are often difficult, uncomfortable and time-consuming to surmount, and the quality of the relationship between Inuit and a particular species is lost in the much easier task of uncovering how many of that species were sighted at specific places and times. While the latter may be of interest to the researching biologist, it is of little interest to the Inuit hunter, who knows that animals are always on the move, responding not only to environmental phenomena but also to the thoughts and actions of respectful or disrespectful humans.

By reducing animals to mere statistics to be manipulated through restrictions on numbers hunted, the relationship between Inuit and these animals will continue to be overlooked. This will be to the detriment of both Inuit culture and animal populations. This is not merely a concern for the future of Arctic biodiversity. Across the world, conservationists tend to overlook the extensive and diverse knowledge possessed by local peoples whose lives are intimately entwined with the environments in which they live. While scientifically-informed approaches to conservation are imperative to the future of biodiversity, so too are non-scientific, relational and culturally-informed approaches. On the surface, customary conservation practices are often difficult to understand, as people with different cultural perspectives do not frame their practices in the same way. But, as this chapter has highlighted, the outcomes of concepts such as respect, animals as neighbours and so on, frequently lead to the protection of species and ecosystems. Conservation scientists should no longer ignore the vast knowledge possessed by non-scientific experts, and the relational and semiotic roles of animals among non-Western peoples. Not only will the cultural diversity of the world diminish, taking with it infinite ways of knowing and understanding the world, but so too will opportunities to protect and enhance biodiversity. If, however, we can learn to open our eyes to other ways of knowing and being in the world, then not only will the uniqueness of indigenous cultures continue to thrive, so too will the health and vitality of the diverse environments that these peoples call home.

Notes

1 Throughout this chapter I refer collectively to Esk-Aleut speaking peoples as Inuit. However, across Arctic homelands, these people refer to themselves variously as Yup'ik, Iñupiat, Eskimos, Inuit and Kallalit. These peoples share a common language (distinguished by dialect), a common culture, a common history, and a common relationship with the environment and Arctic animals (see Burch, 1988; Sonne, 1988; Hoffecker, 2005 for discussion of Esk-Aleut prehistory, history and culture).

2 Historically, and in the present day, animals are harvested not only for immediate consumption by the household, but also for preservation (freezing, drying, fermenting) and consumption by household and extended kin, later in the season or year. Wildlife biologists have in the past misrepresented these practices as 'wanton slaughter' (Foster and Hammond, 1998; Usher, 2004).

3 Inuit continue to own vast numbers of sled dogs. Inuit in eastern Greenland still rely almost exclusively on sled dogs for winter and spring excursions to the floe edge to hunt seals, while individual Inuit in Canada continue to keep sled dogs for cultural reasons, and also to engage in small-scale tourism, to accommodate polar bear trophy hunting by non-Inuit, and to participate in the many sled dog races that are organized across the Arctic each year. In one typical example, in 2006 in Arviat, Nunavut, a hamlet of 2000 people, 27 dog teams were registered with Arviat Racing Club. With approximately 10–15 dogs per team (including pups) this amounted to around 400 dogs.

4 Each autumn Inuit also take advantage of the profusion of berries (e.g. cranberries, cloudberries, crowberries) that grow on the tundra, consuming some immediately and freezing the rest for use throughout the year.

5 *Qablunaat*: Euro-North Americans; non-Inuit people. This spelling is in the Kivalliq dialect from the west coast of Hudson Bay.

6 It is easy to imagine the potential cultural difficulties that can arise when southern researchers conduct harvest surveys.

7 While soapstone carving is very often done for commercial purposes, I refer here to the whittling and carving that many men engage in simply to pass the time as they await the arrival of animals.

8 See Kalland and Sejersen (2005), who argue that the management of marine mammals is, in effect, the management of marine mammal hunting practices.

9 Minutes from a 1953 meeting representing the interests of the Canadian Wildlife Service read as follows: the solution to the caribou 'problem' lay 'in educating the Eskimos themselves to realize the necessity and reasons for sound conservation practices'. The same minutes called for 'the movement of Eskimos ... to places where they can be assured of being able to make a better living' (Sandlos 2007).

10 Many authors (Brody, 1975; Tester and Kulchyski, 1994; Kulchyski and Tester, 2007) frame this within part of the overall colonial endeavour of the Canadian state. Throughout the 20th century, Canada has sought to stamp its sovereignty over its Arctic territory. An important part of this process has been the attempted transformation of the Arctic into a 'productive' place, where both Inuit and settlers are engaged in productive endeavours such as mining and industry. Diverting Inuit off the land and away from their traditional hunting lifestyles, and into 'modern' ways of living were seen (and in some cases are still seen) by government as a welcome improvement for Inuit, but also a means of making the north truly Canadian.

11 See Tyrrell (2008) for a descriptive account of the cultural, social, economic and nutritional importance of the beluga whale for Nunavik Inuit.

12 Nunavimmiut: the people of Nunavik.

13 The phrase 'hunting complex' (after Wenzel 1991, and others) expresses the complexity and social interconnectedness of all aspects of the hunt, from the preparation by women and men of the clothing and tools essential to the hunt, the sociality of the hunt itself, and the preservation and preparation of meat and pelts. It conveys the sense that hunting is an ongoing process involving extended families and entire communities, and is so much more than the mere stalking and killing of animals.

14 There are obvious reasons for this: predation by other species cannot be managed; the Greenland halibut fishery is outside of Canada's jurisdiction; and the impact of habitat degradation on belugas has, for the most part, only been discovered after the massive hydro-electric projects of the 1970s and 1980s had been completed.

15 For more detailed discussion of the history and philosophy behind beluga management in Nunavik, see Kishigami (2005) and Tyrrell (2007, 2008).

16 Arviarmiut: the people of Arviat.

17 See Fikkan et al (1993), Tyrrell (2006) and Freeman and Foote (2009) for detailed accounts of polar bear management.

18 Inuit *qaujimajatuqangit*: Inuit knowledge.

19 These policy-makers comprise a board of Inuit hunters and southern wildlife biologists (the latter permanently employed by the government of Nunavut).

20 In early November, this 48-hour hunting window amounts to a total of only 12 hours daylight, and there are few hunters willing to go in pursuit of polar bears during the hours of darkness.

21 Northern Quebec has a much smaller polar bear population than that on the west coast of Hudson Bay.

22 I recently submitted an article to a peer-reviewed journal. The title, 'Belugas are our neighbours', was a direct quotation from an interview with a respected Inuit elder who was one of my research informants. One of the reviewers, an eminent beluga whale biologist who works in Nunavik, suggested that my title was inappropriate and would make more sense if I altered it to 'Belugas are our food'.

References

Barkham, S. de L. (1980) 'A note on the Strait of Belle Isle during the period of Basque contact with Indians and Inuit', *Études/Inuit/Studies*, vol 4, no 1–2, pp51–58

Bird-David, N. (1990) 'The giving environment: Another perspective on the economic systems of gatherer-hunters', *Current Anthropology*, vol 31, no 4, pp189–196

Bodenhorn, B. (1990) 'I'm not the great hunter, my wife is: Inupiat and anthropological models of gender', *Études/Inuit/Studies*, vol 14, no 1–2, pp55–74

Borré, K. (1991) 'Seal blood, Inuit blood and diet: A biocultural model of physiology and cultural identity', *Medical Anthropology Quarterly*, vol 5, no 1, pp48–62

Bourdages, H., Lesage, V., Hammill, M. O. and de March, B. (2002) 'Impact of harvesting on population trends of beluga in Eastern Hudson Bay', DFO, Canadian Science Advisory Secretariat, Research Documents, 2002/036

Brody, H. (1975) *The People's Land*, Douglas and McIntyre, Vancouver, Canada

Brody, H. (2000) *The Other Side of Eden: Hunter-gatherers, Farmers and the Shaping of the World*, Faber and Faber, London, UK

Burch, E. S. Jr. (1988) 'Knud Rasmussen and the "original" inland Eskimos of Southern Keewatin', *Études/Inuit/Studies*, vol 12, no 1, pp81–100

COSEWIC (Committee on the Status of Endangered Wildlife in Canada) (2004) 'Assessment and update status report on beluga whale, *Delphinapterus Leucas*, in Canada', COSEWIC, Ottawa, Canada

Damas, D. (2002) *Arctic Migrants, Arctic Villagers: The Transformation of Inuit Settlement in the Canadian Arctic*, Mc-Gill-Queen's University Press, Montreal, Canada

DFO (2005a) 'Nunavik and adjacent waters beluga management plan 2005', DFO, Canada

DFO (2005b) 'Stock assessment of northern Quebec (Nunavik) beluga (Delphinapterus Leucas)', DFO, Canada, Science Advisory Report 2005/020

Doidge, D. W., Adams, W. and Burgy, C. (2002) 'Traditional ecological knowledge of beluga whales in Nunavik: Interviews from Puvirnituq, Umiujaq and Kuujjuaraapik', Report 12-419 of the Nunavik Research Centre submitted to Environment Canada's Habitat Stewardship Program for Species at Risk, Project PH-2001-2-2002, Makivik Corporation, Kuujjuaq, Canada

Downie, D. L. and Fenge, T. (eds) (2003) *Northern Lights Against POPs: Combatting Toxic Threats in the Arctic*, Mc-Gill-Queen's University Press, Montreal, Canada

Fienup-Riordan, A. (1988) 'Eye of the dance: Spiritual life of Bering Sea Eskimo', in W. W. Fitzhugh and A. Crowell (eds) *Crossroads of Continents: Cultures of Siberia and Alaska*, Smithsonian, Washington, DC, pp256–270

Fikkan, A., Osherenko, G. and Arikainen, A. (1993) 'Polar bears: The importance of simplicity' in O. Young and G. Osherenko (eds) *Polar Politics: Creating International Environmental Regimes*, Cornell University Press, London, pp96–151

Foster, J. and Hammond, L. (1998) *Working for Wildlife: The Beginning of Preservation in Canada*, University of Toronto Press, Toronto, Canada

Freeman, M. M. R. and Foote, A. L. (eds) (2009) *Inuit, Polar Bears and Sustainability*, Canadian Circumpolar Institute Press, Edmonton, Canada

George, S. (2006) 'Beluga hunters face cuts if quotas exceeded', *Nunatsiaq News*, 9 June

Guemple, L. (1986) 'Men and women, husbands and wives: The role of gender in traditional Inuit society', *Études/Inuit/Studies*, vol 10, no 1–2, pp9–24

Hammill, M. (2001) *Beluga in northern Quebec: Impact of harvesting on population trends of beluga in eastern Hudson Bay*, DFO, Canadian Science Advisory Secretariat, Research Document 2001/025

Hammill, M. O., Lesgae, V., Gosselin, J. F., Bourdages, H., de March, B. G. E. and Kingsley, M. C. S. (2004) 'Evidence for a decline in northern Quebec (Nunavik) belugas', *Arctic*, vol 57, no 2, pp183-195

Hardin, G. (1968) 'The tragedy of the commons', *Science*, vol 162, pp1243–1248

Hoffecker, J. F. (2005) *A Prehistory of the North: Human Settlement of the Higher Latitudes*, Rutgers University Press, New Brunswick, Canada

Ingold, T. (2000) *The Perception of the Environment: Essays in Livelihood, Dwelling and Skill*, Routledge Press, London

Johannes, R. E., Freeman, M. M. R. and Hamilton, R. J. (2000) 'Ignore fishers' knowledge and miss the boat', *Fish and Fisheries*, vol 1, pp257–271

Jolles, C. Z. (2002) *Faith, Food and Family in a Yupik Whaling Community*, University of Washington Press, Seattle

Kalland, A. and Sejersen, F. (2005) *Marine Mammals in Northern Cultures*, Canadian Circumpolar Institute, Edmonton, Canada

Kishigami, N. (2005) 'Co-management of beluga whales in Nunavik (Arctic Quebec), Canada', in N. Kishigami and J. M. Savelle (eds) *Indigenous Use and Management of Marine Resources*, Senri Ethnological Studies 67, The National Museum of Ethnology, Osaka, Japan, pp121–144

Kuhnlein, H., Receveur, O. and Chan, H. M. (2001) 'Traditional food systems research with Canadian indigenous peoples', *International Journal of Circumpolar Health*, vol 60, no 2, pp112–122

Kuhnlein, H., Chan, L. H. M., Egeland, G. and Receveur, O. (2003) 'Canadian Arctic indigenous peoples, traditional food systems, and POPs', in D. L. Downie and T. Fenge (eds) *Northern Lights Against POPs: Combatting Toxic Threats in the Arctic*, McGill-Queen's University Press, Montreal, Canada, pp22–40

Kulchyski, P. and Tester, F. J. (2007) *Kiumajut (Talking Back): Game Management and Inuit Rights 1900–1970*, University of British Columbia Press, Vancouver, Canada

Laugrand, F. (2002) *Mourir et Renaître: La Réception du Christianisme par les Inuit de l'Arctique de l'Est Canadien*, Les Presses de l'Université Laval, Montreal, Canada

Lesage, V. and Doidge, D. W. (2005) 'Harvest statistics for beluga whales in Nunavik, 1974–2004', DFO Canada Science Advisory Secretariat, Canada

Lowenstein, T. (1999) *Ancient Land: Sacred Whale*, The Harvill Press, London

McEachern, J. (1978) 'A survey of resource harvesting, Eskimo Point, N.W.T.', Quest Socio-Economic Consultants, Delta, BC, Canada

Nelson, R. K. (1986) *Make Prayers to the Raven: A Koyukon View of the Northern Forest*, Chicago University Press, Chicago

Northern Contaminants Programme (2003) 'Canadian Arctic contaminants assessment report', Ministry of Indian and Northern Affairs, Ottawa, Canada

Nuttall, M. (1997) 'Nation-building and local identity in Greenland: Resources and the environment in a changing north', in S. A. Mousalimas (ed) *Arctic Ecology and Identity*, International Society for Trans-Oceanic Research, Los Angeles, pp69–84

Pars, T., Osler, M. and Bjerregaard, P. (2001) 'Contemporary use of traditional and imported food among Greenlandic Inuit', *Arctic*, vol 54, no 1, pp22–31

Pelly, D. (2001) *Sacred Hunt: A Portrait of the Relationship Between Seals and Inuit*, Greystone Books, Vancouver, Canada

Richard, P. (2005) 'An estimate of the Western Hudson Bay beluga population size in 2004', CSAS Research Document 2005/017, Canada

Saladin D'Anglure, B. (1990) 'Nanook, super-male: The polar bear in the imaginary space and social time of the Inuit of the Canadian Arctic', in R. Willis (ed) *Signifying Animals: Human Meaning in the Natural World*, Routledge, London, pp178–195

Sandlos, J. (2007) *Hunters at the Margins: Native People and Wildlife Conservation in the Northwest Territories*, University of British Columbia Press, Vancouver, Canada

Sandlos, J. (2009) 'NiCHE: Network in Canadian history and environment', Pod-cast, 16 February, http://niche.uwo.ca/naturespast

Searles, E. (2002) 'Food and the making of modern Inuit identities', *Food and Foodways*, vol 10, no 1, pp55–78

Siku News (2007) 'Beluga hunt ends in northern Quebec's Hudson Strait', *Siku News*, 16 November

Sonne, B. (1988) 'In love with Eskimo imagination and intelligence', *Études/Inuit/Studies*, vol 12, no 1, pp21–44

Stevenson, M. G. (1996) *Inuit, Whalers and Cultural Persistence*, Oxford University Press, Toronto, Canada

Stirling, I., Lunn, N. J. and Iacozza, J. (1999) 'Long-term trends in the population ecology of polar bears in western Hudson Bay in relation to climate change', *Arctic*, vol 52, no 3, pp294–306

Tester, F. J. and Kulchyski, P. (1994) *Tammarniit (Mistakes): Inuit Relocation in the Eastern Arctic 1939–1963*, University of British Columbia Press, Vancouver, Canada

Turner, E. (1990) 'The whale decides: Eskimos' and ethnographers' shared consciousness on the ice', *Études/Inuit/Studies*, vol 14, no 1–2, pp39–52

Tyrrell, M. (2006) 'More bears, less bears: Inuit and scientific perceptions of polar bear populations on the west coast of Hudson Bay', *Études/Inuit/Studies*, vol 30, no 2, pp191–208

Tyrrell, M. (2007) 'Sentient beings and wildlife resources: Inuit, beluga whales and management regimes in the Canadian Arctic', *Human Ecology*, vol 35, no 5, pp575–586

Tyrrell, M. (2008) 'Nunavik Inuit perspectives on beluga whale management in the Canadian Arctic', *Human Organization*, vol 67, no 3, pp322–334

Tyrrell, M. (2009a) 'Western Hudson Bay polar bears: The Inuit perspective', in M. M. R. Freeman and A. L. Foote (eds) *Inuit, Polar Bears and Sustainability*, Canadian Circumpolar Institute Press, Edmonton, Canada, pp 95–110

Tyrrell, M. (2009b) 'Guiding, opportunity, identity: The multiple roles of the polar bear trophy hunt', in M. M. R. Freeman and A. L. Foote (eds) *Inuit, Polar Bears and Sustainability*, Canadian Circumpolar Institute Press, Edmonton, Canada, pp 25-38

Usher, P. (2004) 'Caribou crisis or administrative crisis? Wildlife and Aboriginal policies on the Barren Grounds of Canada, 1947–1960', in D. G. Anderson and M. Nuttall (eds) *Cultivating Arctic Landscapes*, Berghahn Books, Oxford, pp172–199

Wein, E. E., Freeman, M. M. R. and Makus, J. C. (1996) 'Use of and preference for traditional foods among the Belcher Island Inuit', *Arctic*, vol 49, no 3, pp256-264

Wenzel, G. (1991) *Animal Rights, Human Rights: Ecology, Economy and Ideology in the Canadian Arctic*, University of Toronto Press, Toronto

Wiig, O. (2005) 'Are polar bears threatened?' *Science*, vol 309, p1814

Williamson, R. G. (1974) *Eskimo Underground: Socio-cultural Change in the Canadian Central Arctic*, Institutionen För Allmän Och Jämförande etnografi vid Uppsala Universitet, Uppsala, Sweden

8
Challenging Animals: Project and Process in Hunting[1]

Garry Marvin

Introduction

In his essay 'Why Look at Animals?', John Berger (1980) suggested that with industrialization and urbanization animals have become, and will continue to become, marginal to everyday human lives. In some senses he is correct – for example, in the industrialized and urban world, the processes of the living and dying of animals that become food are scarcely experienced by those who eat them – but animals are still present in our lives and even our cities teem with animal life. The number of people who keep pets increases, millions of people go to zoos every year, our television channels are full of animals in documentaries, they appear in movies, cartoons, as main characters in children's books, and they still inhabit the toy sections of department stores and supermarkets. These are all animals that are incorporated into our lives. Adrian Franklin's (1999) study of human-animal relations suggests that although fewer and fewer people live in close proximity to what might be categorized as wild animals, such animals still seem to draw a wide range of people to the spaces they inhabit in order to attempt to see them there. He points to animal watching trips, photographic safaris and forms of ecotourism as examples of the continuing allure of wild animals in wild places and the possibility of encounters with them (see also Bulbeck, 2005; Peace, 2005; Servais, 2005). These kinds of touristic practices have often been labelled as 'non-consumptive' wildlife practices, in which the encounter between the human and the non-human usually remains at the level of observation, captured only by photographic images. In this chapter my focus will be on a different, 'consumptive' set of practices that are underpinned by the desire to encounter and engage with wildlife – hunting as a recreational activity.[2]

Recreational forms of hunting are the most contested of events and practices in which humans seek an engagement with the natural world. Much has been written about hunting from the perspectives of morality and ethics, and it has often been condemned as an unacceptable form of engagement with wild animals today. The charge of the immorality of hunting as a recreational activity not only relates to the fact that it results in the deaths of wild animals, but to the idea that such deaths are unnecessary except for the pleasure that they afford to those who hunt. The pleasure of hunting, from this condemnatory perspective, is centred on the supposed enjoyment of killing animals. In this chapter my aim is to suggest that recreational hunting is a complex cultural practice and that to focus solely on killing, on the deaths of animals, obscures the complexities of the purposes, processes and practices of hunting. Not only is hunting something worth investigating as a cultural practice, but the results of such an investigation may well offer a significant contribution to other debates. Recreational hunting intersects with many forms of wildlife management and environmental management issues, and I believe that hunting needs to be more fully understood in and of itself in order to assess its relationships to, impact on and contribution to such issues.

The focus here is on the nature of this hunting, how it is practiced and experienced, and how it is thought about and reflected on from the inside, from the points of view and the perspectives of hunters themselves.[3] As an anthropologist, I seek to understand social and cultural worlds from the perspectives of those who live in these worlds. My research into hunting has therefore been one of ethnographic, anthropological fieldwork by spending as much time as I can with hunters. This form of research offers the opportunity to observe actual hunting practices and to listen to hunters' conversations and commentaries during the course of hunting. My interest is not so much in why people hunt – their personal motivations or why they choose to hunt – rather it is in the construction of hunting; how it is to hunt; what is going on in hunting in terms of the relationships between hunters and the hunted, and between hunters and the spaces in which they hunt; the nature of the experiences that hunters seek; and how the conditions and situations are created for such experiences.

Although there is a rich, historical literature that consists of individual accounts of hunting, there is little academic work that explores the nature of modern, recreational, hunting practices. This is in marked contrast to the ethnographic material that examines hunting in those societies in which it forms an important part of the procurement of food. This literature, particularly relating to indigenous peoples and their hunting practices, suggests that the relationships between human hunters and animal prey are multifaceted. In these societies the acquisition of meat is both an intention and end of hunting, but the hunted animal is far from being a utilitarian object for the hunters and hunting is underpinned by complex beliefs concerning the appropriate, proper and necessary relationships between hunters and potential prey. Although such hunting requires acts of violence against animals as they must be killed, the beliefs about the nature of that violence certainly do not seem to be related to notions of aggression or domination.

It is impossible here to review all the social and cultural specificities of these beliefs and their accompanying practices, but it is worth highlighting some aspects that can be traced across a wide range of hunting cultures and which are at the core of forms of modern recreational hunting. A fundamental aspect is the notion of an essential and necessary respect that the hunter holds for the prey, and the manner in which the hunter orientates himself (in such cultures it is usually a male activity) towards his potential prey. Hunted animals are not regarded as mere animate objects but rather as creatures possessing immensely valuable lives and often as having personhood. Such creatures deserve to be treated with the greatest respect, and without it, hunting is unlikely to be successful. In many such cultures there are beliefs that being hunted is not something that is imposed on animals. Rather, hunting is something that they willingly participate in, so long as the hunter approaches them in the appropriate ritual and respectful manner. Such cultures often incorporate even more complex beliefs that hunted animals offer themselves to the hunters and that, rather than simply taking, hunting is based on receiving animals.[4]

I would like to stress here that hunting is not an incidental activity: it is a directed, purposeful and serious activity that depends on and requires both an attitude of respect and the practice of respect. Although I do not want to argue for any direct connection between indigenous hunting and modern recreational hunting, I suggest there are continuities, in terms of notions of seriousness, respect and responsibility, at the heart of both forms of hunting, that are the focus of this chapter.[5]

The Nature of the Hunter

I have been using the term 'hunter' rather unspecifically but it is important to emphasize that hunters do not constitute a homogeneous group, even within recreational hunting. Although there is not the space in this chapter to discuss how one might distinguish different types of hunters and different forms of recreational hunting, I do believe that a full account of hunting would need a careful delineation of both hunters and practices. Again, I am not speaking about the psychology of hunters but rather how different sets of hunters might be identified in terms of their orientation towards aspects of hunting.

Stephen Kellert's (1978) study of 'Attitudes and Characteristics of Hunters and Antihunters' relating to the US is a useful starting point. The research for the study involved both in-depth interviews and a set of highly structured questionnaires. Kellert distinguished three categories of hunters: utilitarian/meat hunters, dominionistic/sport hunters and nature hunters. Here I will offer extracts of Kellert's defining characteristics of each, but my emphasis will be on 'nature hunters' because I believe the hunters I have been working with most closely relate to this category.

Kellert (1978) suggested that:

> *Utilitarian/meat hunters appeared to perceive animals largely from the perspective of their practical usefulness. Of primary consideration was the animal's utility and its potential for returning*

some tangible material reward. The utilitarian/meat hunter was not typically concerned with the virtues of the living animal, but rather with those of the dead... The utilitarian/meat hunter viewed hunting as a harvesting activity and wild animals as a harvestable crop not unlike other renewable resources.

For the dominionistic/sport hunters:

The hunted animal was valued largely for the opportunities it provided to engage in a sporting activity involving mastery, competition, shooting skill and expressions of prowess... They were not items of food but trophies, something to get and display to fellow hunters... A frequent accompaniment of the dominionistic/sport hunting focus was a stress on prowess and masculinity ... as a way of expressing strength, aggressiveness and physical endurance.

For the nature hunters:

The desire for an active, participatory role in nature was perhaps the most significant aspect of the nature hunter's approach to hunting. The goal was the intense involvement with wild animals in their natural habitats. Participation as a predator was valued for the opportunities it provided to regard oneself as an integral part of nature... The primary focus was on wildlife, and the opportunities hunting provided for first-hand contact with wild animals in natural settings. Even the prey animals [sic] was perceived as an object of strong affection and respect.

Kellert alerts us not to read these as exclusive categories. Although hunters tended 'to be orientated toward one *primary* attitudinal relation to hunting', they were 'typically characterized by more than one attitude'. I also suggest that what is important for understanding hunting is that it is not simply a matter of 'attitudinal relation' but a matter of attitude informing and guiding hunting practice on different occasions and for different purposes. Rather than using these types as a classification or categorization of *hunters*, I find them more useful for thinking about modes of *hunting*, or as different orientations to hunting in any particular hunting event. Different modes or orientations may come to the fore on different occasions. For example, a hunter interested in hunting for meat can also be a naturalistic hunter, while being a nature hunter does not preclude that person from seeking a trophy that will later be displayed. A utilitarian/meat hunter might be hunting in order to have meat in the freezer but may also seek the experience of intense involvement and hunt in a manner to achieve this. A hunter whose primary orientation is to obtain a trophy, and so apparently a dominionistic hunter, might only pursue this goal in terms of self-imposed restrictions based on how they believe it is acceptable to hunt. If it is important to consider the place of hunting in the context of wildlife management and environmental management issues, then it is of great value to understand the nature of these different modes of hunting.

In the remainder of this chapter I explore the orientations of hunters who would closely identify themselves with Kellert's description of 'nature hunters'. For these hunters the most important element of hunting is an active, participatory role in nature. It is thus the notion/nature of participation they seek that is crucial to understanding this form of hunting and this mode of *being* a hunter. It is the desire for, and experience of, intense involvement with wild animals in their natural habitat that constitutes the 'challenge' in the title of this chapter.

Project and Process

I have taken the terms 'project' and 'process' from Allen Jones' evocatively titled *A Quiet Place of Violence* (1997). Jones is a hunter and writer, and this work is a meditation on the experiences, practices and ethics of hunting. By 'project' Jones refers to the intention of a hunter, the intention of finding, pursuing or waiting for, and killing a wild animal. The 'process' is how the hunter accomplishes this project: 'The *project* may be to kill an animal, but the project is useful only to the extent that it allows us to orientate ourselves to the process.' Jones offers, by way of clarification, another example of a recreational pursuit in the natural world. The project of a mountain climber would be that of reaching the summit of a mountain but this is not the climber's purpose. If simply standing on a summit is the purpose of the climber then they could be deposited there by helicopter but it would hardly count as an achievement. What is important is the process of achieving the intention, the process of climbing, the engagement or, as Jones puts it, the orientation, with the mountain. The process in hunting is how hunters orientate themselves towards both the hunted animal and the environment that constitutes the habitat of the animal. For an event to be hunting, in Jones' terms, a hunter should not set out to find, pursue and kill an animal by any means possible: 'The animal must never be a means to an end. If you want to acquire a trophy so badly that you're willing to do anything to get it – if you let the project take precedence over the process – then the animal has been reduced to an object, a means to an end.' There is an end to any particular hunting project but the means is far more significant for nature hunters than that end. As David Petersen (2000) writes, in comment on Jones' position: 'In all true hunting, process outranks project.'

In order to explore this participation and the mode of being a hunter I start from the apparent end of hunting – and I am using the term 'end' to refer to both purpose and completing or finishing – the killing of animals in hunting. This end is only of significance to the nature hunter in terms of that which precedes it. When I embarked on this hunting research, it appeared clear to me that I was going to attempt to understand why hunters killed animals, because it seemed, from the outside, that this was their most obvious intention. If that were not their intention then why would a hunter set off into the woods or the mountains carrying a shotgun or a rifle? However, it soon became apparent that I needed to study not why but how hunters killed animals and, furthermore, that it was impossible to understand or explain the cultural practice that is hunting simply by focusing on killing as the hunter's reason for hunting. A useful perspective is a passage by the Spanish philosopher José Ortega y Gasset in his essay on hunting:

> *In utilitarian hunting, the true aim of the hunter, that which he looks for and values, is the death of the animal. Everything that goes before is purely a means to achieve this end which is its proper objective. But in sports hunting this order of means and ends is reversed. The sportsman is not interested in the death of the animal, that is not what is sought. What interests him is everything that he has had to do to bring it about, that is, to hunt. In this way the actual objective is that which was previously only a means. Death is essential because without it there is no authentic hunting: the death of the animal is its natural end and finality: that of the hunt in itself, not that of the hunter. The hunter endeavours to achieve this death because it is the sign which gives truth to the whole hunting process, nothing more. In summary, one does not hunt in order to kill, but rather the reverse, one kills in order to have hunted* (Ortega y Gasset, 1968, translated by author).[6]

Once again this is a relationship between project and process. This elliptical point that the hunter does not hunt in order to kill an animal but only kills in order to have hunted opens up a complex set of issues. I will shortly come to the issue of hunting being that which occurs before the kill but the notion of killing demonstrating that hunting has taken place, and suggest it is a little more complex than Ortega y Gasset suggests. Instead, hunters that have spoken to me say that killing should only come about after hunting. That is to say there is a notion of proper killing in hunting – a legitimate or acceptable kill only occurs at the end of a legitimate/acceptable hunting event. Reversing the perspective, killing a wild animal at the end of a hunting event does not mean that the event was a legitimate or acceptable hunt; even if the hunter does not break the law. I later also explore how a hunting event can be regarded as successful, even though an animal is not killed at the end of it.

What is Proper Hunting?

There are several elements to the notion of properness. A key element is that a wild, potentially huntable, animal should be able to escape the attention and intentions of the hunter.[7] The huntable or hunted animal should remain a free, unrestricted and active agent in its own environment. Hunters do not want the animals they seek to hunt to conform to human demands or to be artificially restricted. The wild animals they seek to engage with should be able to resist the human. They should be able to exercise their natural, animal abilities – to remain hidden, to flee, to be in inaccessible places and, in some cases, even to attack. Those hunters with whom I have spoken were all of the opinion that if an animal really could not escape from the hunter then, even if what went on before the hunter killed looked like hunting, it was not for them true hunting.

Many of them mentioned game ranches on which wild animals are bred for hunting. In the opinion of these hunters, however extensive the ranch, this was not a truly wild space, and however long the pursuit, this was not hunting because, in the end, the desired animal would be found and it could never escape the enclosure. In such an event there would be no challenge and hence it could not be called hunting. Killing is guaranteed from the beginning and hence there is no value in the process. An interpretive point can be suggested here. The fact that such animals are specially bred and raised on ranches moves apparently wild animals towards being domesticated animals, animals that are controlled and cared for by humans; domesticated animals are rarely regarded as huntable animals. Hunters also commented that such pseudo-hunting events were unnatural in that game ranches often raise non-native species, for example, African species raised on game ranches in the US, and that huntable animals should be both wild and hunted in their natural habitat. One of my Spanish hunting friends described how he had shot a trophy quality elk in a game park – he valued the trophy antlers 'as a natural work of art' but there was no challenge in shooting it – 'there were 250ha but however many turns around it I made it was always going to be there'. He confided that, 'it is the most shameful animal I have here [in his trophy collection]' because it was not a properly hunted animal.

The same hunter, attempting to explain his views on the nature of hunting, commented that not all huntable wild animals could or should be hunted. On an evening walk in the mountains of northern Spain we were searching for Spanish Ibex, wild mountain goats. We came across a group feeding on a patch of grass near the path. I took out my camera and my hunting friend directed me to a large male which had trophy quality horns. I asked him whether this creature was worthy of being hunted as a trophy animal. His response was that although Spanish Ibex in other parts of the country were huntable, the animals in this particular area were not; they were too accustomed to humans and, although wild creatures, they were effectively 'tame'. 'I have taken my daughters to see them and they have eaten bread out of our hands, that's why I couldn't shoot one of them, I wouldn't like that at all.' Here the appropriate relationship between wild animals and humans was configured as one of distance and the approaches of these wild animals to humans were unnatural. They refused to flee in the presence of humans and so, as in the case of game ranch animals, shifted towards being like domesticated and, hence, unhuntable animals.

A second element of the properness of hunting relates to the abilities of the hunters compared with those of the hunted. Nature hunters are very clear that hunters should voluntarily restrict their ability to find and kill prey if they are interested in, what some have termed, the 'fair chase'. A clear state of this is Posewitz's (1994) declaration: 'Fundamental to ethical hunting is the idea of fair chase. This concept addresses the balance between the hunter and the hunted. It is a balance that allows hunters to occasionally succeed while animals generally avoid being taken.' This voluntary restriction applies in particular to an excess of technology. Hunters must make use of instruments of technology, shotguns, rifles, even bows, and all of these allow hunters to kill at a distance. While accepting that they must make use of such weapons, and they also accept that binoculars or field glasses are acceptable to enable hunters to see at a distance, they are critical of hunters who rely on technology to get themselves into a position where they can use their weapons. They are particularly critical of hunters who use vehicles to drive close to huntable animals or who use vehicles to move animals into a position where they can be shot. For many the greatest sin is committed by the hunter who shoots from a vehicle. Many also criticize the use of items like walkie-talkies or mobile telephones by which hunters can pass on information about the location of animals.

David Peterson (2000) is enraged by technologies that are, 'encouraging humans to hunt with their butts and wallets – cruising roads and trails, searching for easy targets – rather than their boots, brains and hearts... The net effect of all of this *stuff*, of course, is to reduce the need to develop traditional woodcraft skills, exercise patience, or exert physical effort and to help assure consistent kills – in effect, to take the *hunt* out of hunting; to *buy* success rather than earn it'. On one occasion, while I looked at a taxidermized head of a mountain sheep, its hunter said to me, 'I killed that beast with my legs and feet'. Enjoying my obvious confusion for a moment or two, he then explained that he had only been able to take a shot because he had gone into the mountains, walked and then climbed for many hours before he found the creature. For these hunters a respectful relationship between the hunter and the hunted should include the possibility of the animal becoming alert to the human and outmanoeuvring him. Too great a reliance on technology, or too great an ease of access, creates too great an advance for the hunter. As Allen Jones (1997) puts it, 'the elk, as prey, is constantly aware of your capacities as a hunter, unless your technology exceeds its expectations'.

Such concerns about the appropriate relationships between the hunter and hunted is at the heart of what many hunters believe ought to be ethical hunting. The ethics of hunting is not an ethics that seeks to justify the event itself, rather it is an ethics of how hunting should be carried out, an ethics that is woven into the process itself. Hunters have commented to me that there are rules and regulations relating to what animals may be hunted, at what times and in what spaces, and there may even be regulations governing what weapons may be used. There is, however, very little in the way of regulation of how a hunter should comport him or herself in relation to huntable animals. Those who spoke with me about this pointed out that most hunters are alone while they are stalking or waiting for their prey, they are usually unwitnessed, and how a hunter behaves is, in the end, a question of personal judgement and personal ethics.

Another fundamental element in the making of proper hunting is that it cannot be guaranteed that a hunt ends in the killing of an animal. As Ortega y Gasset (1968) puts it, 'the special quality of the hunt is that it is always problematic'. This is a point echoed in a comment about hunting by American ex-president Jimmy Carter, 'success, when it comes, must be difficult and uncertain. The effortless taking of game is not hunting – it is slaughter' (in Petersen,

1996). For nature hunters then, the challenge is not that of being able to hit a wild animal with a bullet – that is merely the challenge of marksmanship (although taking a long shot in very difficult conditions would be commented on as an example of being a skilled hunter) – rather the challenge is all that comes before the moment of taking a fatal shot. As one deer hunter remarked, 'getting a deer is not important; going hunting is what is important' (Petersen, 1996).

Being in the Natural World

Having offered some thoughts on the end of hunting, I will now turn to the beginning. Fundamental to this 'going hunting' is a mode of being in the natural world and it is the intensity of this being that seems to be the core experience sought by nature hunters. In commentary on their experiences, they stress the sense of connectivity, of being a part of nature when they are hunting and that hunting itself is a natural process. Although I, as an anthropologist, can comment on recreational hunting as a cultural practice, it is important to pay attention to how hunters express their hunting as a natural practice configured as a form of predator/prey relationship with wild animals.[8] This gives sense to their emphasis that they feel they are *in* the natural world in very different ways from other users of that world, and suggests that they are participants in nature, rather than simply users of it.[9] So, for example, their view suggests that activities such as hiking, climbing, canoeing, or even animal watching are cultural activities in the natural world but are not activities that fully connect with, or form part of, processes of the natural world.[10] As Jones asserts: 'If I'm hunting, I am no longer *using* nature; I *am* nature' (1997).

In conversations with me, hunters have contrasted their connectedness when hunting with someone walking in the countryside. Such a person might be enjoying intently the sights, sounds and scents of nature, but they are always one step removed from it. Such a person is *in* the natural world, *in* the countryside, but they are not *of* it; they are observers rather than participants because there is nothing in which they can participate. From this point of view hunters do participate because they take part in the processes of life and death in the natural world. Ortega y Gasset (1968) puts this point of view rather well – 'the countryside of the tourist ... is a "picture" and its existence depends on the lyrical qualities that man wants from it and can draw from it' – this countryside of 'landscape' and 'mere spectacle' is something 'that is before us but which we cannot enter'. He also says: 'It is only in the hunting field, which is the primary and "only" countryside, that we are able to migrate from our human world to an authentic "outside".'

What is suggested here, and this is something with which many nature hunters seem to concur, is that hunting is part of, or essential to, the condition of being human, and that being in hunting mode within the countryside returns humans to a more authentic way of being. This is not a position I will attempt to explore or discuss here because it demands a fuller engagement with both philosophical perspectives and issues relating to how hunters construct their self image. Once again, as with my consideration of hunters' notions of hunting as a natural practice, my responsibility is not to challenge this view of what it is to be human, rather it is to take this view seriously and to report on it ethnographically. For nature hunters it seems that hunting brings to the fore that which they regard as an authentic relationship with the natural world that is obscured in or by modern culture: this is something they cherish.

What these nature hunters seek is a way of inserting themselves into Ortega y Gasset's 'outside' so that they are within it and part of it. As one American hunter put it: 'Think of the natural world as a great play, an incredible drama held on the world stage in which all creatures play a part. When I carry binoculars, I stand with the audience, an omniscient observer to all that goes on around me, I enjoy this very much. But when I carry a gun, I become an actor *become part of the play itself*. This I relish too' (Petersen, 1996).

Although no hunter with whom I have spoken has ever used the image of a play in which they seek to have a part, they have spoken about the drama of hunting, particularly of participating in life and death, of connecting with wildlife, and the need to become part of the natural world. Hunters have suggested to me that people taking a walk through the woods are walking as humans who do not need to be aware of their presence and how their presence might be responded to by that world around them. This is not their concern; they are outsiders. The hunter has to be present, but the skill of hunting requires an intense awareness of self so that that presence is not registered by others. This is essential if hunters are going to have any chance of the engagement – the engagement of an insider, of a predator – that they seek. When accompanying hunters on their trips I have always been struck by the concentrated effort that is needed in order to even begin hunting. This is particularly so when such hunting trips involve getting into a hunting area in darkness in order to begin hunting in the first light of dawn or, at the end of the day, hunting in the faintest of available light before nightfall. For deer hunts these are preferred times because before dawn animals can be found browsing beyond cover, before the light pushes them back and, as the light fades at the end of the day, they once again emerge, but darkness soon ends any possibility of hunting. This hunter captures the sense of the initial absorption into the woods:

> *When you go into the woods, your presence makes a splash and the ripples of your arrival spread like circles in water. Long after you have stopped moving your presence widens in rings through the woods. But after a while it fades, and the pool of silence is tranquil again, and you are either forgotten or accepted – you are never sure which. Your presence has been absorbed into the pattern of things, you have become part of it, this is when hunting really begins.*

> *You can always feel it when those circles stop widening; you feel it on the back of your neck and in your gut, and in the awareness of other presences. This is the real start of the hunt, and you'll always know when it happens and when you are beginning to hunt* (Petersen, 1996).

In order to attempt to be a non-presence when hunting, hunters need to be totally aware of the world – alert to and mindful of the world around them and how any creatures present will be alert to and mindful of them if they do not manage their presence most carefully. Hunters must pay attention to the terrain, to sound, sight, scent, wind direction, stillness and movement from the perspectives of the animals that might be there. In my experience, hunters are particularly occupied by wind and air movement because these might expose their presence to animals through scent – the detection of which is a sense far more acute in hunted animals than in humans. It was of particular concern because, if human hunters do not orientate themselves carefully to the probable movement of air then the hunted animals were likely to be aware, at a great distance, of the hunters well before the hunters could see, and hence be aware, of the animals. Movement and stillness are crucial. One hunter, emphasizing the importance of learning to be still, used as an example a hunting encounter I had witnessed. At a fairly short distance from us there seemed to be a deer obscured by low cover in the woods, we were still but the animal seemed concerned and unsure of what was in front of it. The deer could not be seen clearly, but my hunting friend thought that he had seen a movement. He waited and waited, the deer relaxed and moved, my friend saw it and shot it. His comment was succinct: 'In hunting he who moves first loses.' In general the necessary peaking of the senses (Swan, 1995) is an experience that the hunter must have the skills to create and sustain if there is to be any chance of an engagement with a huntable animal.

The challenge presented by the animal is that it will be able to remain invisible or distant. The challenge for the hunter is the slow movement through the habitat of the animal – seeing while remaining undetected – closing the distance so that there is the possibility of a lethal shot. All of this for the hunter *is* hunting – all that goes before the successful shot, the missed shot or the failure to get into a position to take a shot. This links back to killing as the intention and end of hunting. In conversations after hunting, I have heard hunters expressing their pleasures of the day's hunting despite not having killed, and sometimes not even having seen, the species of animal they were seeking. Listening to hunters' conversations at the end of a hunting day, I have been struck by how much of the talk has been about how it was to have been there. The comments were more about the weather, terrain, movements of the air, glimpses, sightings, rustling of vegetation, than of shooting. They had been out there, they had been in it, and they had hunted despite not having killed.

It would seem that there must be a balance between never and always with regard to shooting. If there is never any doubt about the outcome then the act of hunting loses an essential element, but if there is never a chance to kill then there is no point to hunting at all. The hunters I know best speak about particular hunts as being 'interesting', and what makes a hunt interesting is its level of difficulty; the more difficult the hunt, the more effort that was required of them, the more interesting the hunt. One hunter, telling me of hunting buffalo in Australia, commented that the moment of killing was like 'shooting against a wall' because, although the hunt took some skills, in the end the animals were too 'stupid' to flee even when they saw a human in front of them. What seems to be important for hunters is that shooting a huntable animal should not be something that is guaranteed.

I have been out on eight occasions with a Spanish hunting friend who has come to England specifically to hunt muntjac. During those hunts he has never had the opportunity to shoot, and I am not sure he has even seen, a muntjac. One evening I asked him if he wasn't disappointed after spending so much money coming to England and spending so long in the woods. 'No', he replied, 'that's hunting'. As I was completing this chapter I had the opportunity to accompany him on another weekend hunting trip in England. Twice we went out at dawn and dusk and there was not a muntjac to be seen. On the last morning, well after dawn, he spotted a shadowy shape in the woods, he stalked it and shot a muntjac. A short while later, as we were about to leave the woods, he saw and shot two more. Three animals hunted in under an hour. I reminded him of his lack of success on his previous visits and asked how he felt about everything happening in the last hour of this visit. His response was exactly the same: 'That's hunting'.

Concluding Comments

In this chapter I have focused on what are some of the defining features of one form of, or orientation to, recreational hunting. I regard this as a contribution to what I believe ought to be more wide-ranging research into recreational hunting in the modern world. Such hunting might not seem a significant activity compared with other leisure activities, but the number of people involved is significant. It is difficult to obtain reliable information about the demographics of hunting, but the 2006 National Survey of Fishing, Hunting and Wildlife-Associated Recreation in the US contains an estimate of 12.5 million registered hunters nationally, and related surveys suggest that the population of hunters worldwide totals anywhere between 23 million and 43.7 million.[11] Many millions of dollars, euros and pounds move through economic markets as payment for the experience of hunting. Each one of these hunters will be going into spaces of the natural world in order to hunt a wide variety of species of wild animals; this constitutes an intense and in-depth engagement with the natural world.

As I suggested earlier, studies of hunting as a recreational activity are not simply interesting from an academic perspective; although they are that. Hunting activities intersect with wildlife management and environmental management policies and practices, and might also overlap and perhaps come into conflict with other leisure activities in the natural world. In these contexts it is essential to understand what it is that hunters seek in and from hunting. What do they look for in their pursuit of huntable animals? What are the engagements they create? What do they want from the spaces of the natural world in which they hunt? How do they behave in such spaces?

This research relates only to a small group within Kellert's tripartite classification. From his survey he suggested that of the total numbers of hunters, utilitarian/meat hunters constituted 44 per cent, dominionistic/sports hunters 38 per cent and nature hunters 18 per cent. He also notes that this latter group 'comprised the largest proportion of persons who hunted often' (1978). I believe that we need to know more about those hunters who have rather different orientations to hunting from the nature hunters. I also believe it is important to understand how hunters shift their orientations on different occasions and in different spaces.

There are quantitative studies of hunters and studies of their attitudes to hunting, but what is generally lacking are qualitative, ethnographic, studies of hunters; studies that would generate a greater, cultural and social, understanding of their hunting. In the rapidly developing academic field of human-animal studies, one of the key interests is understanding the grounded specificities of human-animal relations. Studies of hunting should be of importance to this field because hunting is underpinned by a rich and complex spectrum of attitudes and activities ranging from treating animals as mere objects, targets for shooting, to the most respectful pursuit of lifefull creatures. Hunters have a passionate interest in wild animals and much of their lives are orientated to such animals and the natural world. Recreational hunting is a vibrant and modern activity. Hunters treat what they do very seriously and, in turn, what they do requires serious understanding.

Notes

1   Intriguingly, in England, the terms 'hunt' and 'hunting' are only used to refer to events which use packs of hounds to pursue quarry as in hare, stag, mink and foxhunting. 'Hunt' is not used to refer to hunting-type practices in which shotguns and rifles are used to kill prey. So, for example, 'hunting' deer is 'deer stalking', 'wildfowling' refers to the pursuit of waterbirds and 'shooting' refers to events based on game birds such as pheasant, partridge and quail.
2   Various terms – for example, sports, hobby, leisure – are used to describe or define hunting of this type and I find all of these unsatisfactory. A suitable term would need to capture the sense that for many hunters the practice is akin to a vocation, for others it is the defining element of their lifestyle. However, this chapter is not the place to discuss the significance of terminology. Despite my reservations about the term, I will use 'recreational'. This captures some of the elements of stepping outside everyday life and the sense that hunters are, in significant ways, re-created through this activity, that they re-create themselves as persons in a different way when they are in hunting mode.
3   I am a social anthropologist and hence much of the material in this chapter has been generated from anthropological perspectives. In particular it has been generated from an ongoing, ethnographic, participant-observational study of the cultures of hunting. My initial foray into this terrain was a study of foxhunting in England. Since then I have been working on the literature of hunting (particularly North American literature) and through ethnographic work with hunters in England and Spain.
4   Although I am unable to discuss these issues, I offer the following references for any reader interested in the complexities of indigenous hunter–prey relationships, the notions of animal personhood and how animals offer themselves to hunters: Tanner (1979); Nelson (1980 and 1983); Brightman (1993); Viveiros de Castro (1998);

Ingold (2000); Nadasdy (2007); Willerslev (2007); Henriksen (2009). Although only an indicative list, these authors – especially Ingold, Nadasdy and Willerslev will guide the reader into the issues and towards other literature.

5   As a further research project I am interested in investigating whether there is a connection between what has been written about indigenous hunting practices and what has been written about modern recreational hunting. My initial thoughts, although these would have to be demonstrated, are that the anthropological accounts of hunting may have been influential in the development of the thinking of modern hunter/ writers, particularly in North America, who have attempted to describe and explain their activities and to set an agenda for what they regard as ethical hunting.

6   As I suggested earlier, what Ortega y Gasset refers to here as 'utilitarian' hunting, is more complex than simply bringing about the death of a hunted animal. His notion of 'purely a means' does not adequately capture the processes that underpin many of the forms of such hunting. This, however, is a discussion for another occasion.

7   In all the cases of hunting that I have examined the hunted animal is always what we might term a 'wild animal'. However, this term is a culturally specific term and form of classification and will not apply in all cultures. How different cultures configure the nature of hunted animals would have to be examined in each case. For the sake of simplicity I will use the term to refer to animals that are not in the regular care of humans. Again, simplistically, that they are not domesticated animals.

8   For a fuller discussion of issues relating to predation, hunting and natural and cultural processes see Ingold 1986 and Marvin 2006.

9   Here I do need to stress that what I am offering are hunters' views of how other people use the natural world. These are not my personal views, which would be more complex, of how other people use the natural world for their own projects and processes of recreation.

10 It is worth noting that many nature hunters spend a great deal of time walking through, or sitting quietly, in the environments where they hunt to observe wildlife. In this sense they might seem to be doing something similar to other animal watchers. These activities however, actually form part of the larger process of hunting, learning about where animals are in that environment and how they behave there, so as to hunt them more effectively on other occasions.

11 US Department of the Interior, Fish and Wildlife Service, and US Department of Commerce, US Census Bureau. 2006 National Survey of Fishing, Hunting, and Wildlife-Associated Recreation.

References

Berger, J. (1980) 'Why look at animals', in J. Berger, *About Looking*, Writers and Readers, London

Brightman, R. (1993) *Grateful Prey: Rock Cree Human-Animal Relations*, University of California Press, Berkeley

Bulbeck, C. (2005) *Facing the Wild: Ecotourism, Conservation and Animal Encounters*, Earthscan, London

Franklin, A. (1999) *Animals and Modern Cultures: A Sociology of Human-Animal Relations in Modernity*, Sage Publications, London

Henriksen, G. (2009) *I Dreamed the Animals: Kaniuekutat – The Life of an Innu Hunter*, Berghahn Books, Oxford and New York

Ingold, T. (1986) *The Appropriation of Nature; Essays on Human Ecology and Social Relations*, Manchester University Press, Manchester

Ingold, T. (2000) 'Hunting and gathering as ways of perceiving the environment', in
T. Ingold, *The Perception of the Environment: Essays in Livelihood, Dwelling and Skill*, Routledge, London, pp40–60

Jones, A. (1997) *A Quiet Place of Violence: Hunting and Ethics in the Missouri River Breaks*, Bangtail Press, Bozeman, Massachusetts

Kellert, S. (1978) 'Attitudes and characteristics of hunters and antihunters', *Transactions of the North American Wildlife Resources*, vol 43, pp412–423

Marvin, G. (2006) 'Wild killing: Contesting the animal in hunting', in Animals Study Group *Killing Animals*, University of Illinois Press, Urbana and Chicago, pp12–15

Nadasdy, P. (2007) 'The gift in the animal: The ontology of hunting and human-animal sociality', *American Ethnologist*, vol 34, no 1, pp25–43

Nelson, R. (1980) *Shadow of the Hunter: Stories of Eskimo Life*, University of Chicago Press, Chicago

Nelson, R. (1983) *Make Prayers to the Raven: A Koyukon View of the Northern Forest*, University of Chicago Press, Chicago

Ortega y Gasset, J. (1968) *La Caza y Los Toros*, Revista de Occidente, Madrid

Peace, A. (2005) 'Loving leviathan: The discourse of whale-watching in Australian ecotourism', in J. Knight (ed) *Animals in Person: Cultural Perspectives on Human-Animal Intimacies*, Berg, Oxford

Petersen, D. (1996) *A Hunter's Heart: Honest Essays on Blood Sport*, Henry Holt and Company, New York

Petersen, D. (2000) *Heartsblood: Hunting, Spirituality, and Wilderness in America*, Island Press/Shearwater Books, Washington, DC

Posewitz, J. (1994) *Beyond Fair Chase: The Ethic and Tradition of Hunting*, Falcon Publishing, Helena, Massachusetts

Servais, V. (2005) 'Enchanting dolphins: An analysis of human-dolphin encounters', in J. Knight (ed) *Animals in Person: Cultural Perspectives on Human-Animal Intimacies*, Berg, Oxford

Swan, J. (1995) *In Defense of Hunting*, Harper San Francisco, California

Tanner, A. (1979) *Bringing Home Animals: Religious Ideology and Mode of Production of the Mistassini Cree Hunters*, St Martin's Press and Hurst, New York and London

Viveiros de Castro, E. (1998) 'Cosmological deixis and Amerindian perspectivism', *Journal of the Royal Anthropological Institute*, vol 4, pp469–488

Willerslev, R. (2007) *Soul Hunters: Hunting, Animism and Personhood among Siberian Yukaghirs*, University of California Press, Berkeley

Part IV
Agriculture

9
Culture and Agrobiodiversity: Understanding the Links

Patricia L. Howard

Introduction

Generations of farmers, herders, pastoralists, hunters, gatherers and fisherfolk have developed and managed complex, diverse and locally adapted livelihoods and agroecological systems using combinations of techniques, knowledge and practices that have frequently led to community food security, sustainable natural resource management, high levels of unique biodiversity and the preservation of cultural identity, human dignity and equity. Such systems usually comprise cooperative traditional social relations, cultural beliefs and rituals, local languages, knowledge, and technologies that have developed over long time frames and that have become institutionalized or built into the fabric of these societies. These act to manage the inter-related sets of resources and habitats, while ensuring that they meet the cultural and material needs of local peoples.

There is considerable consensus that biodiversity provides the basis for many livelihood activities of rural subsistence populations across the world. A single community often purposefully manages and exploits many hundreds of plant and animal species. However, the degree to which rural subsistence societies depend on agrobiodiversity for livelihoods is generally under-estimated by policy makers even though evidence for the direct use values of a wide range of environmental resources for rural people continues to mount (Vedeld et al, 2004).

Together, these biologically and culturally diverse social and ecological systems (or 'socio-ecological systems') are estimated to support some two billion people – just under a third of the global population (Howard et al, 2009). These systems provide cultural and ecological services to humankind as a whole, such as the management and maintenance of unique landscapes and biodiversity, a wealth of traditional knowledge, local crop and animal species, a myriad of other ecological services, and forms of social organization that have evolved to ensure adequate management of local resources and social welfare. These systems thus represent both human heritage and sources of hope for the future as we struggle to live harmoniously with nature in the present. Today, many rural subsistence societies inhabit regions where extinction rates are among the highest in the world and where climates are expected to alter dramatically. Together with strong policies to address climate change and resource depletion, supporting such societies should be seen as a global insurance policy.

Agrobiodiversity includes all biological resources that directly and indirectly contribute to human well-being – wild and domesticated, above and below ground (FAO, 1999). It embraces the full range of environments in which agriculture is practised, including fields, homegardens, grazing lands, and forests. Agrobiodiversity can be understood as an emergent property of the intended and unintended effects of human actions that lead to modifications or transformations of landscapes and ecological relationships. Scientists agree that possibilities for *in situ* conservation of agrobiodiversity depend on the extent to which it continues to meet the needs of local people and that conservation needs to be undertaken by local communities. It thus must reflect these communities' values and concerns, and be based not only on ecological and economic priorities but also on local social, cultural, ethical and spiritual values (Heywood, 1999; Hodgkin and Ramanatha Rao, 2002; Sastrapradja and Balakhrisha, 2002; Wolff, 2004). There is growing realization that the drivers which erode cultural diversity (in the form of subsistence practices, dietary habits, knowledge, values, institutions and languages) also erode agrobiodiversity. In many subsistence societies, the main reasons for maintaining agrobiodiversity are related not to environmental selection pressures or to markets, but rather to non-monetary, cultural factors. Despite this, there is often little understanding in scientific or policy circles about such cultural factors or the ways in which culture is related to agrobiodiversity.

In this chapter, an overview is presented of key concepts for understanding how cultural values and institutions give rise to agrobiodiversity and how, in turn, agrobiodiversity gives rise to cultural diversity. The central proposition is that the unique ability of humans to develop language, values and belief systems, and to accumulate knowledge, gives us the capacity to adapt and change our own behaviour and relations, as well as to change the ecosystems of which we are a part. Cultures and ecosystems thus co-evolve. This does not mean that cultural traits are simply adapted to specific ecosystems: this would imply that all cultures within a particular ecosystem would be similar. Rather, cultures evolve in relation to an array of dynamic factors, which gives rise to cultural diversity in terms of worldviews, beliefs and institutions even where ecological circumstances and production strategies are similar (Ellen, 1982). Nor are all cultural adaptations positive: some are clearly maladaptive, which can lead to ecological degradation and have negative effects for human welfare and social stability.

Richerson and Boyd (2005) defined cultural evolution as the accumulation of 'complex, highly adaptive tools and institutions that ... have allowed people to expand their range to every corner of the globe. By cumulative cultural evolution, we mean behaviours or artefacts that are transmitted and modified over many generations'. Five dimensions of culture have been shown to be important to the co-evolution of cultural and biological diversity (biocultural diversity) in subsistence societies: cultural identity, cosmology and religious beliefs, traditional knowledge, social position and status, and natural resource tenure regimes. Cultural institutions have evolved to ensure certain types of behaviour among individuals in a given society, for example, by creating a system of sanctions and rewards or by socializing people according to certain values. In many modern societies, cultural institutions – for example, relating to law, education and religion – are often separated from each other, whereas in many traditional societies they may be incorporated within a single institution. Such institutions regulate access to and management of agrobiodiversity, conserve and transmit information about agrobiodiversity, determine who does what, as well as who receives the benefits and pays the costs. Each of these dimensions is discussed in turn, and their interrelationships and embeddedness within cultural institutions are highlighted.

Cultural Identity and Agrobiodiversity

The sense of attachment and belonging to a specific group or society constitutes our cultural identity. Shared language, territory, forms of worship, dress, architectural styles, culinary traditions, art forms and status symbols knit people together in communities, collective action groups and households. Cultural identity gives people a sense of what distinguishes them from others or 'outsiders'. Cultural identity is expressed in many ways that depend upon and in turn affect local agrobiodiversity. For example, dress styles are traditionally based on local climatic conditions, needs for protection and comfort, religious considerations, social status, aesthetics, and culturally important botanical and animal resources. Architecture and the design and engineering of homes and other structures for animals, storage, transport, irrigation and religious observances are likewise adapted to local environments, and are derived from locally available building materials, aesthetics, social and religious principles, and traditional knowledge.

Culinary traditions provide a good illustration of the strong link between cultural identity and agrobiodiversity. The decision about what to eat is closely tied to cultural concepts of nutrition, well-being, aesthetics and spirituality, and culinary traditions have been strongly influenced by local agrobiodiversity. In the Andes, the cradle of the world's potato diversity, many researchers have tried to understand why farmers maintain a large number of potato and maize varieties and have sought the answer in their need to adapt the crop to fit diverse environments and agronomic conditions. However, Zimmerer (1991, 1996) showed that these factors alone could not explain the great diversity maintained. Rather, specific varieties are cultivated in accordance with local culinary traditions and post harvest processing and storage needs (freeze-drying, soup-making, boiling). This is also true of maize, where farmers use culinary distinctions as the basis for planting separate fields in different habitat niches. Besides parching and boiling maize, women make hominy, crushed maize, popcorn, mush, corn-on-the-cob, soup-thickener, pudding and tamales.

Another puzzle is why so many people across the tropics prefer to grow 'bitter' (toxic) varieties of manioc rather than 'sweet' (non-toxic) varieties. Much labour is needed to remove the toxin from bitter varieties, most of which could be avoided if only sweet varieties were grown. Wilson (1997) spent time with the Tukanoan peoples of the Colombian Amazon and concluded that, 'while yield and damage suffered by the plant may be factors which influence cultivar selection, it is the foods which can be made from each cultivar which is the most important consideration'. Descola (1994) wondered why the Achuar peoples living in Amazonian Ecuador have not developed maize cultivation as a substitute for manioc, since maize contains far more protein and requires less labour. He found that it is, once again, cultural preferences that are key: manioc is often used as a synonym for *yurumak*, or food in general, and the beer that is made from cassava is essential to both domestic and social life.

The importance of agrobiodiversity to the maintenance of cultural identity in new places becomes obvious when people migrate. Many studies show that agrobiodiversity is diffused across communities, regions and even the globe, together with its curators. Greenberg (2003) researched Mayan migrants who moved from a subsistence-based agricultural economy in the Yucatan Peninsula to a wage-labour, cash economy in Quintana Roo, Mexico. Immigrant homegardens have become sites of conservation, not only of traditional Yucatec crops but also of elements of traditional Yucatec cuisine (some 140 plant species were found in 33 gardens, the majority of which were used in traditional cuisine). In this way, gardens and their plants help to preserve the cultural identity of immigrants in their new environment (Niñez, 1987; Airriess and Clawson, 1994; Gladis, 2001; Corlett et al, 2002; Vogl et al, 2002).

Traditional Belief Systems in Agrobiodiversity Management

Traditional belief systems (cosmologies, religions) provide theories as well as value frameworks for how humans should relate to other humans, to other life forms, and to their environments. It is still a subject of debate to what extent, and under what conditions, religious beliefs provide a stimulus for conservation of agrobiodiversity, but there is substantial evidence that many societies entertain strong theories about ecological relationships, including notions that lead to conservation. Another way in which traditional belief systems are closely linked to agrobiodiversity is through traditional ecological knowledge (TEK), which is often indistinct from traditional belief systems or religions. Both TEK and conservation ethics are widely considered to be essential to the management of agrobiodiversity.

Anderson (1996) argues that it is often difficult to distinguish 'science' from 'religion'. Before Western societies had laboratories and other technical means to test their theories, many of their theories were similar to those found in traditional societies today. Even within science today, there are many 'black boxes' where theories, rather than facts, are established to explain aspects of the world. Most anthropologists would agree with Anderson when he notes, 'people naturally assume human-like agency as a default, whenever explaining anything. It makes evolutionary sense for people to look for someone causing things that happen in the world. When traditional peoples infer similar "black box" mechanisms, we label those supernatural'. It is also true that supernatural beliefs are often used to promote ethics, and that religion is an emotional force that can be mobilised to ensure adherence. In traditional societies, knowledge, ethics and the belief in supernatural beings are often integrated in a single system. Many traditional belief systems are 'animist' or 'pantheist', where personalized, supernatural beings (souls or spirits) are endowed with reason and intelligence, and inhabit ordinary objects such as landscapes or food items, as well as living things (such as animals and plants), and govern their existence. Such supernatural beings and the things they inhabit must thus be respected.

Much research demonstrates the strong connections between cosmological beliefs, cultural identity, and the maintenance of agrobiodiversity (Condominas, 1986; Patterson, 1992; Piper, 1992; Anderson, 1993; Rimantha Ginting, 1994; Trankell, 1995; Alexiades, 1999; Steinberg, 1999). The process of rice production, post-harvest processing and consumption over much of Asia illustrates the relationship between spiritual beliefs and agrobiodiversity (Box 9.1). Asian farmers have developed some 120,000 rice varieties, each of which has become adapted to specific agroecological conditions, and many of which were brought into being as an expression of spiritual beliefs (Hamilton, 2003a, b). Many rituals are associated with the agricultural cycle and take place at sacred sites within or near the fields where the rice spirits may be harboured, and such rituals may involve a special rice type or other sacred plants.

Harvesting of the first grain often involves important cultural rituals. In Bali, a household's senior woman will make a trip to the rice fields before harvest Beginning at the corner where irrigation water enters, a place considered to be sacred, she cuts the first rice grains. The stalks are placed into two bundles that are bound together and decorated with flowers and ornaments of palm leaves, which represent the Nine Pantun, or Rice Mother. Another important task that women then go on to perform is to select rice seed that will be saved for the following rice crop. Senior Ifugao women make their way through the fields at dawn on the day selected for the harvest. Only senior women can carry out this task, since they must be able visually to distinguish between dozens of different varieties, as well as know their desired characteristics. Thus, the genetic heritage of the community's rice varieties is in the hands of these women and relies upon the performance of these rituals (Hamilton, 2003c).

Some rice diversity is also destined for ritual and religious use. Among the Pahang Malay, certain varieties of glutinous rice are bred for their colour, each of which 'has specific ritual uses, or is required for making special confections or cakes' (Lambert, 1985). Rice wine is also very important in rituals, ceremonies, and festivities. Iskandar and Ellen (1999) showed how religious beliefs could also lead to the rejection of modern varieties. In the Javanese uplands, despite the mass introduction of modern varieties, the Baduy people have

Box 9.1 Twenty tenets of rice cultures[1]

Each rice-growing society has its own unique set of values and beliefs associated with rice. Listed here are 20 of the most common tenets that are widespread in many parts of Asia. Though no single society holds to every tenet, many do follow a large number and, taken together, they can be said to amount to a creed of rice culture.

1 Rice is a special sacred food, divinely given to humans.
2 A rice plant has a living spirit or soul comparable to that of humans, and the life cycle of the rice plant is equated with the human life cycle. Rice spirits must be honoured and nurtured through rituals in order to ensure a bountiful harvest.
3 The stages of rice agriculture determine the annual cycle of human activity, including the conduct of proper rituals at each phase of the rice crop's cycle.
4 The work involved in growing rice is the ideal form of human labour, reflecting a well-ordered, moral society.
5 The mythological origin of rice is attributed to a Rice Mother or Rice Goddess; in many versions of the story the goddess is killed and the first rice grows from her body.
6 The fertility of the rice crop is metaphorically equated with the fertility of the Rice Goddess and with the fertility of human females; rice is thus often regarded as female and in exchange systems it functions as a female good.
7 Rice must always be treated with respect to avoid offending the rice spirits or Rice Goddess; at harvest time rice may be cut with a special type of knife to avoid harming them.
8 The granary is the home of the rice spirits and is often built to resemble a small human house. After the harvest, the grain is ritually installed in its home.
9 Special objects may be placed in the granary to accompany the rice; these include anthropomorphic figures made of rice stalks symbolizing the Rice Goddess, or in other cases carved wooden figures or even copies of religious texts.
10 The spirit of the rice remains alive at least until the rice is milled; thus the rice that is set aside before milling to serve as seed rice perpetuates the rice spirit, keeping it alive until the rice is planted again in the following agricultural cycle.
11 The maintenance of special ancestral genetic strains of rice is a primary link between living humans and their ancestors.
12 The daily milling of rice by pounding it in a mortar is traditionally one of the most characteristic activities of village life. Only after it is milled can the rice be brought into the house.
13 The daily milling, cooking and eating of rice determine the daily schedule of human activity.
14 Language reflects the special nature of rice as the primary food of humans; often there is no general word for 'food' other than the word for rice and an invitation to 'eat' implies the eating of rice.
15 The household or family unit is defined as those who eat rice together, especially the rice that is produced through the joint efforts of the family members.
16 Cooked rice is the ultimate human food and only rice is capable of properly nourishing humans; other foods are regarded as condiments to accompany the rice or as snacks and do not constitute a meal if rice is not served.
17 Because humans live by eating rice, their bodies and souls are made from rice.
18 Rice and rice alcohol are quintessential offerings made to spirits, deities and ancestors.

19 The offering of a portion of daily cooked rice to spirits, deities or ancestors sanctifies the remainder of the rice, which becomes sacred food to be eaten by humans. The living humans, the ancestors and the deities are united through the daily sharing of food.
20 Because rice ties humans to their ancestors, defines the family unit and provides the ultimate human nourishment, the growing and eating of rice defines what it means to be human.

maintained their swidden farming system and folk rice varieties mainly because their agriculture has great religious significance, and folk varieties are essential to the performance of rituals and feasts. The Baduy maintain their cultural identity in part through their religious beliefs and rice varieties.

But it would be misleading to assume that religious beliefs always lead to agrobiodiversity conservation. There are also many cases where certain religious beliefs lead to unsustainable practices, both because they are not consciously developed to guide practical action and because these beliefs do not change at the same pace as ecological, economic and social change. There is, however, a substantial body of evidence demonstrating that spiritual beliefs and conservation ethics are closely inter-related in societies that have proven to be sustainable over time (Turner and Berkes, 2006), since they are often inseparable from ways in which traditional people value and understand agrobiodiversity and ecosystems.

Traditional Knowledge: Its Transmission, Distribution and Erosion

Most knowledge in traditional or subsistence societies is not derived from books or through formal education: it is generally unwritten, culturally and ecologically contextualized, and transmitted through a multitude of means (Richerson and Boyd, 2005). Here, the term TEK is used to refer to the nexus of knowledge, practices and beliefs that are related to ecological components and processes (Berkes, 1999). Turner (2003), for example, describes the breadth of indigenous women's TEK in the Pacific Northwest of the US and Canada (Table 9.1), where much of this knowledge is related to the use and management of agrobiodiversity. But there is no research that thoroughly documents the breadth or depth of TEK for any society, not only because there are so many domains of knowledge, but also because knowledge is not distributed evenly within communities.

The transmission of TEK depends on the continuity of social relations over time, and often upon specific groups and individuals and their social networks. Considering that knowledge is essential to human survival, it is not surprising that its transmission does not occur randomly: social mechanisms ensure not only that knowledge is transmitted from generation to generation, but also that innovations are fed back into a culture (Davidson-Hunt, 2003;

Table 9.1 *Pathways by which TEK may have been accrued using examples from North American indigenous people*

Lessons from the past	Stories of positive and negative experiences remembered by individuals, recounted within families and communities, or embedded in art, place names and ceremonies.
Language	Terms that embody conservation concepts, understandings and teachings, e.g. the Heiltsuk word *mnaquels*, which refers to 'selectively collecting things outside', and *miaisila*, which refers to someone whose responsibility it is to be a guardian of certain fish-bearing rivers, or the Nuu-chah-nulth word *uh-mowa-shitl*, to keep some and not take all.
Metaphorical sayings and narratives	Symbolic and metaphorical stories also teach lessons about conservation (e.g. Nlaka'pamux story of Old One and the Creation of the Earth).
Lessons from other places	Technologies, products, names and ideas relating to conservation for environmental stewardship (e.g. the use of fire for clearing; digging and propagation techniques) passed from one community to another through intermarriage, potlatches, trade.
Learning from animals	Observations of animal foraging strategies, populations, browsing and predation; behaviours that might engender understandings of kinship and reciprocity (e.g. grizzly bears foraging for edible roots; birds' egg-laying habits; pack and leadership relationships in wolves; bears 'pruning' berry bushes).
Monitoring – building on experience and expectations	Routine observation of seasonal changes, animal migrations, plant life cycles, and berry production brings recognition of expected patterns and ability to detect variation from the norm.
Observing ecosystem cycles and disturbance events	Relative abundance and productivity of plants and animals in particular circumstances, both temporal and spatial, can guide people's land and resource management strategies (e.g. successional stages following fire; effects of flooding on salmon migration patterns; relation between moisture and berry productivity).
Trial and error experimentation and incremental modification	Observing the results – positive and negative, intentional or incidental, short term and long term – of people's own activities, such as selective harvesting or of emulating natural disturbance (e.g. harvesting cedar bark and planks).
Learning by association, extension, and extrapolation	If a practice works in one place at one time, it might work in another place at another time; conversely, if a practice or activity results in negative consequences in one circumstance, it might be avoided at another time or place (e.g. knowledge about harvesting or conserving one type of shellfish, berry or root might be extended to other, similar types).
Elaborating and building sophistication	Gained from all of these pathways, and building up knowledge, practices and beliefs into complex systems of land and resource management (e.g. Heiltsuk berry gardens; Saanich reefnet fishery).

Source: Turner and Berkes (2003)

Puri, 2005; Turner and Berkes, 2006). For example, elders often have the responsibility to pass on knowledge to younger members of a tribe or clan; they are seen as the wise ones who people turn to for guidance and who can encourage, accept or reject innovations.

Knowledge is not distributed evenly within communities. It varies by age, sex, livelihood strategies, positions of power and authority, specializations, individual competency, geographical origin, residence, ethnicity, religion, occupation, educational level and wealth status (Boster, 1985a; Alexiades, 1999; Howard, 2003; Pfeiffer and Butz, 2005). Boster (1985a, b) argues that knowledge distribution and social structure are interrelated. While a body of knowledge exists within a community that could be likened to a body of Western scientific knowledge in a discipline, not everyone has equal access to this body of knowledge: it depends on an individual's position in the social structure. 'An individual can be considered as having a number of identities: a member of a society, an actor in a sex role in that society, a member of a household, and an individual. Different amounts of knowledge are shared at each of these layers.' It is clear that some knowledge is shared, but not all; the more important a set of knowledge is to peoples' everyday lives, the more likely it will be shared.

Boster (1985b) studied Aguaruna manioc cultivators in northern Peru, where women are responsible for manioc cultivation and hold the most knowledge about manioc diversity. Women in Aguaruna communities are expected not only to cultivate manioc, but also to maintain manioc diversity and transmit knowledge about manioc to other women through kinship networks. The wives of community chiefs are expected to have the greatest knowledge and hold the most manioc varieties, which can be used to replenish lost planting stock or to share. Underlying this division of labour are values and beliefs that prescribe the types of activities and responsibilities that are appropriate for different groups of people, including with whom people of different social positions can appropriately interact. In the Peruvian Andes, Zimmerer (1991) found that Quechua women have the greatest knowledge about potato and maize diversity. The explanation for their superior knowledge is found in their role as the managers of seed (men are not permitted to manage seed or to enter seed storage areas), which has its roots in Andean cosmology:

> The feminine and divine element represents the fertile mother: in Quechua, all of the plants that are useful to humans are venerated under the names of Mother: Mama sara (maize), Mama acxo (potato), Mama oca (Mama cocoa). The belief in this relation between women and seeds is coherent with the tradition of Andean thinking in terms of a dual concept of reality: the duality defined by the principles of masculine and feminine ... 'seed' also refers to ... semen... [providing a metaphor] between the 'seed' that the male deposits in the womb and the seed that is sown in the field, collected, and later deposited in the home (Tapia and de la Torre, 1993).

Just as in all societies, knowledge in traditional communities is used to confer status, manipulate social relations, gain material advantage, and maintain control over certain aspects of one's life. Descola (1994) found among the Achuar in Ecuador that religious and practical knowledge are a closely guarded secret, and 'probably the most precious possession a mother can transmit to her daughter'. It is commonplace, therefore, for certain types of knowledge to be kept secret or shared only with very specific groups in such societies (Howard 2003). The distribution of TEK within societies thus cannot be understood without reference to belief systems that legitimize and mediate relations of prestige, power and authority. These social relations and the ways in which they are manifest may be relatively transparent and clearly demarcated by exclusive occupation of prestigious social positions (e.g. shamans), or they may be much more subtle, embedded in language and in rules about the production and exchange of goods, such as those relating to healing or to access to gardening resources. Different people also have differing access to formal and exogenous knowledge. For example, women members of lower castes, and ethnic minorities, often have less access to exogenous knowledge sources since they receive less formal education, have less contact with extensionists and other government agents, and may be physically isolated or unable to leave their communities. But, for these reasons, they may be the holders of significant and diverse local TEK bases.

Yet across the globe, TEK is rapidly being lost. Research on TEK relating to homegardens provides some context-specific reasons for this. Assimilation into a dominant culture through education and migration commonly leads younger generations to learn less about plants and homegardening (Benjamin, 2000; Angel Peréz and Mendoza, 2004). Keys (1999) observed that young Guatemalan women working in textile factories have no time for homegardening, and so know less. Hoffman (2003) stressed that off-farm employment, in addition to migration and participation in formal educational systems, denigrates women's traditional gardening knowledge, leading to loss of TEK. While formal education can bring many benefits, it is rarely oriented towards the realities of traditional or indigenous peoples' lives, taught in the local language, or respectful of traditional ways of life (Turner et al, 2000; Hoffman, 2003).

Language, aesthetics and agrobiodiversity Language is the principle vehicle through which knowledge is accumulated and transmitted. In traditional societies, culture is largely orally transmitted; TEK is 'written' only in the active use of native language terms. Kokwaro (1995) showed that, even though Ethiopians have had their own written language for more than 2000 years, 'knowledge of traditional medicines is [still] handed down orally by medical practitioners, parents, elders, or priests and priestesses'. This oral tradition is continually enriched but, because it is oral, it is also vulnerable. The erosion or disappearance of language may mean the loss of capacity to classify, understand and talk about important plant and animal resources and practices. Many subsistence societies today continue to preserve extremely detailed knowledge of agrobiodiversity in their own languages, in spite of rule by others speaking different languages (Hunn, 2001).

Language is also used to convey cultural identity, history, social relations, spirituality, humour and aesthetics. Sadiki et al (2007) reviewed the literature, noting that 'the names farmers give to their traditional varieties or folk varieties are fundamental to their very essence and use'. Studies investigating the development and maintenance of folk varietal diversity have found that, while agroecological adaptation and utility (such as culinary and storage characteristics) play a role in landrace selection and classification, linguistic and aesthetic qualities are often more important in determining the vast quantities of diversity present in agricultural fields. Shigeta (1996), for example, shows how the creation of folk varietal diversity in ensete in Ethiopia is related to both language and aesthetics. The Ari people have more than 100 names for local ensete varieties, 78 of which refer to distinguishable types. Most of the names seem to have meanings, including tastes of food, locations, means of propagation, or the names of people who are considered to have developed or introduced the variety. They classify different varieties first by the colour of the plant and then by secondary characteristics, including outer appearance, maturation period, means of propagation, and use and taste. The Ari do not select the 'best' varieties based on practical use. If only practical uses were considered, 'these folk varieties would be reduced to a few superior ones'. The Ari say that they maintain great diversity of ensete folk varieties because it is 'custom' or 'good' rather than for utilitarian reasons.

Referring back to Boster's work (1985b) with the Aguaruna Jivaro of northern Peru, who developed an astonishing diversity of folk manioc varieties, Boster found that women identify folk varieties on the basis of leaf shape, petiole colour and stem colour. The capacity to distinguish varieties on the basis of sight, taste and smell is necessary for selection to occur. Women of the Aguaruna Jivaro have bred folk varieties until they have the entire 'rainbow' of leaf colours: this enables them to fulfil their social duties and achieve status within their communities.

Social Status and Agrobiodiversity

Social status provides another strong motivation for developing and maintaining agrobiodiversity. Status is culturally, historically and contextually defined, and thus must be understood for each society in question. Social status determines an individual's place in society and influences access to resources and capacity for decision-making. People are constrained in their ability to make decisions and to act in accordance with their social status. In many traditional societies, where livelihoods depend overwhelmingly on agrobiodiversity and agrobiodiversity represents the majority of people's wealth, it is clear that well-being and social status are related to access to this diversity. Moreover, different social groups make differing contributions to subsistence and the perpetuation of agrobiodiversity-related TEK.

Status can be related to agrobiodiversity in subtle ways that are not readily observed. Descola (1994) highlighted how researchers often mistakenly assume that the agrobiodiversity of Achuar swidden gardens arises out of utilitarian need (ecological or economic). Instead, he found that it is related to the ways in which people gain social status in their communities. For the Achuar, it is a point of honour for women to cultivate large swidden gardens. The garden diversity evident, particularly in tubers, cannot be attributed to nutritional or culinary needs since 'men – whose attitude openly encourages their wives' agronomic capacities – recognise by taste alone only a very low proportion of the varieties of manioc, yams, or sweet potatoes'. Nor can it be attributed to the need to reduce plant diseases, since only one serious manioc disease is recognized and only a few plants are usually affected. Rather, 'a woman who successfully grows a rich pallet of plants thereby demonstrates her competence as a gardener and fully assumes the main social role ascribed to women by proving her agronomic virtuosity'. Among the Aguaruna, Boster (1985b) similarly reported that women select variation partly to demonstrate their agronomic virtuosity. Indeed, among Amazonia Amerindians in general, male prestige is often related to ceremonial exchange of food products such as manioc beer, and women gain prestige as producers of the crops that men exchange (Descola, 1994; Heckler, 2004).

Homegardens are also visual means to convey social status. In the Ecuadorian Andes, people observe each other's homegardens and make assumptions about the owners' wealth status, occupation and market orientation, as well as health status. Here, the diversity of a household's garden (which often contains hundreds of species with perceived medicinal properties) is an important source of status for women. A woman with a species rich garden earns a reputation as a skilled gardener, and with it, social standing. Women's homegardens thus reveal:

> The extent of the owner's commitment to family well-being. The presence of a garden rich in ... [medicinal plants] epitomizes her exertions on behalf of kin, and her proficiency as primary health provider; a spacious and productive garden filled with medicinal plants suggests that the family, too, is prosperous and fit... Gardens themselves [are] a manifestation of the community's most deeply held values: autonomy, status, religious piety, and personal investment in family... A garden demonstrates a woman's freedom from dependence on products from neighbors and commercial vendors; her fiscal standing evidenced by her ability to expend valuable land on a garden; her faith displayed by a sacrifice of resources to adorn the church; and her industriousness and devotion to family exhibited by her investment in plant cultivation' (Finerman and Sackett, 2003).

It is clear that the status provided through agrobiodiversity management is not confined to visible characteristics of agricultural fields or gardens, or to the skills of their owners. Many studies show that much produce is not consumed but given as gifts or exchanged with others (Zimmerer, 1996; Shrestha, 1998; Wright and Turner, 1999; Nuijten, 2005). Such exchanges are not only important in terms of the products or planting materials that people access: they are just as important as a means to create and maintain social networks. Gift giving through the exchange of planting materials often helps people to maintain kinship and neighbourly ties with people, both within their communities and in distant places, and provides opportunities to accumulate knowledge or to seek help in times of crisis.

Property Rights and Management of Agrobiodiversity

Another cultural institution that is of great importance to human welfare and agrobiodiversity conservation is 'law', or systems of property rights. Scientists agree that such institutions are among the most important in determining whether resources can be managed sustainably. One of the most important negative drivers that has repeatedly been found to undermine natural resource management is the disruption or breakdown of customary property rights systems. This can be through the imposition of colonial legal regimes, state-led land reform programmes, forestry regimes, water management systems, conservation initiatives, forced privatization or collectivization of resources, or any other intervention that imposes a new set of rules about who can use and manage common natural resources. Property rights are also linked to the ability of people to meet their livelihood needs. Like all cultural institutions, property rights institutions vary from culture to culture, context to context, and over time. Indeed, they are so dynamic in cases that some have argued that property rights need to be seen not as static regimes but as processes. Struggles and conflicts over property and the benefits to be derived from property are one of the major driving forces of social and ecological change.

Property rights systems structure the way in which wealth (in whatever form) can be acquired, used, managed and transferred. Biological resources are an important form of wealth. As is the case with all other property rights, claims to use, manage, control and transfer valuable resources are best conceived as a collection of rights and obligations. Property relations, as a basic organizing constituent of social relations, are highly complex, multi-layered, context specific, and are continually contested and changing.

Property is always multifunctional. It is a major factor in constituting the identity of individuals and groups. Through inheritance, it also structures the continuity of such groups. It can have important religious connotations. And it is a vital element in the political organization of society, the legitimate command over wealth being an important source of political power over people and their labour, no matter whether we think of domestic or kinship modes of production, capitalism or communism. Property regimes, in short, cannot easily be captured in one-dimensional political, economic or legal modes (Benda-Beckmann et al, 2007).

What is considered to be a 'resource' varies from society to society, and from group to group within societies. There is not one uniform set of property relations: legal pluralism prevails not only because formal (e.g. state) and customary tenure systems often exist side by side, but also because reforms in these systems create many layers of often contradictory rules, which are interpreted and applied in different ways by different people. Even in Europe, where property rights are codified and where landowners are considered to be autonomous actors with respect to their property, private property rights have been limited and overlain with many different public legal regulations in relation to environmental protection, land use planning and historic conservation.

Because natural resources (land, and also water, trees, animals, plants and air) are often not privately owned, many scientists classify them as common property and such property relations are the subject of much scholarly discussion and debate. Economic theory indicates that, when access to natural resources is 'open' (where anyone can use a resource without limit), such resources will be exploited as long as there is immediate benefit to be derived. Open access to common property resources leads to over-exploitation and resource degradation, or the Tragedy of the Commons (Hardin, 1968). Open access may exist because there are no owners but, even with formal ownership, *de facto* open access may prevail because owners are unwilling or unable to exclude access to the resource. For some time, common property was assumed to be open access and, thus, common property-based systems were considered to be both inefficient and ineffective. This idea has been largely overturned, since it has been found that many traditional common property resource management regimes have been largely effective and sustainable over the long term. Certain principles have been suggested to explain why common property regimes succeed or fail (Table 9.2).

Another set of factors relates to the physical and biological characteristics of the resources themselves and their many uses. Rights to agrobiodiversity are illustrated here using the example of rights to plants. An assumption is often made that those who own the land also have exclusive rights to plants growing on that land, and plants found on common land are also common property. Nevertheless, rights to plants are often different from land rights (Cleveland and Murray, 1997; King and Eyzaguirre, 2005; Howard and Nabanoga, 2007). It is well known that rights to trees (tree tenure) are complex, different from, yet related to, rights to land, and that different groups of users within cultures have different sets of entitlements to trees and tree products (Conklin, 1954; Fortmann and Bruce, 1988; Acharya, 1990; Nugent, 1993; Robbins, 1996; Kundhlande and Luckert, 1998; Meinzen-Dick and di Gregorio, 2004; IWMI, 2005; King and Eyzaguirre, 2005; Benda-Beckmann et al, 2007). Rights to plants are, however, different from rights to trees, but like rights to trees, they are mainly regulated by customary institutions and norms, and are generally informal.

In 1975, Howe and Sherzer demonstrated that systems of rights to plants are culturally rather than legally embedded. Among the San Blas Cuna Indians of northeast Panama, each household cultivates many swidden fields that are dispersed throughout the landscape, so people cannot monitor them against theft. The Cuna believe that generosity is a strong virtue and food sharing is a major way in which such generosity is expressed. Howe and Sherzer found that each plant or tree, and/or its products, is identified by terms in the local language that indicate whether the species has an owner (*ipet nikka*) or not ('belongs to God' – *tios kati*). Of those that have an owner, there is a subcategory that shows whether its products can be obtained as a gift from the owner or not (coconuts can only be obtained from one's parent), and another that distinguishes what must be asked for and what can be taken without

Table 9.2 *Design principles derived from studies of long-enduring institutions for governing sustainable resource use*

Principles	Definitions
Clearly defined boundaries	The boundaries of the resource system (e.g. irrigation system or fishery) and the individuals or households with rights to harvest resource units are clearly defined.
Proportional equivalence between benefits and costs	Rules specifying the quantity of resource products that a user is allocated are related to local conditions and to rules requiring labour, materials, and/or money inputs.
Collective-choice arrangements	Most individuals affected by harvesting and protection rules are included in the group who can modify these rules.
Monitoring	Monitors, who actively audit biophysical conditions and user behaviour, are at least partially accountable to the users and/or are the users themselves.
Graduated sanctions	Users who violate rules-in-use are likely to receive graduated sanctions (depending on the seriousness and context of the offense) from other users, from officials accountable to these users, or from both.
Conflict resolution mechanisms	Users and their officials have rapid access to low-cost, local arenas to resolve conflict among users or between users and officials.
Minimal recognition of rights to organize	The rights of users to devise their own institutions are not challenged by external governmental authorities, and users have long-term tenure rights to the resource.
For resources that are part of larger systems: nested enterprises	Appropriation, provision, monitoring, enforcement, conflict resolution and governance activities are organized in multiple layers of nested enterprises.

Source: Anderies et al (2004)

permission being sought. Since some of the Cuna's crops are now sold for cash, which contradicts traditional ethics that ensure gift giving and reciprocity, plant rights have been adapted. Today, the Cuna differentiate among cash crops of their own accord, so that they may be ungenerous with some crops while they are still expected to be generous with others.

A general framework for understanding rights to plants attempts to capture how the sets of rights and obligations may correspond to social structures (Howard and Nabanoga, 2007) (Box 9.2). The allocation of rights to plants bears a relation to the distribution of plant knowledge within traditional and indigenous communities, where both are determined by the importance of particular plants to particular people. In subsistence communities, the most commonly used plant resources are likely also to be those where access rights and obligations extend to all members of a community, or where the obligation to produce such plants is pervasive. Gender is one of the major factors influencing rights to plants: being a woman implies sharing certain rights and obligations with other women, but not necessarily with men. However, unlike knowledge, plants are physically tied to land, and different plant resources have different biological

---

Box 9.2 FORMULA AND FACTORS AFFECTING PLANT RIGHTS REGIMES

Part Z of species A in resource area X can be used by person Y if the use is for B and Y abides by rule M, during season C

Patterns thus emerge in relation to:

- Resource areas (variable X) since this is also governed by land tenure (*de facto* and *de jure*; customary and formal);
- Particular uses (variable B – e.g. for medicine, for consumption on the spot, for household subsistence, for sale);
- Trees (a subset of variable A - tree tenure);
- Groups of persons (variable Y – e.g. children, herbalists, poor women, cattle owners);
- Particular periods or seasons (variable C – e.g. when species become available; when competition with livestock does not occur; during religious holidays);
- Harvesting limits or management practices (variable M) e.g. the requirement to use a digging stick to cultivate roots at the same time that a wild species is being harvested; to not uproot plantlets, which are also associated with the species, the resource area, the uses and the persons involved.

Source: Howard and Nabanoga (2007)

---

characteristics that create quite different sets of conditions around access and use. For example, certain uses of plants will destroy them whereas others will not, and rights regimes may proscribe uses that destroy plants or stipulate that plants be managed in such a way as to stimulate reproduction or production of particular parts, such as fruit (see Peluso, 1996; Pfeiffer and Butz, 2005). The occurrence of particular species in a particular landscape niche may lead to the creation of rules applicable only to that species and niche that may be quite different from rights regimes that apply more generally.

Among the Baganda, in Uganda, women's homegardens are regarded as their property and anyone wanting to harvest for any use is expected to obtain permission. But there are important exceptions. For instance, elderly women can harvest vegetables for their own consumption from any homegarden. When asked why there are no restrictions on these rights, Mrs Katumba, a Bagandan woman, said: 'We get all our vegetable seed from the elderly women, therefore all vegetables belong to them. They even have the power to curse vegetable yields if you don't let them harvest' (Howard and Nabanoga, 2007). Closely related kin are allowed to harvest enough vegetables to consume at one meal without permission as long as this does not become routine. Anyone can harvest a handful of the vegetable from another's garden without permission as long as the vegetable is meant to feed a sick person, since it is then seen as a medication and not a food. No one can sell what is harvested from others' gardens. It is disgraceful for men to be seen harvesting vegetables. Although men can harvest from any tree found in their wives' gardens, they do not have the right to harvest vegetables for any use other than for medicine. Single men must rely on their female relatives to harvest vegetables. The right to harvest plants depends not only on the ownership and control of the resource area, but also on one's relationship to the plants' owners, the use to which the product is put, and the resource area in which the product is located (Howard and Nabanoga, 2007).

Customary rights regimes have helped to ensure sustainable management of both resources and landscapes over the long term, as well as a distribution of resources that permits all members of a society to meet subsistence needs. Such rights are not only threatened by legal changes and changes in land use, but also by increased social differentiation, migration and population growth (Howard and Nabanoga, 2007). Most traditional societies have already experienced erosion of such rights as well as conflicts over access. Such threats are likely to accelerate in the future, so that reinforcement of local rights to plants, trees, animals, water and land are essential for biodiversity conservation and local livelihoods.

Traditional Societies, Agrobiodiversity, Adaptation and Resilience

This chapter has highlighted the importance of agrobiodiversity, indicated how the livelihoods of some two billion people who live in subsistence socio-ecological systems depend on it, and identified the cultural institutions that have co-evolved over time to bring this diversity to life, manage it and give it meaning. There are multiple global drivers that seriously threaten both biodiversity and related cultures, which together constitute the basis of our planet's richest heritage. These drivers include climate change, biodiversity loss due to deforestation, land fragmentation, invasive species, pollution and the expansion of the global agroindustrial food system (Howard et al, 2009). These will continue to interact with national and local drivers, leading to loss of indigenous knowledge and values, and changing property rights regimes.

Humans have a tremendous capacity to cooperate, understand, learn, manage and innovate – to affect their environments as well as their cultures – both to induce change and to respond to it. These capacities comprise the foundations of human resilience and they need to be harnessed and empowered in order to permit societies to confront the enormous challenges that the future holds. Policy support for traditional societies would represent a reversal of centuries of exploitation and devaluation of such systems, as well as disregard for the rights of local and indigenous peoples to maintain their ways of life. The loss of traditional societies, on the other hand, with their wealth of knowledge, technologies and forms of social organization, will be irreversible and can only reduce the resilience of local, national and global populations, at a time when socio-ecological resilience is required more than at any other time in human history. It is thus imperative that traditional socio-ecological systems, and their sense of stewardship and adaptability, be considered as a globally significant resource to be protected, so that they may continue to adapt and evolve to future local and global challenges.

Acknowledgements

This chapter represents an extract from a document prepared for the Globally Important Agricultural Heritage Systems (GIAHS) Programme of the Food and Agriculture Organization (FAO) of the UN. Thanks go to Parviz Koohafkan and David Boerma of the FAO for commissioning and guiding this work. The full document is referenced as Howard et al (2009).

Note

1   Roy W. Hamilton (2003b) 'Introduction' in *The Art of Rice: Spirit and Sustenance in Asia*, Fowler Museum at UCLA, Los Angeles. Reproduced with permission.

References

Acharya, H. (1990) 'Processes of forest and pasture management in a Jirel community of highland Nepal', PhD thesis, Cornell University, New York

Airriess, C. and Clawson, D. L. (1994) 'Vietnamese market gardens in New Orleans,' *The Geographical Review*, vol 84, no 1, pp16–31

Alexiades, M. (1999) 'Ethnobotany of the Ese Eja: Plants, health and change in an Amazonian society (Plant Medicinals)', PhD thesis, City University of New York, New York

Anderies, J., Janssen, M. and Ostrom, E. (2004). 'A framework to analyze the robustness of social-ecological systems from an institutional perspective', *Ecology and Society,* vol 9, no 1, p18

Anderson, E. F. (1993) *Plants and People of the Golden Triangle: Ethnobotany of the Hill Tribes of Northern Thailand*, Dioscorides, Oregon

Anderson, E. N. (1996) *Ecologies of the Heart: Emotion, Belief, and the Environment,* Oxford University Press, New York

Angel-Peréz, A. and Mendoza, B. (2004) 'Totonac homegardens and natural resources in Veracruz, Mexico', *Agriculture and Human Values,* vol 21, pp329–346

Benda-Beckmann, F. von, Benda-Beckmann, K. von and Wiber, M. (2007) 'The properties of property', in F. von Benda-Beckmann, K. von Benda-Beckmann and M. Wiber (eds) *Changing Properties of Property*, Berghan, Oxford.

Benjamin, T. (2000) 'Maya cultural practices in Yucatecan homegardens: An ecophysiological perspective', PhD thesis, Purdue University, West Lafayette, Indiana

Berkes, F. (1999) *Sacred Ecology: Traditional Ecological Knowledge and Management Systems,* Taylor & Francis, Philadelphia

Boster, J. (1985a) 'Requiem for the omniscient informant: There's life in the old girl yet', in J. Dougherty (ed) *Directions in Cognitive Anthropology*, University of Illinois Press, Illinois

Boster, J. (1985b) 'Selection for perceptual distinctiveness: Evidence from Aguaruna cultivars of Manihot esculenta', *Economic Botany*, vol 39, no 3, pp310–25

Cleveland, D. and Murray, S. (1997) 'The world's crop genetic resources and the rights of indigenous farmers', *Current Anthropology*, vol 38, no 4, pp477–515

Condominas, G. (1986) 'Ritual technology in Mnong Gar swidden agriculture', in I. Nørlund, S. Cederroth and I. Gerdin (eds) *Rice Societies: Asian Problems and Prospects*, Curzon Press and The Riverdale Co, London and Riverdale

Conklin, H. (1954) 'The Relation of Hanunoo Culture to the Plant World', PhD thesis, Yale University, New Haven

Corlett, J., Dean, E. and Grivetti, L. (2002) 'Hmong gardens: Botanical diversity in an urban setting', *Economic Botany*, vol 57, no 3, pp365–79

Davidson-Hunt, I. (2003) 'Journeys, plants and dreams: Adaptive learning and social-ecological resilience', PhD thesis, University of Manitoba, Winnepeg

Descola, P. (1994) *In the Society of Nature: A Native Ecology in Amazonia*, Cambridge University Press, Cambridge

Ellen, R. (1982) *Environment, Subsistence and System: The Ecology of Small-scale Social Formations*, Cambridge University Press, Cambridge

FAO (1999) 'What is agrobiodiversity?' FAO, Rome, ftp://ftp.fao.org/docrep/fao/007/y5609e/y5609e00.pdf

Finerman, R. and Sackett, R. (2003) 'Using home gardens to decipher health and healing in the Andes', *Medical Anthropology Quarterly*, vol 17, no 4, pp459–482

Fortmann, L. and Bruce, J. (1988) 'Why land tenure and tree tenure matter: Some fuel for thought', in L. Fortmann and J. Bruce (eds) *Whose Trees? Proprietary Dimensions of Forestry*, Westview Press, Boulder Colorado

Gladis, T. (2001) 'Ethnobotany of genetic resources in Germany: Diversity in city gardens of immigrants', in J. Watson and P. Eyzaguirre (eds) *Home Gardens and In Situ Conservation of Plant Genetic Resources in Farming Systems,* Proceedings of the Second International Home Gardens Workshop, 17–19 July, International Plant Genetic Resources Institute, Federal Republic of Germany and Rome, Witzenhausen

Greenberg, L. (2003) 'Women in the garden and kitchen: The role of cuisine in the conservation of traditional house lot crops among Yucatec Mayan immigrants', in P. Howard (ed) *Women and Plants: Gender Relations in Agrobiodiversity Management and Conservation*, Zed Press and Palgrave-Macmillan, London and New York

Hamilton, R. (ed) (2003a) *The Art of Rice: Spirit and Sustenance in Asia,* Fowler Museum at UCLA, Los Angeles

Hamilton, R. (2003b) 'Introduction', in R. Hamilton (ed) *The Art of Rice: Spirit and Sustenance in Asia*, Fowler Museum at UCLA, Los Angeles

Hamilton, R. (2003c) 'Labor, ritual and the cycle of time', in R. Hamilton (ed) *The Art of Rice: Spirit and Sustenance in Asia*, Fowler Museum at UCLA, Los Angeles

Hamilton, R. (2003d) 'The goddess of rice', in R. Hamilton (ed) *The Art of Rice: Spirit and Sustenance in Asia*, Fowler Museum at UCLA, Los Angeles

Hardin, G. (1968) 'The tragedy of the commons', *Science*, vol 162, pp1243–1248

Heckler, S. (2004) 'Tedium and creativity: The valorization of manioc cultivation and Piaroa women', *Journal of the Royal Anthropological Institute*, vol 10, pp241–259

Heywood, V. (1999) 'Trends in agricultural biodiversity,' in J. Janick (ed) *Perspectives on New Crops and New Uses*, ASHS Press, Alexandria

Hodgkin, T. and Ramanatha Rao, V. (2002) 'People, plants and DNA: Perspectives on the scientific and technical aspects of conserving and using plant genetic resources', in J. Engels, V. Ramantha Rao, A. Brown and M. Jackson (eds) *Managing Plant Genetic Diversity*, International Board of Plant Genetic Resources (IPGRI), Rome

Hoffman, S. (2003) 'Arawakan women and the erosion of traditional food production in Amazonas Venezuela', in P. Howard (ed) *Women and Plants: Gender Relations in Agrobiodiversity Management and Conservation*, Zed Press and Palgrave-Macmillan, London and New York

Howard, P. (2003) 'Women in the plant world: An exploration', in P. Howard (ed) *Women and Plants: Gender Relations in Agrobiodiversity Management and Conservation*, Zed Press and Palgrave-Macmillan, London and New York

Howard, P. and Nabanoga, G. (2007) 'Are there customary rights to plants? An inquiry among the Baganda (Uganda) with special attention to gender', *World Development*, vol 35, no 9, pp1542–1563

Howard, P., Puri, R. K., Smith, L. and Altierri, M. (2009) *Globally Important Agricultural Heritage Systems: A Scientific Conceptual Framework and Strategic Principles*, FAO, Rome, www.fao.org/nr/giahs/documents1/papers-publications/en

Howe, J. and Sherzer, J. (1975) 'Take and tell: A practical classification from the San Blas Cuna', *American Ethnologist*, vol 2, no 3, pp435–460

Hunn, E. (2001) 'Prospects for the persistence of "endemic" cultural systems of traditional environmental knowledge: A Zapotec example', in L. Maffi (ed) *On Biocultural Diversity: Linking Language, Knowledge and the Environment*, The Smithsonian Institution Press, Washington DC

Iskandar, J. and Ellen, R. (1999) '*In situ* conservation of rice landraces among the Baduy of West Java', *Journal of Ethnobiology*, vol 19, no 1, pp97–125

IWMI (2005) *Proceedings of the African Water Laws Plural Legislative Frameworks for Rural Water Management in Africa, 26–28 January*, IWMI, University of Greenwich, University of Dar-es-Salaam, and South African Department of Water Affairs and Forestry, Johannesburg, South Africa

Keys, E. (1999) 'Kaqchikel gardens: Women, children, and multiple roles of gardens among the Maya of Highland Guatemala', *Yearbook of the Conference of Latin American Geographers*, vol 25, pp89–110

King, A. and Eyzaguirre, P. (2005) 'Intellectual property rights and agricultural biodiversity: Literature addressing the suitability of IPR for the protection of indigenous resources', *Agriculture and Human Values*, vol 16, pp41–49

Kokwaro, J. (1995) 'Ethnobotany in Africa', in R. Schultes and S. von Reis (eds) *Ethnobotany: Evolution of a Discipline*, Dioscorides Press, Oregon

Kundhlande, G. and Luckert, M. (1998) 'Towards an analytical framework for assessing property rights to natural resources: A case study in the communal areas of Zimbabwe', Staff Paper 98-05, Department of Rural Economy, University of Alberta, Edmonton, Canada

Lambert, D. (1985) *Swamp Rice Farming: The Indigenous Pahang Malay Agricultural System*, Westview Press, Boulder, Colorado

Meinzen-Dick, R. and di Gregorio, M. (eds) (2004) *Collective Action and Property Rights for Sustainable Development*, CGIAR System-wide Programme on Collective Action and Property Rights (CAPRi), International Food Policy Research Institute, Washington, DC

Niñez, V. K. (1987) 'Household gardens: Theoretical and policy considerations. Part I', *Agricultural Systems*, vol 3, pp167–186

Nugent, D. (1993) 'Property relations, production relations, and inequality: Anthropology, political economy, and the Blackfeet', *American Ethnologist*, vol 20, no 2, pp336–362

Nuijten, E. (2005) 'Farmer Management of gene flow: The impact of gender and breeding system on genetic diversity and crop improvement in the Gambia', PhD thesis, Wageningen University, The Netherlands

Patterson, S. D. (1992) 'In search of a Mesoamerican floricultural tradition: Ceremonial and ornamental plants among the Yucatecan Maya', PhD thesis, University of California, Los Angeles

Peluso, N. (1996) 'Fruit trees and family trees in an anthropogenic forest: Ethics of access, property zones, and environmental change in Indonesia', *Comparative Studies in Society and History*, vol 38, no 3, pp448–510

Pfeiffer, J. M. and Butz, R. J. (2005) 'Assessing cultural and ecological variation in ethnobiological research: the importance of gender', *Journal of Ethnobiology*, vol 25, no 2, pp240–278

Piper, J. M. (1992) *Bamboo and Rattan: Traditional Uses and Beliefs*, Oxford University Press, Singapore

Puri, R. K. (2005) *Deadly Dances in the Bornean Rainforest: Hunting Knowledge of the Penan Benalui*, KITLV Press, Leiden, Netherlands

Richerson, P.J. and Boyd, R. (2005) *Not by Genes Alone: How Culture Transformed Human Evolution*, University of Chicago Press, Illinois

Rimantha Ginting, J. (1994) 'Plants that cool and clear the mind: The symbolism of rice cultivation, holy places and rituals among the Karo Batak of North Sumatra (Indonesia)', PhD thesis, Leiden University, Netherlands

Robbins, P. (1996) 'Negotiating ecology: Institutional and environmental change in Rajasthan, India', PhD thesis, Clark University, Worcester, Massachusetts

Sadiki, M., Jarvis, D., Rijal, J., Bajracharya, N. N., Hue, T., Camacho-Villa, C., Burgos-May, L. A., Swadogo, M., Lope, D., Arias, L., et al (2007) 'Variety names: An entry point to crop genetic diversity and distribution in ecosystems?' in D. Jarvis, C. Padoch and D. Cooper (eds) *Managing Biodiversity in Agricultural Ecosystems*, Columbia University Press, New York

Sastrapradja, S. D. and Balakrishna, P. (2002) 'The deployment and management of genetic diversity in agroecosystems', in J. Engels, V. Ramantha Rao, A. Brown and M. Jackson (eds) *Managing Plant Genetic Diversity*, IPGRI, Rome

Shigeta, M. (1996) 'Creating landrace diversity: The case of the Ari people and ensete (Ensete ventricosum) in Ethiopia', in R. F. Ellen and K. Fujui (eds) *Redefining Nature: Ecology, Culture and Domestication*, Berg, Oxford

Shrestha, P. K. (1998) 'Gene, gender and generation: Role of traditional seed supply systems in the maintenance of agrobiodiversity in Nepal', in T. Partap and B. Sthapit (eds) *Managing Agrobiodiversity: Farmers' Changing Perspectives and Institutional Responses in the Hindu Kush-Himalayan Region*, International Center for Integrated Mountain Development (ICIMOD), Kathmandu

Steinberg, M. K. (1999) 'Maize diversity and cultural change in a Maya agroecological landscape', *Journal of Ethnobiology*, vol 19, no 1, pp127–139

Tapia, M. and de la Torre, A. (1993) *Women Farmers and Andean Seeds*, FAO and IPGRI, Rome

Trankell, I. (1995) *Cooking, Care, and Domestication: A Culinary Ethnography of the Tai Yong, Northern Thailand*, Acta Universitatis Upsaliensis, Uppsala Studies in Cultural Anthropology 21, Uppsala University, Sweden

Turner, N. (2003) '"Passing on the news": Women's work, traditional knowledge and plant resource management in indigenous societies of North-western North America', in P. Howard (ed) *Women and Plants: Gender Relations in Agrobiodiversity Management and Conservation*, Zed Press and Palgrave-Macmillan, London and New York

Turner, N., Ignace, M. B. and Ignace, R. (2000) 'Traditional ecological knowledge and wisdom of aboriginal peoples in British Columbia', *Ecological Applications*, vol 10, no 5, pp1275–1287

Turner, N. and Berkes, F. (2006) 'Coming to understanding: Developing conservation through incremental learning in the Pacific Northwest', *Human Ecology*, vol 34, pp495–513

Vedeld, P., Angelsen, A., Sjaastad, E. and Kobugabe Berg, G. (2004) 'Counting on the environment: Forest incomes and the rural poor', *Environmental Economics*, Series Paper no 98, World Bank, Washington, DC

Vogl, C. R., Vogl-Lukasser, B. N. and Caballero, J. (2002) 'Homegardens of Maya migrants in the district of Palenque, Chiapas, Mexico', in J. Stepp, R. Zarger and F. S. Wyndham (eds) *Ethnobiology and Biocultural Diversity: Proceedings of the Seventh International Congress of Ethnobiology*, University of Georgia Press, Athens

Wilson, W. (1997) 'Why bitter cassava (Manihot esculenta Crantz)? Productivity and perception of cassava in a Tukanoan indian settlement in the Northwest Amazon (Colombia)', PhD thesis, University of Colorado and University Microfilms International, Boulder, Ann Arbor

Wolff, F. (2004) 'Legal factors driving agrobiodiversity loss', *ELNI Review*, vol 1, pp1–11

Wright, M. and Turner, M. (1999) 'Seed management systems and effects on diversity,' in D. Wood and J. M. Lenne (eds) *Agrobiodiversity: Characterization, Utilization and Management*, CABI Publishing, UK

Zimmerer, K. (1991) 'Seeds of peasant subsistence: Agrarian structure, crop ecology and Quechua agriculture in reference to the loss of biological diversity in the southern Peruvian Andes', PhD thesis, University of California, Berkeley

Zimmerer, K. (1996) *Changing Fortunes: Biodiversity and Peasant Livelihood in the Peruvian Andes*, University of California Press, Berkeley

10
Food Cultures: Linking People to Landscapes

E. N. Anderson

Evolving Food Cultures

Throughout almost all human existence, people ate what their immediate environment provided. For millions of years, ancestral humans lived solely by hunting and gathering. Agriculture arose some 11,000 years ago, but it remained very local and produced little surplus. Only in the last 2500 years have large segments of humanity farmed, traded food and relied on food produced far from their homes. Now a majority of the world's people live in cities, producing virtually no food. Many of the rest live in areas dominated by monocrop or low-diversity agriculture. Subsistence farming is far from extinct, but has been pushed into marginal areas. Hunting-gathering has all but gone from many environments. Few people now eat what they produce. Urbanites have little chance, but even many rural residents are often landless farm workers or specialized producers of one or a few commodities for sale. Very few are diversified farmers who live off their lands.

Modern agribusiness centres on a few varieties of a small number of crops, mainly wheat, rice, maize and potatoes, with barley, sorghum, millet, manioc and sweet potatoes making up much of the rest. Protein comes largely from beans, cattle, sheep, goats, pigs and chickens. Sugar is enormously important (Mintz, 1985); coffee, tea and chocolate provide stimulation (Anderson, 2003a), while grape wine and grain alcohol preparations provide relaxation. The worldwide food industry tends to centre on those items that are easy to grow and process, and that store without spoiling. Markets and human needs have driven agriculture to produce vast quantities of sugars and starches. These have become increasingly cheap relative to other nutrients. The problem of long-term shelf storage limits staple foods to the least nutritious, such as white flour, white sugar, alcohol and cooking oil. As processing became cheaper and storage more necessary, these former luxuries – such as white flour and polished rice – became the preferred and cheapest forms of food. At the same time as foodstuffs travel the world, so natural resources are used up. National soil fertility is effectively exported, and virtual water is shipped from countries of production to importing countries (Smil, 2008).

This represents an enormous shift over the past century. British and American seed catalogues in the early 20th century contained hundreds of varieties of beans and peas, along with dozens of tomatoes and other species variants. More than 7000 varieties of apples were known (Ripe, 1994). Besides this, peasants and homegardeners had countless unique or local varieties and strains growing on their lands. People selected and propagated their own strains, and homegarden evolution was a living force, as it is today among subsistence cultivators. The rapid implosion of diversity reached a peak around 1970, but has reversed since. We have become more conscious of heirloom varieties and heritage crops, and groups like Native Seed Search (of Tucson) work to preserve and propagate these. Supermarkets stocked only three or four varieties of apples 30 years ago, even in peak season. They now often have 20. Still, that is a long way from 7000.

In the US, tastes have expanded in many areas, but contracted in others. Organ meats such as liver and tongue that were popular in my childhood are widely unacceptable now. Squeamishness even denies us many fish and other marine life. Insects and other small fauna were never acceptable to Euro-Americans, in spite of the deserved popularity of such nutritious animal foods in most other parts of the world (Schwabe, 1979). Clearly, our relationship to food has changed, but perhaps too fast for us to adjust. Psychologically and phenomenologically, we humans are still hunter-gatherers. We are cognitively and emotionally tied to landscapes. We bond with place and setting. Some have suggested that we bond most easily to places that look like the high savannahs in which we evolved (Orians and Heerwagen, 1992), but ethnography shows that people feel most connected with the places and environments in which their cultures have evolved. As hunter-gatherers, we were adapted to scan homelands for predators as well as for food, and to react with stress to the former and delight to the latter. A modern life of constant stress, with only rare exposure to green places, is not good for human health (Pretty, 2007). The hunting-gathering lifestyle paved the way to good physical health: walking many miles every day to hunt or gather, foraging for plants, game, fruits and leaves. We evolved to require the balance of exercise and nutrients that such a lifestyle provides. Modern processed foods and their associated lifestyles do quite the opposite.

Vanishing Food Systems: The Yucatec Maya

It is well to look back at what we have lost. For 20 years, I have been studying the Yucatec Maya of western Quintana Roo, Mexico (Anderson, 2003b, 2005a; Anderson and Tzuc, 2005). In this area I have recorded some 150 wild or domestic plant species used for food, as well as 7 domestic bird and mammal species and 15 important game species, plus several small birds and mammals occasionally taken. Food plants range from the staple food, maize, to minor forest species used only in famine periods or for casual snacks while in remote forest areas. Maize is the most important calorie source in traditional households, but beans, sweet potatoes, manioc, squash (including the seeds), tomatoes, chillies and fruit, as well as meat from chickens, ducks (two species), turkeys and pigs can all be important food sources. Until recently, game, ranging from deer to gophers, was important, but overhunting has almost eliminated this resource, causing major hardship to the poor.

Still today in rural areas, every household has its own fields growing maize and other staples, and its dooryard garden can contain 90 or more species of food and medicinal plants. The complex of small fields, gardens and forest plots in varying stages of regrowth (including some old-growth tracts) has created what Fedick (1996) felicitously calls a 'managed mosaic'. This has led to the maximization of biodiversity, preserving organisms dependent on all different stages of succession in one ecosystem. It also provided natural protection against fires. The Maya cut firebreaks, or in wetter areas avoid burning in the wet season. The results of this management was demonstrated after Hurricane Gilbert in 1988, when vast fires devastated forests that were uninhabited, but stopped short at the edges of Maya cultivated lands where firebreaks, open tracts, wet gardens and other features of this mosaic suppressed the flames.

Until very recently in this region, and still today in outlying hamlets, people produced almost all their own food, buying only salt and occasionally a few spices, or some maize if the harvest had been poor. Today, white flour and white sugar are common, and this has had a detrimental effect on health. Maya who have migrated to cities have access to a far more restricted range of foods, in spite of the lavish and colourful markets for which Mexico is so famous. Even the largest markets do not stock traditional forest foods or the rarer bean and squash varieties. Mexico is succumbing to mass production (Stanford, 2008) and so urban Maya may have even less access to fresh foods in the future.

Modern agriculture does not succeed across the Yucatan Peninsula. The searing dry season, torrential rainy season and thin soil over limestone rock foil most attempts. A staggering array of pests and diseases take care of the rest. In fertile areas, citrus and other fruits prosper, and cattle flourish on savannah grasslands, but otherwise the land must be left to Maya cultivation or it quickly degrades. The Maya cope with these challenges by having a profound knowledge of the ecology of the peninsula (see Arellano Rodriguez et al, 2003), and by investing enormous labour in using that knowledge to their advantage. Because of this, they thrive where others fail. Thus the Maya system of agriculture has been left largely intact, though urbanization and dense populations have eliminated or degraded it around cities.

This Maya system of mosaic land management has been sustained by traditional religious teachings that involve ceremonies and intrinsic respect for the spirits of the forest, the fields, the gardens and the weather. Communities pray for rain (*ch'a chaak*) just before the rainy season. A long cycle of ceremonies, collectively known as *loh*, provides food for the spirits to ask their help in the future and thank them for their support in the past. The *jaanlikool* 'food of the fields', for instance, thanks spirits for the harvest. Guardian spirits of the game are believed to punish anyone who takes too much, in other words, more than needed for one's family or more than can be harvested sustainably. As in John Locke's philosophy of natural resource use, one had to leave 'enough, and as good' for others (Locke's *Second Treatise on Government*, 1690, sect 33; Locke, 1924). But again, this religious complex that has helped to maintain biodiversity is eroding in the face of advancing modernization.

Burning Landscapes for Food

On the northwest coast of North America, native peoples (First Nations) traditionally used hundreds of species of marine organisms, dozens of plants, and dozens of land animals and birds. These species were collected from the immediate area by hunting, fishing, gathering and foraging. Extracted resources ranged from whales to seagrass harvests. Tobacco was the only domesticated crop but many edible roots (from clover to wild lilies) were cultivated without being truly domesticated. Many areas were regularly burned to maximize the number of berries and other foods. Game was managed by strict rules enforcing sustainable extraction. Some groups trapped mountain goats in spring, removed their shedding fur and turned them loose again. This suggests taming and thus a complex strategy of animal management. Some groups stocked salmon in streams or carefully managed clam-digging grounds (Williams, 2006). A comprehensive and complex ideology, with many detailed management rules enforcing sustainability, underlay the economy (Turner, 2005).

Native peoples of California created a food-rich landscape in spite of the lack of agriculture covering most of the state (Blackburn and Anderson, 1993; Anderson, 2005). Almost all parts of the state that would burn were subject to fire regimes that maximized the production of food plants. A dramatic testimony is Father Juan Crespí's record of his travels in 1769–1770 through what is now the southern half of the state (Crespí, 2001). He found that in all well-populated areas the land was burned over, and was either bare from recent fires or covered with pasturage (*pastizal*) of grasses and annual forbs. Most of the latter were used for seeds or greens. Species like tansy-mustard (*Descurainia* sp., described by Crespí as a 'cumin-like' herb), redmaids (*Calandrinia ciliata*) and chia sage (*Salvia columbariae*) are fire-following annuals that produce great quantities of nutritious seeds. Crespí also described grass high enough to hide a horse, presumably giant wildrye (*Leymus condensatus*), which produces edible seeds and its growth is stimulated by burning. Through ethnographic research, we now know of abundant accounts of controlled and carefully timed burning, pruning of bushes and trees, transplanting and even sowing of wild crops, irrigation of wild meadow and streambank lands, and other types of landscape management for food and fibre crops.

There is every reason to believe that burning has been used by California native peoples since the earliest settlement more than 12,000 years ago, in which case indigenous management has shaped the local evolution of species and the creation of whole landscapes with varying vegetation types. California's peoples also impacted game and marine animal populations (Kay and Simmons, 2002), periodically depleting some resources while facilitating the recovery of others. The degree of deliberate management involved in this latter process is unknown, but, as on the northwest coast, strict rules against taking too many animals applied in at least some cases. Animals were also managed so as not to become too wild. Hunting deer, for instance, was done by trapping or by using disguises and very close approach. When Latta (1977) asked Yokut hunters how far their bows could shoot and bring down a deer, they explained that they always sneak up close to the deer before shooting. The assumption that California's indigenous peoples had no significant impact, or left no footprint, on their landscapes is clearly a misnomer. They shaped the land by their actions and 'management' activities.

In some parts at least, First Nation peoples are likely to have produced more food per acre, and certainly produced it more sustainably, than we do today. In the lake-rich Klamath River system on the California-Oregon divide, replacement of ducks, fish, antelope, water lilies, nuts, edible roots and shoots, and other life forms, with cattle, barley and potatoes, is likely to have led to a net loss of food production and certainly a loss of diversity. Early descriptions of an abundance of foods contrast strikingly with the desolate and unproductive landscape of today, where the rich lake bottoms produce low-value crops and most of the land is unused. Nowhere is this more obvious than at Tulare Lake, California. It previously produced high densities of fish and marsh plants (Latta, 1977). Drained for cotton cultivation, its basin soils rapidly became too saline for cultivation, and most of it is abandoned today. This may be compared with the situation on the High Plains further east, where many have speculated that the land produced more meat per hectare in the days of buffalo and elk, than it does under cattle today.

Even the vast 'virgin forests' encountered by the European settlers, in what is now the eastern US, were not virgin at all. They consisted largely of oak, hickory and chestnut, all intensely cropped for nuts by local indigenous communities. There may too have been carefully managed burning to maintain these forests (Delcourt and Delcourt, 2004), which thrive on long-cycle large-scale burning, and annual or near-annual ground-cover burns. These cyclic fires also support berry production, as well as deer and other game species. The whole landscape was thus managed for a maximum yield of a highly diverse range of foods. The human race may well have evolved as a fire-following species and certainly has co-evolved with fire (Anderson, 2005b).

Key to all this was a religious worldview that centred on respect and care for non-human beings and sustainable use of natural resources. Among the Maya, it is not uncommon for individuals to ask a tree's permission to take a piece of bark and then thank the tree afterwards. Traditional ceremonies further reinforce this worldview, which centres on taking no more than necessary and leaving some for others. Variants of what is essentially the same worldview persisted throughout the region. This was not the idyllic 'harmony with nature' alleged in some accounts (condemned by Kay and Simmons, 2002, among others) but rather a pragmatic, though religiously represented, set of management plans. Some resources were run down, others enhanced, others merely cropped. Plans were grounded in a belief system that was based on a highly spiritual and divinized conception of nature. In short, even without agriculture, traditional peoples can, and often do, create enormously rich, diverse, complex and carefully managed landscapes of food. Ironically, many of these landscapes were set aside as 'wilderness' areas (considered pristine environments, untouched by humans) after disease and military action eliminated or displaced the First Nation communities that created them. Furthermore, subsequent departure from traditional management practices led to the widespread loss of biodiversity in these areas (Kay and Simmons, 2002; Anderson, 2005b).

Landscapes for Food

In the Americas, there are other human-created landscapes more modified than those of the Yucatec Maya (Doolittle, 2000; Denevan, 2001; Whitmore and Turner, 2001). One of the most spectacular is the terraced slopes of Peru, which are often modified from top to bottom over a vertical mile or more with stonework so careful that it is almost a texture of rock. With fairly minimal maintenance, these terraces have endured for hundreds or even thousands of years (Denevan, 2001). Attempts to modernize them, such as by converting to concrete, have failed. The system was fine-tuned and hard to improve. Similar landscapes for food exist in Europe (Grove and Rackham 2001). Spain's system of growing grain on flatlands, olives and almonds on hills, grazing on high pastures, and pig-rearing in oak forests is an ancient system that has created beauty and stability. Italy's mountain fields, which preserve ancient crops like *farro* (emmer wheat) as well as local wolf populations, persist today. Switzerland's high pastures (Netting, 1981), France's vineyards and Ireland's hedged fields are justly famous.

Perhaps the greatest array of anthropogenic landscapes created for food production exists in Asia. My previous work with Chinese households in China, Hong Kong, Malaysia and Singapore revealed a level of landscape intensification unseen elsewhere (Anderson, 1988). South China was characterized by complex agroecosystems involving intensive rice agriculture in the valleys, tree and vegetable crops around houses, rough grazing on hills and marshes, and production of medicinal herbs and wild foods on left over marginal lands. Every patch of the landscape was intensively used, and until recently, the diversity and stability of the system were considerable.

In the more remote valleys of the New Territories of Hong Kong (Anderson and Anderson, 1973; Anderson, 1988), the high hills comprised wild brushland and grassland, which was burned regularly to drive away destructive animals, making both grazing and walking easier. The ash from the fires was washed downhill in seasonal streams, the water from which was directed onto vegetable terraces or lower rice terraces within the watersheds. Groves of mature trees were carefully conserved above and around villages to provide shade, timber and erosion protection. These, in turn, were protected by religious sanctions –

large trees were worshiped and respected. Vegetables were grown near the villages, easily accessible to care for and later for harvest. Rice was grown where water could be directed into the paddy fields during growing season and drained away for harvest and winter fallow. Lower down, areas that flooded deeply and erratically were used for duck farming and water buffalo pastures. Then came permanently flooded areas which produced lotus and other aquatic crops. Lakes and swamps were turned into fishponds. Beyond these were salty stream deltas, and these were used for shrimp and fish trapping. Even the sea was cultivated, producing oysters in beds laid out in long strips. The whole system was maintained by recycling all possible nutrients, even old ropes and sandals were composted, and sometimes old roof tiles were ground up and spread on the land. Nothing was wasted. Dozens of species of vegetables were grown. Wildlife ranging from frogs and insects to wild ducks were eaten. But modern practices in the later 20th century brought many such systems to near collapse. Pesticides, monocropping and clearing of sacred groves have undermined these sustainable and diverse systems, at the expense of the many ecosystem services and functions they provide.

The Roles of Knowledge and Religion in Food Systems

The common focus of local eating is diversity. Traditional people around the world know every significant plant and animal within a large radius of their homes. However, it is not only the edible and useful ones that are known to them. The Maya, for instance, recognize more than 100 species of birds, most of which are not eaten or otherwise used. Often, even intricate details of the nesting and calls of rare and obscure birds are widely known. Clearly, though, it is the edible species that are best known – where to find them, when to harvest and how to prepare. Even the most poisonous plants are used if they can be processed. Amazingly complex systems have developed for detoxifying plants like cycads and wild yams. Under the constant and imperative pressure to obtain food, such knowledge and skill was important. This had many effects, the most obvious being to preserve local landscapes. If everything is used, people have every incentive to maintain their landscapes in a steady state. All peoples today should at least consider the benefits of designing systems that make use of the widest possible range of resources, and thus provide the greatest possible justification for maintaining the diversity of natural systems.

The role of religion in all this has been unquestionably influential. Resource management systems were maintained by religious rules and taboos, not otherworldly mysticism but pragmatic rules enforced by sanctions. Anyone taking too many game animals or wasting too many resources would attribute any subsequent misfortune to divine punishment. Often, sheer fear from the prick of conscience was sufficient to enforce community compliance. Religion, therefore, gives communities the means to enforce long term, wide reaching strategies and regulations, on people who may otherwise be short-sighted. Failing a strict sanction, even the most well-meaning person is likely to take too much if individual needs are strong. However, it is not only local religions of small societies, but most world religions incorporate good ecological management in their teachings (see the Religions of the World and Ecology series, edited by Tucker and Grim, including Hessel and Ruether, 2000; Girardot et al, 2001). Religion is not inflexible, but does encourage people to do their best. We often fail to understand such systems today, because of opposition to religion and the evolution of secular life (including science) that has developed in the last 200 years.

Can Some Balance be Restored?

Religion no longer has the force it once had, and in many cases, has tended to become divorced from local ecosystems and local ecosystem maintenance. Yet there has been an emergence of understanding about human love of nature and food traditions. But disconnections may now be too profound. Rural residents in many regions today may be too far removed, not only from subsistence production, but from any kind of diversified, small-scale farming of the sort that could support a local community. In this case, what hope is there for urbanites?

In Europe and in much of Asia and even the US, there is, however, a great deal of hope (cf. Nestle, 2002). Cities have public markets and street markets where local farmers sell their best and freshest produce. Farmers' markets, that have long been a fixture in most places in the world, are propagating with amazing speed in the US and Canada. Most European cities have open green spaces assigned to local people for gardens and allotments. Cuba's agricultural revolution has centred on several thousand urban gardens. In China, any dooryard, sidelot or roadside strip is cultivated in some way. Even fence space is not wasted, for thorny but fruit-bearing bushes are used as hedges. Many gardeners cultivate windowsill and veranda gardens. Across Latin America, families grow vegetables and herbs such as cilantro, oregano and mint in containers. In Tegucigalpa, Honduras, this has been developed into a food and herb programme, encouraged and assisted by agencies because of the huge nutritional and medical benefits that can be realized from even a tiny container (Barbara Anderson, pers comm). Gardening in powerline rights-of-way is common in Los Angeles but often leads to harassment of gardeners by power companies or public agencies. Urban agriculture remains an important source of both food and identity in many parts of the world.

The slow food and locavore movements have made people much more conscious of freshness and quality in ordinary foods. The latter movement is particularly prevalent in Seattle. Locavores pledge themselves to eat only foods produced within a certain distance, e.g. 100 miles, of their homes. This idea may be limited by location – one gardener Willie Galloway put it this way: 'I developed a lot of recipes for winter kale'. But locavores are seeking to establish diversity in food systems by making choices that may affect whole landscapes. By combining gardening and farm visits, a modern urbanite or suburbanite can regain much of the contact with food and landscape that we have lost. However, it will never be the same as making one's living in the forests and coasts of a once wilder world.

Acknowledgement

I am grateful to my wife Barbara Anderson for assistance with this paper.

References

Anderson, E. N. (1988) *The Food of China*, Yale University Press, New Haven

Anderson, E. N. (2003a) 'Caffeine and culture', in J. Jankowiak and D. Bradburd (eds) *Drugs, Labor, and Colonial Expansion*, University of Arizona Press, Tuscon, pp159–176

Anderson, E. N. (2003b) *Those Who Bring the Flowers*, ECOSUR, Chetumal

Anderson, E. N. (2005a) *Political Ecology of a Yucatec Maya Community*, University of Arizona Press, Tucson

Anderson, E. N. (2005b) *Everyone Eats*, New York University Press, New York

Anderson, E. N. and Anderson, M. L. (1973) *Mountains and Water: The Cultural Ecology of South Coastal China*, Orient Cultural Service, Taipei

Anderson, E. N. and Tzuc, F. M. (2005) *Animals and the Maya in Southeast Mexico*, University of Arizona Press, Tucson

Anderson, M. K. (2005) *Tending the Wild: Native American Knowledge and the Management of California's Natural Resources*, University of California Press, Berkeley

Arellano Rodríguez, J. A., Guido, J. S. F., Garrido, J. T. and Bojórquez, M. M. C. (2003) 'Nomenclatura, forma de vida, uso, manejo y distribución de las especies vegetales de la Península de Yucatán', *Etnoflora Yucatanense*, no 20, Universidad Autónoma de Yucatán, Mérida

Blackburn, T. and Anderson, M. K. (eds) (1993) *Before the Wilderness: Environmental Management by Native Californians*, Ballena Press, California

Crespí, J. (2001) *A Description of Distant Roads: Original Journals of the First Expedition into California, 1769–1770*, San Diego State University Press, California

Delcourt, P. and Delcourt, H. (2004) *Prehistoric Native Americans and Ecological Change: Human Ecosystems in Native North America Since the Pleistocene*, Cambridge University Press, New York

Denevan, W. M. (2001) *Cultivated Landscapes of Native Amazonia and the Andes*, Oxford University Press, New York

Doolittle, W. (2000) *Cultivated Landscapes of Native North America*, Oxford University Press, New York

Fedick, S. (ed) (1996) *The Managed Mosaic*, University of Utah Press, Salt Lake City

Girardot, N. J., Miller, J. and Xiaogan, L. (eds) (2001) *Daoism and Ecology: Ways within a Cosmic Landscape*, Harvard University Press, Cambridge, Massachusetts

Grove, A. T. and Rackham, O. (2001) *The Nature of the Mediterranean World*, Yale University Press, New Haven

Hessel, D. and Ruether, R. R. (eds) (2000) *Christianity and Ecology*, Harvard University Press, Cambridge, Massachusetts

Kay, C. E. and Simmons, R. T. (eds) (2002) *Wilderness and Political Ecology: Aboriginal Influences and the Original State of Nature*, University of Utah Press, Salt Lake City

Latta, F. F. (1977) *Handbook of Yokuts Indians* (2nd edition), Bear State Books, California

Locke, J. (1924, 1st pub 1690) *Two Treatises on Government*, J. M. Dent, London

Mintz, S. (1985) *Sweetness and Power*, Penguin, New York

Nestle, M. (2002) *Food Politics: How the Food Industry Influences Nutrition and Health*, University of California Press, Berkeley

Netting, R. M. (1981) *Balancing on an Alp: Ecological Change and Continuity in a Swiss Mountain Community*, Cambridge University Press, New York

Orians, G., and Heerwagen, J. H. (1992) 'Evolved responses to landscapes', in J. Barkow, L. Cosmides and J. Tooby (eds) *The Adapted Mind*, Oxford University Press, New York, pp555–580

Pretty, J. (2007) *The Earth Only Endures: On Reconnecting with Nature and Our Place in it*, Earthscan, London

Ripe, C. (1994) 'Dying of starvation at the supermarket', Unpublished

Schwabe, C. (1979) *Unmentionable Cuisine*, University of Virginia Press, Charlottesville, Virginia

Smil, V. (2008) 'Water news: Good, bad and virtual', *American Scientist*, vol 96, no 5, pp399–407

Stanford, L. (2008) 'Globalized food systems: The view from below', *Anthropology News*, no 10, 7 October

Turner, N. (2005) *The Earth's Blanket*, Douglas and MacIntyre, and University of Washington Press, Vancouver and Seattle

Whitmore, T. M. and Turner II, B. L. (2001) *Cultivated Landscapes of Middle America on the Eve of Conquest*, Oxford University Press, New York

Williams, J. (2006) *Clam Gardens: Aboriginal Mariculture on Canada's West Coast*, Transmontanus, Victoria

Part V
Reconnection

11
Sacred Nature and Community Conserved Areas

James P. Robson and Fikret Berkes

Introduction

There is increasing recognition that biodiversity and cultural diversity are interconnected through cultural meaning (Posey, 1999), the use of language (Maffi, 2001) and local ecological knowledge (Berkes, 2008). Such biological and cultural diversity is thus connected by the various processes of the integrated social-ecological systems that make up the biosphere (Berkes and Folke, 1998). This chapter assesses Indigenous and Community Conserved Areas (ICCAs), and their associated livelihoods, cultures and spiritual beliefs that centre on nature. Biodiversity conservation is imperative to livelihoods that depend upon nature to produce ecological services for human well-being (MEA, 2005). But humans also have cultural and spiritual needs. In many parts of the world, belief in a sacred nature underpins people's land and resource use while in pursuit of livelihoods. Moreover, traditional cultural and spiritual values provide the context in which environmental stewardship can be nurtured. As Kothari (2009) puts it, the future of conservation lies at least partly in the past.

ICCAs provide examples of both novel and age-old approaches to safeguarding against common threats to biological and cultural diversity, and against the social and environmental consequences of this loss (Borrini-Feyerabend et al, 2004a; Kothari, 2006). In 2009, the IUCN defined ICCAs as 'natural and/or modified ecosystems containing significant biodiversity values, ecological services and cultural values, voluntarily conserved by indigenous peoples and local communities, both sedentary and mobile, through laws or other effective means' (IUCN, 2009). ICCAs comprise a diverse set of designated areas. While older ICCAs are more clearly associated with the sacred and spiritual beliefs of local groups, newer ICCAs tend to involve a complex rationale that combines a subtler sense of the sacred together with a more prominent set of livelihood and resource productivity concerns.

ICCAs are found in both terrestrial and marine settings, and can range in size from <1ha in sacred groves in India to >30,000km² indigenous territories in Brazil (Oviedo, 2006). These areas, however, largely remain unrecognized by most conservation agencies. While they could (and should) be recognized for what they contribute to national and international conservation objectives, there is little documentation of their potential or discussion of their policy implications to date. As the IUCN acknowledges, 'the history of conservation and sustainable use in many of these areas is (often) much older than government-managed protected areas, yet they are often neglected or not recognized in official conservation systems... and many face enormous threats' (IUCN, 2009).

In this chapter, we look at the role that ICCAs could play in managing multiple values (conservation-spiritual/cultural-livelihoods) in multiple-use landscapes where a great deal of the world's remaining biological and cultural diversity is located. We begin by examining the historic and contemporary context of ICCAs, before proceeding to a case study from Oaxaca, southern Mexico. This case illustrates the difficulties in matching official interpretations of the ICCA concept to local, on-the-ground realities, and the challenges facing the integration of ICCAs into national protected area (PA) systems. The chapter concludes by discussing a number of policy issues related to assessing the conservation benefits of ICCAs as part of multifunctional, cultural landscapes. These include threats to community control of ICCAs, finding the right mix of governance regimes to further biodiversity conservation while protecting local rights and values, and integrating traditional knowledge into PA management.

Traditional Systems Evolving into Mixed Strategies for Biodiversity Conservation

While ICCA may be a new term, the idea of natural areas conserved by communities is not. The traditional basis of conservation is older than the modern conservation movement and goes back to the time of temple gardens in Asia and European game preserves (Borgerhoff Mulder and Coppolillo, 2005). Probably the best known form of traditional conservation – sacred forests or sacred groves in India – have been documented in detail, and traditional sacred areas of diverse descriptions are found in all parts of the world (Ramakrishnan et al, 1998). There are most likely more of these sacred areas than appreciated today; a preliminary survey conducted in Ecuador identified 328 sacred sites (Oviedo, 2006) and, in a pilot project in the Russian North, 263 sacred sites were identified, described and mapped from interviews with indigenous elders of just one district of the Yamal-Nenets Autonomous region. These sites were located on some of the best hunting grounds and contained high biodiversity or rare species, migration routes and unique landscapes (AHDR, 2004). This demonstrates the intricate connection between ecological values and resources important for livelihoods. Furthermore, similarities between traditional and modern conservation are greater than many appreciate. Colding and Folke

(1997) found that nearly one-third of species-specific taboos held by indigenous peoples worldwide corresponded to threatened species that appeared on the IUCN Red List.

The World Heritage Sites network of United Nations Educational, Scientific and Cultural Organization (UNESCO) includes many sites dedicated to the integrated conservation of cultural and biological diversity; sacred mountains, sacred forests, temples and shrines, and sacred lakes and springs. Table 11.1 illustrates the broad diversity of sacred natural sites and the range of geographic regions in which they may be found (Schaaf and Lee, 2006). It is therefore no accident that many national parks around the world have been established at the sites of former sacred areas. These include the Alto Fragua Indiwasi National Park, the first national park of Colombia created at the request of indigenous groups. Another example is the Kazdagi National Park in western Turkey, established in an area with centuries-old sacred sites and a high diversity of trees used by local woodworkers for crafting a diversity of wood products since the time of the Ottoman Sultan Mehmed II in the 1400s (Berkes, 2008).

In some parts of the world, traditional conservation co-exists with government conservation. In the Western Ghats of south India, one of the world's recognized biodiversity hotspots, researchers have found high levels of biodiversity in traditional sacred groves comparable to that in PAs (Bhagwat et al, 2005). Threatened tree species were more abundant in sacred groves, but

Table 11.1 *A typology of sacred natural sites as reflected in the UNESCO document*

Site	Examples from the UNESCO document
Sacred mountains	• Sacred sites and pilgrimage routes in the Kii Mountain Range, Japan • Mount Fuji, Japan • Sacred peaks of the Nepali and Indian Himalaya • Adam's Peak in the cultural landscape of Sri Lanka
Sacred landscapes	• Sacred hidden valleys (*beyul*) of Nepali Himalayas • Sacred sites and burial sites (*mazars*) of Kyrgyzstan • Cultural landscape (*tsodilo*) of the Kalahari, Botswana • Sacred sites in Globally Important Indigenous Agricultural Heritage Systems (GIAHS), such as rice terrace systems • Gran Ruta Inca, the ancient route through the Andean Highlands • Sacred islands, such as the Solovetsky Archipelago in the White Sea, Russia
Sacred forests	• Sacred forests in temples and shrines, Japan • *Kaya* forests of coastal Kenya, sacred areas with protective magic • Sacred groves and ritual use, Ghana • Co-managed Bolivian sacred forests and indigenous lands
Sacred water	• American Indian sacred springs and waters of New Mexico • Sacred Sites and Gathering Grounds Initiative, Arizona • Sacred lakes and springs, Huascarán World Heritage Site and Biosphere Reserve, Peru • Rivers of the Ainu people, Japan

Source: Schaaf and Lee (2006)

endemic tree species were more abundant in the government forest reserve. Researchers also found high biodiversity in multi-species plantations dominated by shade-grown coffee. In these locally developed agroforestry systems, annual, perennial and tree crops are grown together by small landholders in species combinations that have evolved over hundreds of years. While national governments and international conservation organizations emphasize the importance of formal PAs, often to the exclusion of other kinds of protection, it seems that the biodiversity of the Western Ghats is the product of a combination of traditional conservation, agroforest-dominated cultural landscapes and government protection. In this integrated conservation area, sacred groves and agroforestry plantations are just as important as formal PAs (Bhagwat et al, 2005).

In contrast to the apparently deliberate conservation of landscapes and species through sacred areas and taboos, high species richness in some areas is the (non-deliberate) product of traditional livelihood practices. For example, in Australia's Western Desert, Bird et al (2008) showed that indigenous burning for purposes of small game hunting results in the formation of small-scale mosaics that increase habitat diversity. In areas where indigenous burning no longer takes place, the mosaics are much more coarse-grained, leading to a loss in habitat diversity and a decline of small mammals. There are many examples demonstrating the role of traditional livelihood practices in generating landscape level diversity, which leads to species and genetic diversity. Large areas of southern Mexico exhibit high species richness despite the absence of official PAs. In the case of Oaxaca, Mexico, Robson (2007) attributed this to local and indigenous practices that result in multi-functional cultural landscapes. These areas exhibit high beta-diversity[1] due to a mosaic of multiple-use forests and small-scale agriculture along environmental gradients. In the Peruvian Andes, the centre of origin of the potato, Quetchua indigenous people maintain a mosaic of agricultural and natural areas. The 8500ha area, now a designated biocultural heritage site, contains some 1200 cultivated and wild potato varieties. The Quetchua do not make a distinction between the cultivated and wild, and instead perceive the two as part of a continuum (Pathak et al, 2004).

These Indian, Mexican and Peruvian cases exemplify mixed systems that respond to contemporary issues and livelihood needs, while retaining historic sacred relations and traditional land use practices. This is an increasingly common scenario in areas where historic and sacred values of nature now combine with a livelihood dimension and/or a 'learned' conservation ethic. These cases also highlight the fundamental difference between formal PAs and ICCAs. The primary objective of the former is biodiversity conservation, whereas the latter are established for family and community well-being, such as the provision of clean water, as well as for spiritual and cultural reasons. However, ICCA management systems and practices often produce similar outcomes to those being strived for by conservationists from industrialized nations, and this is not coincidental. Despite lacking a modern conservation discourse, local resource users often have well-developed concepts for productive landscapes and waterscapes that provide a diversity of ecosystem services and products to meet livelihood needs (MEA, 2005).

What is also clear is that many indigenous and non-industrial peoples make little or no distinction between the biological, economic, spiritual and social objectives of conservation, and tend to regard these aspects as inter-related. In the worldview of many indigenous groups, from the Cree and Dene of northern Canada and the Maori of New Zealand, to the Zapotecs, Mixes and Chinantecs of northern Oaxaca, the use and protection of natural resources go hand in hand. A person has to use a resource in order to respect it and feel a responsibility towards it. According to this view, conservation without use can be damaging because it alienates people from their lands and from their stewardship responsibilities. Biological and cultural diver-sity necessarily go together as part of an integrated social-ecological system (Berkes and Folke, 1998).

Rise of ICCAs: The Contemporary Context

In addition to historic ICCAs, such as sacred groves, new ICCAs have been established in recent years. Most of the marine ICCAs fall into this category and are largely situated in the Asia-Pacific region. These areas are a legacy of the rich heritage of traditional reef and lagoon tenure systems in which the use of closed areas, closed seasons and taboo species is common. More than 500 locally managed marine areas are found in the Philippines and more than 300 in Fiji, reflecting rapidly growing networks resulting from the efforts of leading island nations (LMMA Network, 2009). In terrestrial areas, ICCAs often emerge out of a combination of traditional practices applied to new species, and an evolving consensus on what constitutes environmentally friendly land use practices. For example, shade-grown coffee, now common in agroecolog-ical systems across Asia, Africa and Latin America, is a new 'innovation' in response to growing international markets for green products (Tucker, 2008).

A novel development that has driven the designation of some new ICCAs is the policy of payments for environmental services (PES). Although PES-like systems existed in the US in the 1980s (particularly in watershed management and soil conservation) and in various other countries in the 1990s, the prac-tice is relatively recent. The original principle was to compensate communities for foregoing use and conserving their forests, based on the value of environ-mental services generated by these forests, for instance, watershed protection, carbon sequestration and biodiversity conservation. Since then, the concept has been used extensively in Latin America and southeast Asia through the efforts of organizations like the World Bank and the Centre for International Forestry Research (CIFOR). Certification of products is a mechanism that can be connected to PES and facilitates access to 'green' markets. This innova-tive economic approach spread particularly rapidly through countries such as Costa Rica and Mexico (Wunder et al, 2008). PES policies, therefore, provide a mechanism to integrate the conservation, livelihood and cultural objectives of community PAs, explaining their rapid adoption in many areas.

The multiple objectives of ICCAs compared to government-PAs, and how the two sets of objectives might become integrated, are perhaps best considered through a set of case studies. Table 11.2 lists five relatively recent ICCAs from a range of geographic areas, all of which involve indigenous or tribal groups.

The Namibian ICCA shown in the table is designated under PA status. It borders Namibia's Etosha National Park and is part of a national network of conservancies that devolve wildlife rights, use and benefits to local communities (Hoole, 2008). This provides an example of the diversity of community-based conservation areas in southern Africa that originated with the Communal Areas Management Program for Indigenous Resources (CAMPFIRE) in Zimbabwe in the 1980s and spread to other countries like Zambia and Mozambique (Fabricius et al, 2004).

The Guyana example in Table 11.2 involves the community conservation of the giant Amazonian fish, arapaima, and is located within an existing PA, the Iwokrama Forest. Here the monitoring of the fish population and the enforcement of the fishing ban are carried out by the local Makushi people. These fish are territorial and live in shallow water. Monitoring relies on the ability of local fishers to identify individual fish from the surface disturbance they create when they come up to breathe air, and count them without marking or other intrusive measures. The enforcement does not rely on government regulation

Table 11.2 *Diverse objectives for establishing modern ICCAs*

Cases and designations	Local objectives and priorities	Reference
Ehi-rovipuka Conservancy, Namibia, 1975km², one of Namibia's 50-plus conservancies	Capture economic benefits of wildlife use and ecotourism; employment; meat from wildlife; enhance community organization and empowerment; participation in wildlife management	Hoole (2008)
Arapaima Management Project of the North Rupununi District Development Board, Guyana	Community-based conservation as investment for future use of arapaima (*Arapaima gigas*), the giant Amazonian fish; collateral donor support; empowerment through better organization and participation in multiple-resource management	Fernandes (2005)
Paakumshumwaau-Maatuskaau Biodiversity Reserve, 4259km², Cree Nation of Wemindji, Quebec, Canada	Biodiversity and landscape conservation; security from hydro-electricity development threat; biodiversity and landscape conservation to safeguard traditional lifestyle; reaffirming land and resource rights; community identity, cohesion and cultural values	Quebec (2008)
Nuevo San Juan, Mexico, 18,000ha, community-based forestry enterprise	Economic and social development; multiple-use forest ecosystem for timber and non-timber forest products; grazing; financing of health and social services; control of traditional lands	Orozco Quintero (2007)
Regional Committee for Chinantla Alta Natural Resources (CORENCHI), northern Oaxaca, 26,000ha	Conservation of diverse tropical forests; development of common strategy for PES approaches; strategy to preserve common property within territorial borders; creation of communal statutes to normalize and regulate use of and access to resources	Camacho et al (2008)

but is based on social sanctions used by the whole community ('more eyes watching') (Fernandes, 2005).

The Canadian example is a biodiversity reserve created at the request of an indigenous group, the Wemindji Cree of James Bay. The original objective was to save a heritage river from possible hydroelectric development in a region where all the major rivers have already been dammed. Facilitated by a research group based at McGill University, Montreal, the Cree carried out consultations and began to develop a nomination document. In the process, they found many other reasons why protection should be implemented. The Quebec government supported the proposal as it facilitated meeting the PA quota for the province. Thus, the nomination document was prepared in about a year in a country where the nomination of a new PA under indigenous land claims can easily take a decade.

Of the two Mexican examples, the Nuevo San Juan case, state of Michoacán, is a long-standing ICCA dating back to 1983. The forestry enterprise emerged from the struggles of the 1960s and 1970s with forest concessions, in which local forests were heavily exploited by outsiders. Taking advantage of existing legislation and developing numerous partnerships to build capacity for starting up and running enterprises, community leaders chose to fight privatization by the use of a communal entrepreneurship approach (Orozco Quintero, 2007). The enterprise has become renowned for achieving value-added production, high diversity of products and by-products, and re-investment of profits for community social development. Nuevo San Juan, therefore, has been successful at balancing livelihood needs with conservation objectives, while increasing the land area under forest cover (Castillo and Toledo, 2000).

The second Mexican case is a much more recently designated ICCA in Oaxaca and involves a coalition of six Chinantec communities known as CORENCHI. A planning process took place among the CORENCHI communities between 2000 and 2006 that led to the demarcation of different land use zones, including conservation areas to protect high-biodiversity forest ecosystems. These zones cover more than 26,000ha in total, or approximately 80 per cent of the combined territories of member communities. The CORENCHI example is part of a new generation of ICCAs set up to strengthen communal control of natural resources and obtain greater socio-economic benefits through conservation efforts. Parallel to land planning, the CORENCHI community conservation process was stimulated by PES from the *Comisión Nacional Fore*stral (the National Forestry Commission, CONAFOR) and certification by the *Comisión Nacional de Areas Naturales Protegidas* (the National Natural Protected Areas Commission, CONANP). As such, the CORENCHI experience complements the more detailed Oaxaca case study to follow, which focuses on a Chinantec community that is yet to apply for government recognition of its ICCAs.

Perhaps the most striking feature of Table 11.2 is the wide range of motivations for establishing ICCAs: access to livelihood resources, security of land and resource tenure, improving communal resource management regimes, security from outside threats, financial benefit from resources or ecosystem functions (including PES), certification, provision of critical ecosystem services such as clean drinking water, rehabilitation of degraded resources, empowerment, capacity building, and cultural identity and cohesiveness. These motivations match many of the dominant conservation perspectives of local communities identified by Kaimowitz and Sheil (2007). Furthermore, each of the cases has multiple objectives, often combining economic, ecological and social aspects. Most cases in Table 11.2 are recent ICCAs; as such livelihood needs are often the main drivers. However, ethical and cultural values are still important because they underpin livelihood objectives. Thus, customary attachment to land and tenurial security are major motivations.

In some of the examples, cultural values are implicit. In the Guyana case, for example, the arapaima was once considered by the Makushi people as 'mother and father of all the fishes' and was protected by local taboos. The modern Makushi, however, say that they do not believe in such superstitions, but still their actions support the continuity of traditional conservation (Fernandes, 2005). In many cases, the needs of future generations are an integral part of the ICCA narrative. This comes across most strongly in the Canadian case where a locally managed PA was established 'so our grandchildren can hunt and fish'. This is also emphasized in the Guyana case in which the Makushi are willing to forego current arapaima harvests for enhanced future potential, and in the Oaxaca case where the CORENCHI communities hope to strengthen the local economy through ICCAs in order to reduce rural-urban migration of young people. Although spiritual or sacred values are not always explicitly stated as drivers for the establishment of new ICCAs, they typically have an implicit underlying role, particularly in informing family and community-level discussions that evaluate the merit of establishing PAs.

The rise in ICCAs has been paralleled by recent dramatic shifts in international conservation paradigms and thinking. While the formal conservation movement has long attempted to separate people from so-called pristine ecosystems, and focus its efforts on islands of biological diversity, the 'last five years has seen a remarkable turnaround towards linking protected areas (or conservation more generally) with the traditions and practices, livelihoods and aspirations of indigenous peoples and other local communities' (Kothari, 2009). However, some caution is needed, especially given that the successful integration of ICCAs into national and international conservation systems would first require a range of conditions to be in place, including policy support, on-the-ground capacity and tenure security (Kothari, 2009). We explore some of these ideas further through another case study from Oaxaca. Home to more than half of Mexico's ICCAs, Oaxaca offers an ideal opportunity to see how ICCAs work on-the-ground and the implications of their integration into a national PA system.

ICCAs in Oaxaca

Oaxaca is nationally and internationally renowned for its biological and cultural diversity (García-Mendoza et al, 2004). Its highly variable topography and climate has given rise to a range of landscapes and ecosystems. This natural environment has co-evolved with diverse indigenous groups who trace their origins to the hunter-gatherers that arrived in Mesoamerica up to

15,000 years ago (Smith and Masson, 2000). This biocultural diversity is still evident today. It is estimated that between 80–90 per cent of Oaxaca's tropical and temperate forests are under the management and control of 1400 local communities (Sarukhan and Larson, 2001; Bray et al, 2008). These communities represent 16 of Mexico's 53 indigenous groups. Many have developed intimate relationships with their natural resources and innovative management practices that lead to multifunctional, cultural landscapes (Robson, 2007). Territorial planning is based on a mosaic of land uses that include forest protection, timber extraction, the harvesting of non-timber forest products, and maize or bean cropping systems (Robson, 2007; Hunn, 2008).

Recent years have seen the establishment of ICCAs, managed as part of local common property regimes. The reported number of ICCAs in Oaxaca differs depending on the source consulted and the definition being used. According to Anta and Perez (2004), 44 communities have set aside conservation areas comprising a total of 175,000ha. In a subsequent study, Anta (2007) identifies only 42 certified community reserves (covering 91,318ha) and 90 voluntary conservation areas (covering 265,720ha). Bray et al (2008) refer to 236 'informally protected' community areas in Oaxaca, covering an estimated 240,000ha of forestlands. While many areas relate to local conservation efforts that are explicitly recognized by communities, governmental agencies, non-governmental organizations (NGOs) or academics, others fall under different land uses. Despite this somewhat confusing picture, two things are clear about ICCAs in Oaxaca. Firstly, there are far more of these areas here than in any other Mexican state. Secondly, ICCAs in Oaxaca cover at least half the area afforded protection by state or federal parks and reserves. These community initiatives represent a wide range of target eco-zones and cover a number of the 'priority' and 'extreme priority' sites recently identified by the country's conservation planners (CONABIO-CONANP et al, 2007). For example, of the 191 species of mammal found in Oaxaca, 32 per cent were found in state and federal PAs, 37 per cent were found in community PAs and 55 per cent were found in both classifications of PA. It has also been reported that 30 per cent of Oaxaca's endemic species at risk are found in the state's ICCAs (Anta, 2007).

The rise in the number of ICCAs in Oaxaca and other Mexican states has led to important policy and legislative changes. In May 2008, reforms to the *Ley General de Equilibrio Ecológico y la Protección al Ambiente* (General Law of Ecological Balance and Environmental Protection, LGEEPA) opened the door to the certification of ICCAs under the title of Voluntary Conservation Areas (VCAs). The VCA mechanism certifies participating communities for the establishment, administration and management of PAs that meet national biodiversity conservation goals. These areas are recognized for the provision of environmental services and meeting conventional PA objectives. Furthermore, products from the sustainable harvest of (restricted) forest resources will receive a government-endorsed 'sustainability seal' to facilitate access to 'green' markets. The LGEEPA's newly-modified Articles 46 and 59 mean that VCAs could be officially recognized by federal government and incorporated into the *Sistema Nacional de Áreas Protegidas* (National System of Protected Areas, SINAP).

According to the latest figures, 127 VCAs are listed by CONANP (June 2009). Of these, 63 belong to indigenous communities, 17 of which are located in Oaxaca. The majority of Oaxacan ICCAs, however, have yet to be registered under the VCA certification mechanism. Thus, they are not formally recognized by government as contributing to national biodiversity conservation goals, nor do they appear on any map of national or state PAs. The following case study provides an opportunity to explore how the Oaxacan experience fits our broader understanding of the ICCA concept and the challenges faced by communities who seek official recognition for their conservation efforts.

Santiago Comaltepec, Northern Oaxaca

Located in the Chinantla region of northern Oaxaca, the indigenous Chinantec community of Santiago Comaltepec holds title to 18,366ha of communal lands. These include extensive tracts of tropical dry forest, temperate pine-oak forest, montane cloud forest and tropical evergreen forest. The community's cloud forest covers some 5500ha with little fragmentation and forms part of the largest and best conserved areas of this forest type in Mexico. The community's forests provide a range of vital hydrological services to both the local populace and downstream users.

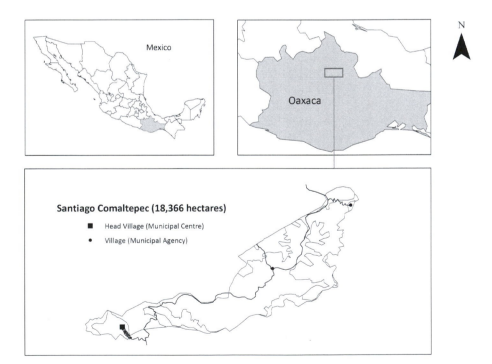

Figure 11.1 *Location and layout of Santiago Comaltepec, Sierra Norte of Oaxaca, Mexico*

Table 11.3 *Land use zones, communal territory of Santiago Comaltepec*

Land Use Zone	Area (ha)
I Forest Production Areas	1436
II Protected Areas	
• Watershed protection	523
• Wildlife protection	4421
• Forest reserve	5068
Subtotal	**10,012**
III Forest Restoration Areas	416
IV Agriculture/Livestock/Urban Use	6206
TOTAL	**18,070²**

Land uses in Santiago Comaltepec include multi-crop production for subsistence and commercial end use, pasturelands for grazing and forests for logging. Ecosystems are also managed for the protection of ecosystem services, the creation of wildlife refuges and the harvesting of non-timber forest products. In this way, territorial use is based on multiple values and needs including subsistence and economic importance, spiritual and sacred significance, and the provision of environmental services (Martin, 1993; Robson, 2009). Table 11.3 shows the community's current territorial plan (2003-13), which divides the communal territory into four main land use categories. ICCAs account for over half of Comaltepec's territory (10,011ha) and are designated for forest, wildlife and ecosystem protection. In such areas, extractive activities are officially restricted by regulations that clearly define (and limit) who has access to and permitted use of forest resources. Village-elected communal authorities supervise and monitor these forest areas. Furthermore, there are at least three sacred areas recognized by local resource users, which play host to important ceremonial activities at the beginning of each year.

The ICCAs were established for a number of reasons. Firstly, local people hold a strong conservationist ethic formed, in part, by the concession period (1957–1982) when forests were taken over and heavily exploited by outside logging interests. Secondly, forests are highly valued for the role they play in the provision of clean drinking water. Thirdly, the establishment of ICCAs in Comaltepec has been incentivized by government PES schemes. The Mexican payment program for hydrological services (for the period 2004–2008) generated approximately US$8000 per year for the community.

Despite setting aside such a large proportion of its communal territory for forest and wildlife conservation purposes, there is still uncertainty and a lack of clarity pertaining to a number of key issues, including the definition of ICCAs, the question of conservation outside of ICCAs, and issues surrounding government recognition of ICCAs.

Defining ICCAs   There are numerous categories of ICCA and so conceptual definitions are subject to interpretation by the different actors involved. In the case of Santiago Comaltepec, the community has established three different ICCAs designated for three specific purposes; forest reserve, wildlife protection and watershed protection. Under the Mexican government's VCA mechanism, it is unclear whether all three areas would be considered for certification or just those protecting forest ecosystems that are of the highest priority to conservation agencies (cloud forest, for example). If the latter is true, then just over half of the community's total designated area (the 'forest reserve') would potentially be recognized by federal government as contributing to biodiversity conservation. This issue points to the problems that could arise when a novel concept is adopted by policy-makers in order to fit pre-existing government conservation objectives.

Conservation outside of ICCAs   The problem of definition is further complicated when we consider that conservation can (and does) take place outside of formal community PAs. Robson (2007, 2009) has shown that conservation benefits in northern Oaxaca are not tied exclusively to the presence of PAs but rather to the multifunctional nature of land use systems. This protects important elements of both forest and agricultural biodiversity across a range of land uses, suggesting that limiting the definition of an ICCA to a 'formally decreed PA' may negate the important contributions made elsewhere in a community's territory. For example, the use of local knowledge and practices to promote a diverse crop selection and, therefore, high levels of agro-biodiversity. Indeed, the community's territory forms part of a wider region considered a centre of domestication, crop evolution, and diversity among and within crop species. Local multi-crop agroforestry systems exhibit important levels of floristic diversity (Bandiera et al, 2005), while the agriculture-forest mosaic provides for a complex, patchy landscape on which a number of bird and mammal species depend (Robson, 2009).

Santiago Comaltepec is also home to a number of sacred natural sites that lie outside of the community's officially designated PAs. Productive activities are typically restricted in these culturally important areas, resulting in biodiversity and landscape conservation (albeit on a small scale). While they are widely acknowledged by the community, their legal status is not clearly established. A large percentage of the community's forestry zones, meanwhile, are certified for sustainable management practices and cutting cycles are employed that seek to protect natural forest processes and functions. Clearings and regeneration cuts imitate the effects of forest fires to help with pine regeneration and reproduce processes of ecological succession. Although the community does not consider these areas to be formal ICCAs, they do contribute to habitat conservation and beta-diversity at the territorial scale.

Government recognition of ICCAs   What would recognition under the VCA scheme imply for a community like Santiago Comaltepec? At the moment, their lands (including community PAs) are managed under customary governance arrangements. This autonomy is very important to local people. With a history of government interference, communities like Comaltepec are wary of most forms of outside intervention. While the community welcomes official recognition of their conservation efforts, they are against any co-management arrangements whereby future ICCA decision-making is shared with external agencies. When interviewed, the village authorities in Santiago Comaltepec were not aware of the new direction taken by Mexican conservation policy, including the development of the VCA mechanism. Most indigenous forest communities were not involved in the discussions that led to recent changes to environmental law. Subsequently, they have little or no knowledge about ICCAs as a new category of official PA and few know how to participate in the VCA mechanism, nor are they aware how involvement would impact their lives in practical terms. The VCA mechanism currently remains an enigma to much of its target audience, and would appear to provide participating communities with few options and a reduced sense of control.

The domination of communal land tenure systems in Oaxaca (and Mexico more widely) means that there is almost no public land for the government to unilaterally set aside as PAs (Robles Gil, 2006; Bray et al, 2008). At the same time, the Santiago Comaltepec case highlights how biodiversity conservation can be compatible with community interests, and that the incorporation of sufficient ICCAs could significantly increase state and national PA coverage. The VCA initiative is an important strategy for the conservation and sustainable use of biodiversity, by which the Mexican government recognizes landowners' voluntary efforts to protect their lands. It does not require local people to relinquish their ownership and management of forest resources, and could lead to official recognition and wide-ranging benefits for participating communities. These potential benefits are especially important for communities that have limited tracts of commercially valuable timber forests or whose forest areas are particularly inaccessible (Bray et al, 2008).

However, the Santiago Comaltepec case also shows that the Mexican government's take on the ICCA concept often fails to match on-the-ground realities, where traditional and contemporary landscape management may achieve conservation in a diversity of ways. Furthermore, there is concern as to what official recognition would imply, in administrative terms, for participating communities. Mexico has yet to fully develop the legal or institutional framework required to stimulate and support the range of voluntary conservation measures. As Anta (2007) explains, novel mechanisms such as the VCA certification scheme are poorly recognized by many of the country's policy-makers and biodiversity specialists. Many key conservation institutions and players have yet to incorporate the idea of voluntary conservation into their agendas, programmes or work plans.

Policy Implications

The findings from Oaxaca and Mexico provide some important policy lessons for other countries that harbour a significant proportion of the world's remaining biological and cultural diversity. They raise a number of questions that have inevitable policy implications. Here we briefly explore four of these:

(i) how to assess the conservation benefits of ICCAs;(ii) what are the perceived threats to community control of ICCAs; (iii) how to find the right mix of governance regimes for ICCAs; and (iv) how to incorporate traditional ecological knowledge into PA management.

In terms of the conservation benefits of ICCAs, policy debate is needed to contextualize the significance of ICCAs. For many signatory countries to the CBD, recent policy has been based on a form of 'systematic conservation planning' (after Margules and Pressey, 2000). This identifies omissions and gaps in national PA systems and selects priority regions for targeting future conservation efforts. Taking this policy to its logical conclusion, many governments will no doubt look to incorporate a significant number of ICCAs into their national PA systems in the future.

There is concern that government ICCA-recognition schemes will not consider some areas of high biodiversity because they are either too small, contain the 'wrong type' of forest, or are considered 'tainted' by existing human disturbance. For example, it is difficult to imagine the VCA mechanism in Mexico certifying large portions of community territory that contain a mosaic of agricultural and natural areas. This would require a progressiveness that is mostly lacking from CONANP, *Comisión Nacional para el Conocimiento y Uso de la Biodiversidad* (National Commission for Use and Knowledge of Biodiversity, CONABIO) and other key government agencies. Rather, government focus is likely to remain on 'wilderness' areas set within community lands that meet pre-existing conservation objectives (Robles Gil, 2006), thereby neglecting the protection of biological and cultural diversity across territorial zones. This would force the ICCA concept away from a 'propoor conservation' approach (Kaimowitz and Sheil, 2007) that is based on finding, developing, maintaining and safeguarding managed landscapes that address basic human needs and values.

With regard to the perceived threat to community control of ICCAs, the degree of government involvement in ICCAs is still unclear. This is something many communities are concerned about because they consider the protection of their natural resources and territory as a key aspect of their cultural identity and reproduction. We know, for example, that many communities in Oaxaca would be strongly opposed to government agencies attempting to formalize autonomous indigenous initiatives. Although the current IUCN PA categories make a distinction between 'co-managed protected areas' and 'community conserved areas' (Borrini-Feyerabend et al, 2004a), all ICCAs would in effect be co-managed by government conservation legislation.

Many indigenous and rural groups, however, still associate 'parks' with 'dispossession'. This is the principal reason why some Indian and Filipino communities with designated ICCAs have been reluctant to take advantage of new legislation (Pathak et al, 2004; Kothari, 2006). The strength of the Australian model for Indigenous PAs (IPAs) is that aboriginal people can decide upon the level of government involvement. The establishment of IPAs thereby creates an arrangement that enables indigenous groups to declare and manage an area, while maintaining control of the resources within their traditional territory (Smyth, 2006; Australian Government, 2009). A similar approach, however, has yet to be adopted in other regions.

In terms of finding the 'right' governance regime for ICCAs, there currently appears to be no single 'correct' governance model. The difficulty is that many existing ICCAs continue to suffer from a range of limitations, including the loss of traditional management capabilities and authority, and insecure land tenure (Borrini-Feyerabend et al, 2004b; Kothari, 2006). On the basis of the cases highlighted in Table 11.2, capacity building can be strengthened through the establishment of partnerships and networks (Berkes, 2007). These typically involve a range of actors such as community, NGOs, government agencies and academic institutions (MEA, 2005). 'Packaged' prescriptions do not work because each ICCA is different. Rather, flexibility and site specific approaches are needed. Strengthening land and resource tenure through government recognition and PES provides incentives for ICCAs to join national systems. This is the major strength of Mexican ICCAs, where power becomes decentralized to communities holding common property rights.

Finally, in terms of the use of local knowledge, ICCAs offer lessons in integrating traditional knowledge and management practices into PA planning. Local and traditional knowledge have only been discussed seriously since the 1990s, and have not to any extent entered mainstream conservation science. Many of the examples mentioned in this chapter, from the Western Ghats to Oaxaca, show that, in many areas, there is in-depth local and traditional knowledge of ecological functions, including how to manage multiple species at multiple scales (Berkes and Folke, 1998). Integrating ICCAs into PA systems means that conservation area managers at all levels need to understand the importance of local knowledge and local institutions (Berkes, 2009). Conventional scientific knowledge and community knowledge operate at distinct spatial scales, and good management requires the use of both (Borrini-Feyerabend et al, 2004b; Berkes, 2008).

Use of local and traditional knowledge in conservation raises questions about the appropriate scale at which conservation should take place. This choice of scale is highly political as it inevitably affects the rights of local people to use and manage resources. Use of biodiversity for livelihoods at the local scale will not necessarily mesh with 'systematic conservation planning' approaches which tend to focus on larger-scale regions. Much of the current debate on community-based conservation (Fabricius et al, 2004; Borgerhoff Mulder and Coppolillo, 2005) focuses on the question of scale, and has no simple answer. It is true that in a complex system all scales are important. Managing ICCAs, and conservation in general, requires taking into account interests at all levels and looking for win-win solutions where possible and trade-offs where not (Berkes, 2007). Indigenous and other rural communities of the world have never been at the centre of the conservation discourse. Getting their voices heard is not going to be easy; rights are won, not given. But there are ways in which local ICCA managers can gain a voice through the actions of community groups, indigenous associations and development NGOs.

Conclusions

The conservation of biological and cultural diversity is intimately intercon-nected, and indeed interdependent (Pretty, 2007). Language, land use prac-tices, traditional ecological knowledge and resource management institutions connect both forms of diversity in social-ecological systems (Berkes and Folke, 1998). We are now at a critical junction in the history of conservation where there is the opportunity to make social–ecological linkages explicit by including people in conservation. ICCAs offer a means to accomplish this, but old paradigms do not change quickly.

A major obstacle facing ICCAs is that these areas do not look much like traditional conservation projects to agencies, officials and policy-makers whose explicit focus remains the protection of threatened and endangered species and their habitats. In fact, in many countries, only a portion of ICCAs are likely to meet the government criteria for recognition and inclusion in national PA systems. Therefore, these recognition schemes risk falling into the trap of restricting their remit to match pre-existing conservation objectives, thereby failing to reflect what is actually happening on the ground.

The issues discussed here have implications for conservation policy nation-ally and internationally. If ICCAs are to play an integral role in future conser-vation policy, the conventional conservation approach would, by necessity, become more inclusive and pluralistic, no longer in the monopoly of conser-vation biologists and government officials. It would broaden the constitu-ency for conservation and make it more real and legitimate for indigenous and rural peoples of the world. Whereas strict preservation will continue to be important, the incorporation of sustainable use and livelihood needs into conservation will contribute to UN Development Programme (UNDP) Millen-nium Development Goals (MDGs) on sustainability and poverty eradication. In particular, ICCAs will help support the poorest families and indigenous peoples that heavily depend on wild resources. In this way, they can and do form a key component of an emerging conservation paradigm that addresses broader, more diversified and more democratically-defined goals.

For ICCAs to work, however, the needs of indigenous people and the threats to these needs must be better recognized, understood and addressed (Kaimowitz and Sheil, 2007). The vast majority of ICCAs have yet to receive recognition from official agencies and the most successful in this regard will likely be communi-ties that are politically perceptive and influential. Weaker and more marginalized groups are likely to find it difficult to lever the required level of institutional and technical support. To help ICCAs fulfil their potential in meeting national and international goals, current government recognition mechanisms need to be backed by a set of supportive legal reforms that are transparent to target commu-nities, and which clearly spell out the costs and benefits of participation. Local and indigenous communities burdened with the costs of conservation generally seek recognition but not intervention, and prefer to receive benefits for their efforts without the imposition of new conditions. Whether national conservation agencies are able to adopt such an approach is yet to be seen.

The conservation of biological and cultural diversity is centred in multiple-use landscapes, where territorial use and protection is informed by multiple values that combine conservation ethics, livelihood needs and long-standing values and beliefs. This complex mix can make it difficult to decide where conservation begins and where it ends. Thus, embracing the diversity and richness of indigenous experiences into the mainstream poses an enormous challenge to conservation. Conventional PAs and conservation science will no doubt continue to be important, but 'next generation' conservation will need to connect cultural diversity with biological diversity through the incorporation of livelihood needs and cultural values into conservation objectives.

Acknowledgements

The authors would like to thank the residents and communal authorities of Santiago Comaltepec for their invaluable contribution to the case study material. Thanks to Francisco Chapela, David Bray and Yolanda Lara for their insights into community conservation initiatives in northern Oaxaca. We also thank the editors for their encouragement and suggestions. This work has been supported by the Canada Research Chairs (CRC) programme in Community-based Resource Management (www.chairs.gc.ca). Robson's work has also been supported by Dr Leticia Merino, Universidad Nacional Autónoma de México (National Autonomoua University of Mexico, UNAM) and a University of Manitoba Graduate Fellowship.

Notes

1 Beta diversity (β-diversity) is a measure of biodiversity that compares the species diversity between ecosystems or along environmental gradients. In Oaxaca, it is common to find a number of different tropical and temperate forests over a relative short distance due to the abrupt topography, aspect and associated climatic variation.
2 The shortfall of 296ha is linked to an area of Comaltepec's territory that was under legal dispute with the neighbouring community of San Pedro Yolox. While this conflict has been resolved, no form of land use is currently permitted in this area.

References

AHDR (2004) *Arctic Human Development Report*, Stefansson Arctic Institute, Akureyri, Iceland
Anta Fonseca, S. and Perez Delgado, P. (2004) 'Atlas de experiencias comunitarias en manejo sostenible de los recursos naturales en Oaxaca', SEMARNAT, Mexico
Anta Fonseca, S. (2007) 'Áreas naturales de conservación voluntaria', Unpublished Report, Iniciativa Cuenca, Oaxaca, Mexico
Australian Government (2009) *National Reserve System*, www.environment.gov.au/parks/nrs/index.html
Bandiera, F. P., Martorell, C., Meave, J. A. and Caballero, J. (2005) 'Floristic heterogeneity in rustic coffee plantations, and its role in the conservation of plant diversity: A case study of the Chinantec region of Oaxaca, Mexico', *Biodiversity and Conservation*, vol 14, no 5, pp1225–1240

Berkes, F. (2007) 'Community-based conservation in a globalized world', *Proceedings of the National Academy of Sciences*, vol 104, pp15188–15193

Berkes, F. (2008) *Sacred Ecology* (2nd edition), Routledge, New York and London

Berkes, F. (2009) 'Community conserved areas: Historic and contemporary context and some policy issues', *Conservation Letters*, vol 2, issue 1, pp19–25

Berkes, F. and Folke, C. (eds) (1998) *Linking Social and Ecological Systems*, Cambridge University Press, Cambridge

Bhagwat, S., Kushalappa, C., Williams, P. and Brown, N. (2005) 'The role of informal protected areas in maintaining biodiversity in the Western Ghats of India', *Ecology and Society*, vol 10, no 1, p8, www.ecologyandsociety.org/vol10/iss1/art8

Bird, R. B., Bird, D. W., Codding, B. F., Parker, C. H. and Jones, J. H. (2008) 'The "fire stick farming" hypothesis: Australian aboriginal foraging strategies, biodiversity, and anthropogenic fire mosaics', *Proceedings of the National Academy of Sciences*, vol 105, pp14796–14801

Borgerhoff Mulder, M. and Coppolillo, P. (2005) *Conservation: Linking Ecology, Economics, and Culture*, Princeton University Press, Princeton, New Jersey

Borrini-Feyerabend, G., Kothari, A. and Oviedo, G. (2004a) 'Indigenous and local communities and protected areas', World Commission on Protected Areas, IUCN, Gland

Borrini-Feyerabend, G., Pimbert, M., Farvar, M. T., Kothari, A. and Renard, Y. (2004b) 'Sharing power: Learning-by-doing in co-management of natural resources throughout the world', IIED/IUCN/CEESP, and Cenesta, Tehran

Bray, D., Duran, E., Anta Fonseca, S., Martin, G. J. and Mondragón, F. (2008) 'A new conservation and development frontier: Community protected areas in Oaxaca, Mexico', *Current Conservation*, no 2.2, pp7–9

Camacho, I., del Campo, C. and Martin, G. (2008) *Community Conserved Areas in Northern Mesoamerica: A Review of Status and Needs*, Global Diversity Foundation and IUCN, Gland

Castillo, A. and Toledo, V. M. (2000) 'Applying ecology to the third world: The case of Mexico', *BioScience*, vol 50, no 1, pp66–76

Colding, J. and Folke, C. (1997) 'The relation between threatened species, their protection, and taboos', *Conservation Ecology*, vol 1, no 1, www.ecologyandsociety.org/vol1/iss1/

CONABIO-CONANP-TNC-Pronatura-FCF/UANL (2007) 'Análisis de vacíos y omisiones en conservación de la biodiversidad terrestre de México: Espacios y especies', CONABIO CONANP, The Nature Conservancy-Programa México, Pronatura, A. C., Facultad de Ciencias Forestales de la Universidad Autónoma de Nuevo León, Mexico City

CONANP (2009) 'Listado de areas certificadas', CONANP, Mexico City, www.conanp.gob.mx/listado_areas.html

Fabricius, C., Koch, E., Magome, H. and Turner, S. (eds) (2004) *Rights, Resources and Rural Development: Community-Based Natural Resource Management in Southern Africa*, Earthscan, London

Fernandes, D. (2005) '"More eyes watching"... Lessons from the community-based management of a giant fish, Arapaima gigas, in central Guyana', Masters thesis, University of Manitoba, Winnipeg

García-Mendoza, A. J., de Jesús Ordóñez, M. and Briones-Salas, M. (eds) (2004) 'Biodiversidad de Oaxaca', Instituto de Biología de la UNAM, Fondo Oaxaqueño para la Conservación de la Naturaleza, WWF, Mexico City

Hoole, A. (2008) 'Community-based conservation and protected areas in Namibia: Social-ecological linkages for biodiversity', PhD thesis, University of Manitoba, Winnipeg

Hunn, E. S. (2008) *A Zapotec Natural History*, University of Arizona Press, Tucson

IUCN (2009) 'Indigenous and community conserved areas', www.iucn.org/about/union/commissions/ceesp/topics/governance/icca/index.cfm

Kaimowitz, D. and Sheil, D. (2007) 'Conserving what and for whom? Why conservation should help meet basic human needs in the tropics', *Biotropica*, vol 39, no 5, pp567–574

Kothari, A. (2006) 'Community-conserved areas: Towards ecological and livelihood security', *Parks*, vol 16, pp3–13

Kothari, A. (2009) 'Protected areas and people: The future of the past', *Parks*, vol 17, no 2, pp23–34

LMMA Network (2009) 'Locally managed marine areas network', www.lmmanetwork.org

Maffi, L. (ed) (2001) *On Biocultural Diversity*, Smithsonian Institution Press, Washington, DC

Margules, C. R. and Pressey, R. L. (2000) 'Systematic conservation planning', *Nature*, vol 405, pp243–253

Martin, G. J. (1993) 'Ecological classification amongst the Chinantec and Mixe of Oaxaca, Mexico', *Etnoecologia*, vol 1, pp17–33

MEA (2005) *Ecosystems and Human Well-being: Current State and Trends*, Millennium Ecosystem Assessment, Island Press, Washington, DC

Orozco Quintero, A. (2007) 'Self-organization, linkages and drivers of change: Strategies for development in Nuevo San Juan, Mexico', Masters thesis, University of Manitoba, Winnipeg

Oviedo, G. (2006) 'Community-conserved areas in South America', *Parks*, vol 16, pp49–55

Pathak, N., Bhatt, S., Balasinorwala, T., Kothari, A. and Borrini-Feyerabend, G. (2004) 'Community conserved areas: A bold frontier for conservation', TILCEPA/IUCN, CENESTA, CMWG and WAMIP, Tehran

Posey, D. A. (ed) (1999) *Cultural and Spiritual Values of Biodiversity*, UNEP and Intermediate Technology Publications, London

Pretty, J. (2007) *Earth Only Endures: On Reconnecting with Nature and Our Place in it*, Earthscan, London

Quebec (2008) 'Proposed Paakumshumwaau-Maatuskaau Biodiversity Reserve Conservation Plan', Quebec Strategy for Protected Areas, Quebec City

Ramakrishnan, P. S., Saxena, K. G. and Chandrashekara, U. M. (eds) (1998) *Conserving the Sacred for Biodiversity Management*, Oxford and IBH, New Delhi

Robles Gil, P. (2006) 'El Carmen: The first wilderness designation in Latin America', *International Journal of Wilderness*, vol 12, no 2, pp36–40

Robson, J. P. (2007) 'Local approaches to biodiversity conservation: Lessons from Oaxaca, southern Mexico', *International Journal of Sustainable Development*, vol 10, pp267–286

Robson, J. P. (2009) 'Out-migration and commons management: Social and ecological change in a high biodiversity region of Oaxaca', *International Journal of Biodiversity Science and Management*, vol 5, no 1, pp21–34

Sarukhan, J. and Larson, J. (2001) 'When the commons become less tragic: Land tenure, social organization, and fair trade in Mexico', in J. Burger, E. Ostrom, R. B. Norgaard, D. Policansky and B. D. Goldstein (eds) *Protecting the Commons: A Framework for Resource Management in the Americas*, Island Press, Washington, DC

Schaaf, T. and Lee, C. (eds) (2006) 'Conserving cultural and biological diversity: The role of sacred natural sites and cultural landscapes', Proceedings of the Tokyo Symposium, UNESCO, Paris

Smith, M. E. and Masson, M. A. (eds) (2000) *The Ancient Civilizations of Mesoamerica: A Reader*, Wiley-Blackwell, Oxford

Smyth, D. (2006) 'Indigenous protected areas in Australia', *Parks*, vol 16, pp14–20

Tucker, C. (2008) *Changing Forests: Collective Action, Common Property and Coffee in Honduras*, Springer Academic Press, New York

Wunder, S., Campbell, B., Frost, P. G. H., Sayer, J. A., Iwan, R. and Wollenberg, L. (2008) 'When donors get cold feet: The community conservation concession in Setulang (Kalimantan, Indonesia) that never happened', *Ecology and Society*, vol 13, no 1, 12, www.ecologyandsociety.org/vol13/iss1/art12

12
Solastalgia and the Creation of New Ways of Living

Glenn Albrecht

Introduction: The Global Transformation of Place and Psychoterratic Dis-ease

Cultures all over the world have concepts in their language that relate psychological states to states of the environment. The Hopi have used the word *koyaanisqatsi* to describe conditions where human life is disintegrating and out of balance with the world. The Portuguese use the word *saudade* to describe a feeling a person has for a loved one, perhaps a loved place, that is absent or has disappeared. People in the front line of environmental change are now telling their stories of distress in the face of unwelcome disturbance to their home environments. As they tell of their experiences, words and concepts in their native languages reveal deep culturally-defined relationships to nature. The Baffin Island Inuit of the Arctic have recently applied the word *uggianaqtuq* to the changing climate and weather. The word means to behave unexpectedly or in an unfamiliar way and has connotations of a 'friend acting strangely' or in an unpredictable way. But now it is the Arctic weather that has become *uggianaqtuq* to them.

Despite the importance of connections between environmental health and human health (physical and mental) in many cultures, we have very few concepts in English that address environmentally induced mental distress and physical illness. In order to rectify that deficiency, I have created two new diagnostic categories: psychoterratic and somaterratic health and illness. These make the connection between the state of the Earth (terra) and mental (psyche) and bodily (somatic) health (Albrecht et al, 2007). Psychoterratic illness arises from a negative relationship to our home environment, be it at local, regional or global scales. The negative relationship involves a loss of identity, loss of an endemic sense of place and a decline in well-being. Conversely, an enduring and positive relationship to a loved home environment delivers the benefits of a strong endemic sense of place and enhanced well-being.

While conventional diagnoses of psychological and psychiatric conditions are possible for some forms of psychoterratic illness, in general such conditions are 'existential' in that they represent a diminution of the quality of existence or well-being. Such existential syndromes can manifest as anxiety, mild distress, forms of depression and in extreme instances, suicide. In order to more fully understand the suite of negative psychoterratic conditions, I wish to revive a pre-existing psychoterratic syndrome, *nostalgia*. This needs to be re-evaluated as a legitimate form of psychoterratic dis-ease to be seen alongside *solastalgia*, a new concept I have introduced to assist our understanding of old and emergent environmentally induced health and illness. Nostalgia was formerly a concept specifically related to physical separation from a loved home environment. Solastalgia, by contrast, I have defined as relating specifically to the place-based distress that is delivered from the lived experience, within a home environment, of unwelcome environmental change.

As a consequence of relentless interconnected global development pressures and global warming, both somaterratic and psychoterratic 'dis-eases' are likely to increase. As ecosystem and climatic health decline, we will see corresponding erosion of the vitality of human psychic and somatic health. Unfortunately, synergistic interactions between biophysical and psychic distress are now implicated in many types of Earth-related pathology.

Cases of Psychoterratic Distress

There is no more graphic illustration of how people respond to a negatively perceived shift or change in the environment than with the case of mining. Exposure to mining activity literally takes people's environment away from them; it undermines their sense of place. Two place-based case studies in environmentally induced psychoterratic distress, the regions of Appalachia in the US and the Hunter Valley of Australia, have much in common. They are now being extensively mined for their coal resources but both places were once seen as exemplifying great beauty where humans could live in harmony with their bioregion. The Hunter Valley landscape was once described as 'the Tuscany of the south' and its beauty has been celebrated by both indigenous and colonial cultures. Appalachia has had its spring beauty celebrated in dance and music. However, in the last few decades, both regions are being desolated by coal mining, with open cuts in Australia creating mountains of spoil and 'voids' full of toxic water, and mountain top removal in the US creating 'valley fills' and smothering streamlines they once contained.

People in both the Hunter Valley and Appalachia are now giving expression to their existential distress about the imposed changes to landscape quality. They are responding to the double pressures of ecosystem distress and a climate that is beginning to act in hostile and unpredictable ways. The victims of imposed place based distress feel that their home is being desolated by forces that they have no power to control (Higginbotham et al, 2006).

Despite the forces of international mining companies and government pitted against them, many in these mining affected regions have been brave enough to acknowledge the distress and still fight the environmental injustice. Maria Gunnoe, who was awarded The Goldman Environmental Prize in April 2009 for her battle against mountain top removal for coal in Appalachia, has given graphic expression to this loss of solace from her home environment:

I'm settin' there on my porch, which is my favorite place in the whole world, by the way – I'd rather be on my front porch than any other place in the world and I've been to a lot of places. As it stands right now, with the new permits I saw last week, they're gonna blast off the mountain I look at when I look off my front porch. And I get to set and watch that happen, and I'm not supposed to react. Don't react, just set there and take it. They're gonna blast away my horizon, and I'm expected to say, 'It's OK. It's for the good of all'.

Am I willing to sacrifice myself and my kids, and my family and my health and my home for everybody else? No – I don't owe nobody nothin'. It's all I can do to take care of my family and my place. It was all I could do before I started fightin' mountain top removal. Now that I'm fightin' mountaintop removal, it makes it nearly impossible. But at the same time, my life is on the line. My kids' lives are on the line. You don't give up on that and walk away. You don't throw up your hands and say, 'Oh, it's OK, you feed me three million tons of blasting material a day. That's fine, I don't mind. It's for the betterment of all' (Gunnoe, 2009).

In the Hunter Valley, open cut or open pit mining for coal has produced similar reactions in people living in what are euphemistically called 'the zones of affectation'. One female grazier had been fighting the coal industry in the Hunter Valley, but the relentless assault on her quality of life with a mine next door to her home finally became too much. In an interview with a research team she describes the psychological and physical pain on her and her farm manager associated with the assault on her rural grazing property from coal mining:

Well, I noticed when this business with [mine name], when I was really fighting here. And my manager would come to me and say he didn't sleep last night. The noise, because they're loading right near the road, he's just across the creek from the road. And you hear a drag line swinging around and dumping rocks into a truck. And then the truck would back away... beep, beep, beep, beep, beep. And then the next one would roar in. He used to say to me 'we just can't cope any longer'... I lost a lot of weight. I'd wake up in the middle of the night with my stomach like that [note: clenched fist], and think, what am I going to do? We're losing money, they won't listen to me, what do I do? Do I go broke? I can't sell to anybody, nobody wants to buy it because it's right next to the mine. What do I do? And I was a real mess ('Eve', as quoted in Connor et al, 2004).

Another resident of the Hunter coalfields reported in an interview that she found the destruction of her home landscape distressing and she went on to say, 'it almost reduces me to tears to think about it [mining]. When the coal is gone, the people ... will be left with nothing but the final void' (Connor et al, 2004). I imagine that mining-affected communities all over the world have similar stories to tell.

An indigenous man in the Hunter Valley, reacting to the massive changes to his traditional lands, expressed his disgust when he was being interviewed about the mining landscape. He explained that he drives hundreds of unnecessary miles to avoid witnessing the desolation of his place:

It is very depressing, it brings you down ... Even [indigenous] people that don't have the traditional ties to the area ... it still brings them down. It is pathetic just to drive along, they cannot stand that drive. We take different routes to travel down south just so we don't have to see all the holes, all the dirt ... because it makes you wild (in Albrecht et al, 2007).

The original people of the Hunter Valley have a documented history of occupation in the Hunter Region going back at least 20,000 years. They were hunters and gatherers, living off the natural productivity of the land. The river valley and the bioregion was seen as an edible landscape, full of emu, kangaroos, fish and fresh water mussels, and many types of edible plants. A 'dreaming story' of the indigenous people at the mouth of the Hunter River tells that coal was originally on the surface of the Earth but that the ancestors covered and buried it. They say that there was a time when a great darkness came over the land and blotted out the sun. The darkness originated from a fire burning from coal deep within the Earth. The elders decided that in order to bring back the light, the darkness had to be stopped from escaping from the hole. All the people gathered rocks, trees and plants and covered the hole and put out the fire. They also covered coal wherever they found it on the surface of the Earth. Then, as thousands of generations of people walked over their country, the coal was rendered safe as it was compressed and remained deep underground.

The Hunter region is famous for the abundance of fossil plants imbedded in the local shale and coal seams, so the indigenous explanation for the presence of these fossils long predates that of Western scientific geology. Despite the Aboriginal dreamtime warnings about the burning of coal, under the impact of coal mining, this 'country' of the indigenous people has been massively transformed and is now a major source of carbon emissions produced in the valley and all over the world. The open cuts emit fugitive methane into the atmosphere and the coal is burnt locally in massive power stations. The great bulk of the coal is exported through the Port of Newcastle at the mouth of the Hunter River. It is now the largest black coal exporting port in the world. The intense reaction of the indigenous interviewee (above) to the coal mining can now be more empathetically understood. Moreover, dreamtime warnings about the dangers of burning coal and the science of global warming seem not that far apart.

Other indigenous peoples worldwide have been in the front line of environmental transformation for centuries. Their identity and sense of place, based on the long term and careful management of renewable sources of sustenance, have been 'sacrificed' so that a different culture, one based on non-renewable resources and the non-sustainable exploitation of renewable 'natural resources', could be developed in its place (Berkes, 2008). The mental and physical health problems of indigenous people worldwide are connected to a significant extent with this severing of personal and cultural identity and sustenance from place. Many indigenous people suffer from psychoterratic or Earth-related mental health syndromes caused by forced separation from a much loved place. They also suffer distress when their remaining home territory is transformed in ways that offends traditional and formerly sustainable connections to place. For example, indigenous people, when confronted by some modern forms of natural resource exploitation such as forestry for paper products, express profound distress responses. David Suzuki reported that a chief of the Haida people in western Canada exclaimed that he 'couldn't breathe' when he came across a recently clear-felled forest and for him, 'it was as if the land had been skinned of life' (Knudtson and Suzuki, 1992).

Other types of unwelcome change to home environments deliver similar anxiety and distress in non-indigenous people. I listened to the testimony of a citizen of Salt Spring Island (west coast Canada) about the noise of float planes as they used Ganges Harbour for takeoff and landing. His life had been made intolerable by the constant noise and the very reason why he chose to live on Salt Spring Island had been overwhelmed by a force that he had no way of controlling. Indeed, as no formal airport has ever been declared on the island, no regulation of the airline industry has been undertaken.

Other forms of disturbance to loved environments cause equally intense forms of human distress. From an Australian website we have the following testimony:

> I heard a bloke say that watching his cherished patch of forest being cleared for development was like losing a child. He was having trouble trying to express the heartbreak. Now more than ever we need a word to describe the pangs of grief and loss felt when witnessing a loved environment being destroyed (Real Dirt, 2008).

In addition to mining and intrusive development, anthropogenic change to our climate is now also having an impact on our physical and mental health. The climate, under the influence of global warming, is becoming less predictable and is shifting. In coastal, eastern Australia, someone would now have to live about 150 kilometres further south from their present location in order to experience a climate similar to that of only 50 years ago. Earlier and warmer springs in the US and Canada have already changed the endemic sense of place and many species are moving their range further north and into higher altitudes. In order to stay in their 'home' comfort zone, the residents of ecosystems are seeking to move further north or south – depending on the hemisphere.

Some species are favoured by warmer temperatures while others are disadvantaged. Insect populations can explode in warmer temperatures with a now classic, but tragic, case being the mountain pine beetle (*Dendroctonus ponderosae*). This beetle, a native to the forests of western Canada (British Columbia) and the US, has emerged from obscurity to international prominence because it is destroying whole forests of lodgepole pine trees *(Pinus contorta)* and other pine species (BC Government, 2009). Its numbers were once controlled by severe winter temperatures and short cool summers. But now that the winters are warmer and summer temperatures are maintained for longer, the beetle and its larvae are completing an annual life cycle in such numbers that they are transforming whole landscapes into graveyards of dead and dying trees. Now the mountain pine beetle epidemic is affecting the US states of Colorado and Wyoming, and many people fear that the forests will be dead within three to five years. In the eastern states of the US, there are unprecedented outbreaks of spruce budworm and hemlock looper, which are also threatening the viability of forests. The anxiety about changing and disappearing forests has been discussed informally by people on the web. One person, when discussing the pine beetle epidemic commented:

> The entire landscape for hundreds of kilometres has been transformed, and now in the last stages of the epidemic vast areas of immature and non-merchantable trees have been killed too. When I finally grasped the extent of this last insult I found it difficult to remain optimistic about the future of our economy, never mind the effects on the scenic values and the recreational experiences we were used to having. It is indeed sad (Torsten, 2007).

What these case studies and the lived experiences of people illustrate is that one of the most powerful relationships we have as humans is to our home environment. Our sense of place is the outcome of the intersecting ecologies of home, head and the heart. Our physical and mental health is tied to this vital relationship and when it is threatened, we can become distressed; when it is broken, we become 'dis-eased'. The changes to home environments that can be the source of threats to our mental health and sense of well-being are often the result of development impacts and now, anthropogenic global warming. Climate change is a particularly powerful driver of distressing change because it will negatively affect those who must move out of increasingly risky and hostile home environments, as well as those who have no choice but to remain within them.

Clearly there is the need for a typology of psychoterratic conditions where we can see how both negative and positive relationships to place are connected to our mental states and sense of existential well-being. I shall start by examining one of the oldest known forms of place-based distress, nostalgia.

Nostalgia

The concept of nostalgia was first created by Johannes Hofer in a dissertation written in Latin in Basel in 1688. The new word was a translation into Greek and New Latin of the German word *heimweh* or the pain for home. Similar concepts have been used in other languages for the feeling of loss when a person is separated from their home environment. Nostalgia (from the Greek *nostos* – return to home or native land – and the New Latin suffix *algia* – suffering, pain or sickness from the Greek root *algos*) or literally, the sickness caused by the intense desire to return home, can be the source of profound psychological and related physiological distress. According to Hofer, the symptoms of nostalgia included 'continued sadness, meditation only on the Fatherland, disturbed sleep either wakeful or continuous, decreased strength, hunger, thirst, senses diminished ... even palpitations of the heart' (Fiennes, 2002).

Found in the English language from 1757 onwards, nostalgia was considered to be a medically diagnosable psycho-physiological disease right up to the middle of the 20th century. In 1905, nostalgia was defined as 'a feeling of melancholy caused by grief on account of absence from one's home country, of which the English equivalent is homesickness. Nostalgia represents a combination of psychic disturbances and must be regarded as a disease. It can lead to melancholia and even death. It is more apt to affect persons whose absence from home is forced rather than voluntary' (Fiennes, 2002).

Nostalgia was particularly evident in soldiers fighting in foreign countries who experienced homesickness to the point where they became ill and unable to perform their duties. The cure for nostalgia was a prescription for afflicted soldiers to be repatriated (sent back to the father's land) to recuperate and restore their well-being and health. According to Feinnes, nostalgia was still being discussed in journals such as *War Medicine* in the 1940s, and Lowenthal (1985) notes that it was 'termed a possibly fatal "psycho-physiological" complaint by an eminent social scientist' as late as 1946.

However, in general, reference to 'nostalgia' as a sickness or melancholia resulting from a longing or desire to return home while away from 'home' is no longer in common use. The more frequent modern use of the term loses its connection to the geographical or spatial 'home' and suggests a temporal dimension or 'looking back', a desire to be connected with a positively perceived period in the past. Typically, there is a longing for a cultural setting in the past in which a person felt more 'at home' than the present. For individuals who see the past as better than the present there is the possibility that nostalgia remains a very real experience that can lead to deep distress. For example, for indigenous people who have been dispossessed of their lands and culture, the nostalgia for a past where former geographical and cultural integration was both highly valued and sustainable is an ongoing painful experience. As explained by Casey (1993), '[n]ostalgia, contrary to what we usually imagine, is not merely a matter of regret for lost times; it is also a pining for *lost places*, for places we have once been in yet can no longer re-enter'. Casey systematically explores the contexts where symptoms of 'place pathology' are presenting problems for both indigenous and colonial culture. He asserts:

> It is a disconcerting fact that, besides nostalgia, still other symptoms of place pathology in present Western culture are strikingly similar to those of the Navajo: disorientation, memory loss, homelessness, depression, and various modes of estrangement from self and others. In particular, the sufferings of many contemporary Americans that follow from the lack of satisfactory implacement uncannily resemble (albeit in lesser degree) those of displaced native Americans, whom European Americans displaced in the first place. These natives have lost their land; those of us who are non-natives have lost our place (Casey, 1993).

The anthropologist, W. E. H. Stanner, writing about the plight of indigenous people in Australia and New Guinea during the colonization process of the 19th and 20th centuries, identified a similar kind of syndrome based on the cumulative distress linked to homelessness, powerlessness, poverty and confusion (Stanner, 2009). He wrote:

> What I describe as 'homelessness', then, means that Aborigines faced a kind of vertigo in living. They had no stable base of life; every personal affiliation was lamed; every group structure was put out of kilter; no social network had a point of fixture left... In New Guinea, some of the cargo-cultists used to speak of 'head-he-go-round-men' and 'belly-don't-know-men'. They were referring to a kind of spinning nausea into which they were flung by a world which seemed to have gone off its bearings. I think that something like that may well have affected many of the homeless Aborigines.

As alluded to above, historically, indigenous people are likely to experience psychoterratic distress as they live through the destruction of their cultural traditions and the transformation of their lands. In rural and remote locations, where a collective memory of a continuous culture such as that of indigenous Australians still exists, there is no idealization of a golden past, but a genuine grieving for the ongoing loss of 'country' and all that entails. The strength of attachment to country is difficult for people in many modern 'European' cultures to fathom, however, Deborah Bird Rose (1996) captures the essence of this attachment when she provides an account of what 'country' means to Aboriginal people in Australia:

> Country is not a generalized or undifferentiated type of place, such as one might indicate with terms like 'spending a day in the country' or 'going up the country'. Rather, country is a living entity with a yesterday, today and tomorrow, with a consciousness, and a will toward life. Because of this richness, country is home, and peace; nourishment for body, mind and spirit; heart's ease.

Both the loss of country and the disintegration of cultural ties between humans and the land are implicated in all aspects of the 'crisis' within many Aboriginal communities in contemporary Australia. The difficulty or inability to find 'heart's ease' is a root cause of the identity problems faced by indigenous Australians. Many authors have identified the social problems experienced by traditional indigenous cultures worldwide and their connections to loss of culture and the support provided by a nurturing home environment. In the Australian context, indigenous people experience physical and mental illness at rates far beyond those of other Australians. Their social problems – unemployment, alcoholism, substance abuse (particularly glue and petrol sniffing in youth), violence against women, and disproportionately high rates of crime and custody including an epidemic of deaths in custody – lead to community dysfunction and crisis. Both loss of home and pathological home environments are implicated in the dysfunctionality.

In general, despite the emplaced distress of indigenous people, the major focus of place-based literature has been on 'lost places' and displaced communities. Yet, in the Hunter Valley it was the distress of those who remained in their home environment that was the focus of my concern. The places that I was interested in were not being completely 'lost'; they were places in the process of negatively perceived transformation. The people I was concerned about were not voluntarily or forcedly being removed from their homes/ places, however, their place-based distress was remarkably close to the feelings of homelessness or nostalgia, indigenous vertigo-in-living and *koyaanisqatsi*. In the Upper Hunter, as is clearly the case in many other contexts worldwide, people were suffering from a negative experience of place transition and felt powerless to stop the process delivering such unwelcome change.

In the light of my experience of distressed people and an empathetic understanding of place-based distress in other communities worldwide, there seemed to be a real need and justification for the creation of a new concept/ term (in English) that captured the conceptual space or territory connected to this existential constellation of factors that define place and identity distress. The situations I have focused on are where people are still within a home environment in a state of transition but feel a similar melancholia as that caused by genuine nostalgia. Such melancholia or sadness is connected to the breakdown of the normal relationship between their psychic identity and their home. What these people lack is solace or comfort derived from their present relationship to 'home' and so I created the neologism, 'solastalgia', to describe this specific form of melancholia.

Solastalgia: The Origins

I had been thinking about the relationship between ecosystem distress and human distress for some time, well before my own direct experience of the distress in people affected by mining in the Hunter Region of New South Wales. The breakdown of a healthy connection between human health, broadly conceived, and the health of the biophysical support environment, or 'home' at various scales, has been carefully considered by many thinkers from Hippocrates onwards (see Albrecht et al, 2008).

American Aldo Leopold and Australia's pioneer environmental thinker Elyne Mitchell were two people who had earlier thought about such relationships. In the US, Leopold, with his ecologically-inspired concept of the 'land ethic' in *A Sand County Almanac* (1949), not only broke new ground in the emergent domain of environmental ethics, he also created the concept of 'land health' defined as 'the capacity of the land for self renewal'. However, he did not see in his contemporaries any connection made between 'sick landscapes' and a sense of shame or pathological psychological states connected to ownership of degraded land.

In Australia, even before Leopold's ideas on land health were published posthumously, Mitchell, in her book *Soil and Civilization* (1946) attempted to explain to Australians the importance of the connection between human mental health and ecosystem health. In the context of the impoverishment of the Australian environment by agricultural activity she wrote:

> But no time or nation will produce genius if there is a steady decline away from the integral unity of man and the Earth. The break in this unity is swiftly apparent in the lack of 'wholeness' in the individual person. Divorced from his roots, man loses his psychic stability (Mitchell, 1946).

As the scale and impacts of post-war industrial society began to enlarge, the transdisciplinary thinker, Gregory Bateson, gave creative expression to the feelings I had about the essential connections between ecosystem and human mental health. He argued in the context of the Great Lakes of North America;

> You decide that you want to get rid of the by-products of human life and that Lake Erie will be a good place to put them. You forget that the eco-mental system called Lake Erie is a part of your wider eco-mental system – and that if Lake Erie is driven insane, its insanity is incorporated in the larger system of your thought and experience (Bateson, 1972).

Despite this prescient insight into the vital eco-mental relationships between ecological and psychological domains, they remained poorly articulated. It was not until I was influenced by David Rapport's concept of 'ecosystem distress syndrome' (Rapport and Whitford, 1999) that I began to contemplate the human conceptual correlate of such distress. In what follows, I shall take old nostalgia and some of the history of 'eco-mental' thinking into the new context for an emergent psychoterratic condition, that of solastalgia.

Solastalgia

Solastalgia (Albrecht, 2005, 2006) has its etymological origins in the concepts of nostalgia, solace and desolation. Solace is derived from the Latin verb *solari* (noun *solacium* or *solatium*), with meanings connected to the alleviation or relief of distress or to the provision of comfort or consolation in the face of distressing events. Solace has connections to both psychological and physical contexts. One emphasis refers to the comfort one is given in difficult times (consolation) while another refers to that which gives comfort or strength. A person or a landscape might give solace, strength or support to other people. Special environments might provide solace in ways that other places cannot. If a person lacks solace then they are distressed and in need of consolation. If a person seeks solace or solitude in a much loved place that is being desolated, then they will suffer distress.

Desolation has its origins in the Latin *solus* (noun *desolare*) with meanings connected to devastation, deprivation of comfort, abandonment and loneliness. It too has meanings that relate to both psychological and physical contexts – a personal feeling of abandonment (isolation) and to a landscape that has been devastated. In addition, the concept of solastalgia has been constructed such that it has a ghost reference or structural similarity to nostalgia thereby ensuring that a place reference is imbedded. Hence, solastalgia has its origins in the New Latin word 'nostalgia' (and its Greek roots *nostos* and *algos*), however, it is based on two Latin roots, 'solace' and 'desolation', with a New Latin suffix, *algia* or pain, to complete its meaning.

I describe solastalgia as the pain or sickness caused by the ongoing loss of solace and the sense of desolation connected to the present state of one's home and territory. It is the 'lived experience' of negative environmental change manifest as an attack on one's sense of place. It is characteristically a chronic condition tied to the gradual erosion of the sense of belonging (identity) to a particular place and a feeling of distress (psychological desolation) about its transformation (loss of well-being). In direct contrast to the dislocated spatial and temporal dimensions of nostalgia, it is the homesickness you have when you are still located within your home environment.

The factors that cause solastalgia can be both natural and artificial. In particular, chronic environmental stressors such as drought can cause solastalgia, as can ongoing war, episodic terrorism, land clearing, mining, rapid institutional change and the gentrification of older parts of cities. Urban transformation is an area of expanding solastalgic experience that many now relate to. I claim that the concept of solastalgia has universal relevance in any context where there is the direct experience of transformation or destruction of the physical environment (home) by forces that undermine a personal and community sense of identity. Loss of place leads to loss of sense of place experienced as the condition of solastalgia.

The most poignant moments of psychoterratic distress occur when individuals directly experience the transformation of a loved environment. Watching land clearing (tree removal), for example, can be the cause of a profound distress that can be manifest as intense visceral pain and mental anguish. With constant and graphic news (with images) of ongoing land clearing in locations such as the Amazon, many people distant from the actual events begin to feel solastalgia as they empathize and identify with distant places and the ongoing destruction that is taking place. I contend that the experience of solastalgia is now possible for people who strongly empathize with the idea that the Earth is their home and that chronic processes that are destroying endemic place identity (cultural and biological diversity) at any place on Earth can be personally distressing to them.

A 'diagnosis' of solastalgia is based on the recognition of the degree of distress within an individual or a community that is connected to the loss of a sense of place, especially an endemic one. All people who experience solastalgia are negatively affected by their desolation and likely responses can include the generalized distress outlined above, but these can escalate into more serious health and medical problems such as drug abuse, physical illness and mental illness (depression, suicide). It is possible to view solastalgia as an existential syndrome with somatic expressions or as a psychosomatic illness with conceptual and empirical dimensions. There is, though, a danger in 'medicalizing' solastalgia as it would lose its philosophical origins and meanings, and might only be seen as a treatable 'illness' under a biomedical model of the human psyche.

With an understanding of the psychodynamics of solastalgia, the problems of a pathological or toxic home and a lack of 'heart's ease' can be explained with greater cross-cultural sensitivity and relevance. I argue that the 'dis-ease' of indigenous people can, in part, be explained as a response to solastalgia. Both social and medical epidemics that afflict some indigenous people can be partly understood as their attempt to relieve themselves of the distress and pain of solastalgia. Perhaps solutions to such problems can come from the diagnosis of solastalgia and its negation by empowered indigenous people being directly involved in the repair and restoration of their 'home'. In areas where people still have strong, direct connections to country, the defeat of solastalgia can come from actions that strengthen the endemic and weaken the exotic. Such actions could, for example, range from indigenous responsibility for the removal and management of exotic species (flora and fauna) to the active promotion of indigenous culture. In urban areas where indigenous links to land are more tenuous, building new culturally creative sources of solace will assist in the creation of 'heart's ease'.

An example of the relationship between a healthy landscape and human health is provided by traditional owners in Arnhem Land in the Northern Territory of Australia. Some indigenous people maintain their connections to their traditional land by living on what are called 'outstations'. These are small family-based communities that allow traditional cultural values to be transmitted and a degree of hunting and gathering on the land. In this context, people have been able to maintain a healthy ecosystem (country) with documented benefits for their total well-being, including physical and mental health. Johnston et al (2007) writing about this issue argue:

> *The indigenous testimony of the nexus between human and landscape health is entirely consistent with previous anthropological research in Australia. Moreover, the identified benefits to both people and country are consistent with the more limited*

epidemiological and ecological research. But policies designed to improve Aboriginal health have barely begun to integrate Aboriginal perspectives, underscoring the inherent Western view that human health is largely decoupled from the natural environment.

The results of decoupling of human health from supportive home environments can also be seen in non-indigenous people under the impact of persistent drought and other forms of rural environmental decline (e.g. dryland salinity) in parts of Australia. In rural and regional Australia in particular, there are reports of increasing rates of depression and suicide, particularly in men. The standard explanations for increasing rates of suicide and depression in the rural context include rural reconstruction in the face of economically rational 'deregulation', high indebtedness and financial problems, unemployment, distress over loss of family owned property and heritage, and easy access to chemicals and firearms. However, rarely is the state of the relationship between humans and their biophysical environment considered as part of the matrix of health.

In Australia in general, the psychological health of farmers is closely related to the health of the environment. In the grip of drought, when stock die of malnutrition and thirst, the dams are empty, the pasture is barren and even the wildlife begins to die, severe depression about such a state of affairs is not uncommon. The impact of prolonged drought on rural gardens can also be a cause of solastalgia with the loss of the house garden a tipping point of distress. In collaboration with colleagues from the Centre for Rural and Remote Mental Health (CRRMH) in New South Wales, research has been conducted on farmers and the stress created by persistent drought. One interviewee, a female farmer explains:

> *Well I guess we're coming into our fifth year of the drought ... our gardens have had to die because our house dam has been dry ... so it's very depressing for a woman because a garden is an oasis out here with this dust ... you know, to come home to a nice green lawn ... that's all gone, so you've got dust at your back door* (Albrecht et al, 2007).

During 'natural' events such as drought, the morale of farmers, their families and the local community decline, and with drought-breaking rains, joy and confidence in the future returns. Solastalgia is negated by the natural restoration of the home environment to something that is full of creative and productive potential. Unfortunately, under global warming, some of the negative changes that are taking place such as decades of declining rainfall, loss of biodiversity and more frequent and severe bushfires cannot be negated by one-off solastalgia-breakers. They have become a permanent feature of many climes and landscapes worldwide. It is therefore possible to make more general and global comments about the existential and psychological implications of the decoupling of humans from supportive home environments. The World Health Organization (WHO) predicts that depression will be the second leading burden of illness worldwide by 2020, and I would suggest that under extreme climate change scenarios, this burden is likely to increase even further.

Soliphilia and the Love of Life

We have seen how solastalgia is a human response to the lived experience of an emerging negative relationship to a home environment. It is instructive to think further about the very opposite response, one of strong identity and positive affiliation with a loved home environment. As with negative land relationships, humans have a long history of expressing their positive feelings, their love of home and landscape in art, prose, poetry and philosophy. However, it is again surprising that few systematic attempts have been made to clearly articulate a particular love of place in the English language.

In 1923, Rudolf Steiner created the concept of 'love life' in relation to the life of bees. He argued:

> *That which we experience within ourselves only at a time when our hearts develop love is actually the very same thing that is present as a substance in the entire beehive. The whole beehive is permeated with life based on love. In many ways the bees renounce love, and thereby this love develops within the entire beehive. You'll begin to understand the life of bees once you're clear about the fact that the bee lives as if it were in an atmosphere pervaded thoroughly by love ... the bee sucks its nourishment, which it makes into honey, from the parts of a plant that are steeped in love life. And the bee, if you could express it this way, brings love life from the flowers into the beehive. So you'll come to the conclusion that you need to study the life of bees from the standpoint of the soul* (Steiner, 1998).

Steiner, as a pioneer in the study of organic interconnections, actually saw life and love explicitly in ecological terms. He argued in his 1923 lectures on bees that to understand life, 'you need to take a deep look into the entire ecology that nature has to offer' (see Matherne, 2002).

The concept of an ecological or interconnected life of love and love of life was first clearly articulated by Eric Fromm in the 1960s. He distinguished between 'necrophilia' which involved the love of destruction and death and 'biophilia' which he described as the 'love of life'. In *The Heart of Man* (1965), Fromm develops the idea of biophilia in the context of human character development, productivity and ethics. In the tradition of Albert Schweitzer he argues:

> *The full unfolding of biophilia is to be found in the productive orientation. The person who fully loves life is attracted by the process of life and growth in all spheres. He prefers to construct rather than to retain.*

He further goes on to state:

> *Biophilic ethics have their own principle of good and evil. Good is all that serves life, evil is that which serves death. Good is reverence for life, all that enhances life, growth, unfolding. Evil is all that stifles life, narrows it down, cuts it into pieces* (Fromm, 1965).

Fromm's pioneering concept of biophilia links love of humanity with love of life and nature, in a nexus that anticipates many themes within late 20th and early 21st century environmental ethics. In his 'Humanist Credo', written in 1965 and first published in English in *On Being Human*, he linked biophilia to a comprehensive ethic:

> *I believe that the man choosing progress can find a new unity through the development of all his human forces, which are produced in three orientations. These can be presented separately or together: biophilia, love for humanity and nature, and independence and freedom* (Fromm, 1994).

To develop Fromm's life-based concept of biophilia even further into an empathetic and ecological understanding of life, we might wish also to talk about 'ecophilia' (Soyinka, 2004), an idea long overdue for further development. E. O. Wilson (1984) uses the term 'biophilia' in a way different from the psychosocial character development of Fromm, in that he argued biophilia was biologically-rooted. Wilson argued for a 'deep conservation ethic' based on innate biological affiliation with all other organisms, as a counter to destructive and exploitative relationships with the rest of nature.

The concept of 'topophilia', or love of place, was developed by the geographer Yi-Fu Tuan (1974), and he highlighted the suite of human 'affective ties with the material environment'. Tuan argued that, on most occasions, topophilia is a mild human experience; an aesthetic expression of joy about connection to landscape and place, but that it can become more powerful when human emotions or cultural values are 'carried' by the environment. He acknowledges the work of the anthropologist Strehlow in providing insight into the depth of positive place attachment held by Aboriginal Australians, and what happens when place attachment is severed (Tuan, 1974). He also considers that topophilia can be a powerful human emotion for humans who are closely connected to the land.

As a response to the likely increase in chronic environmentally induced distress at all place scales, from local to global, I have created a counter to solastalgia, the concept of *soliphilia*. This 'philia' is a culturally and politically inspired addition to the other 'philias' that have been created to give positive biological and geographical conceptions of connectivity and place. Soliphilia, put simply, is the love of the totality of our place relationships and a willingness to accept the political responsibility and solidarity needed between humans to maintain them at all scales of existence.

Soliphilia can be added to love of life and landscape to give us the love of the whole, and the solidarity between humans that is needed to keep healthy and strong that which we all hold in common. In order to negate all the 'algias' or forces that cause sickness and extinction, we require a positive love of place, expressed as a fully committed politics and as a powerful ethos or way of life. Soliphilia goes beyond the left-right politics of the control and ownership of cancerous industrial growth, and provides a universal motivation to achieve sustainability through new ways of symbiotic living that are life-affirming.

Conclusion

It is ironic, but at the very moment globalization and technology enables us to connect as a single species, we begin to see the value of the endemic, and the locally and regionally unique. That great philosopher of place, singer Joni Mitchell, argued 'don't it always seem to go that you don't know what you got till its gone'. She was worried about the erosion of the distinctiveness of rural and regional US as it disappeared under the car parks of supermarkets and junk food outlets. Endemic US was in the process of being replaced by a culture going global. Now, that concern about a loss of the regionally distinctive and endemic applies to the whole planet. We are at risk of losing a sense of place about the whole Earth within a few decades of the first ever images of the living planet, Gaia, taken from space by the Apollo 8 astronauts only 40 or so years ago.

It seems that many people, in a variety of contexts, sense that something is wrong with our relationship to our planet home and their unease just might be an expression of deep-seated solastalgia about non-sustainability. Most of us are now acutely aware of the issues affecting the future of this planet and have no desire to lapse into eco-anxiety or worse, forms of eco-paralysis and the melancholia of solastalgia. We want to do something meaningful right now to put the world back onto a sustainable pathway. All the good philias combine into a new and all-encompassing emotion that is a foundation for hope and the deeply rooted urge to pass on present generations' security and safety to future generations. Soliphilia can be the foundation of expressions of new ways of being that reunite us, as people, and with the foundations of life.

Jonathon Lear in *Radical Hope* (2006) writes powerfully about Plenty Coups, the last great chief of the Crow Nation in the US. Coups' world and culture was almost completely desolated by the invasion of white settlers and the loss of the bison on traditional hunting lands. Right now, in the turmoil of a fiscal, ecological and climate crisis all of us are in the position of the Crow and all other indigenous people who, despite having the norms of their culture and place utterly changed, have been able to find the courage to imagine and work for a viable future. The Crow lost their bison and with them, for a while, everything else. However, with inspired ethical leadership, Plenty Coups was able to reconnect elements of his culture's past with a vision of a future containing hope. What is needed now, in this period of global collapse, is ethical commitment and inspired leadership to usher in a sustainable future that we cannot yet fully see, but is likely to transcend almost everything that is comfortable and familiar to us now.

The love of life, and a life filled with love, is a most profound starting point for a sustainable future. At present, we are certainly not united by a universal sense of our shared fate. We are not bees, but we do now live in a globally-interconnected hive where there is the potential for a human ethic based on love of life to pervade the whole Earth. Our love of the whole, the soliphilia, will stimulate dynamic and sustainable responses to human-induced local, regional and global ecosystem distress (including climate distress) by creative people and their affiliations from all cultures and walks of life.

References

Albrecht, G. (2005) 'Solastalgia: A new concept in human health and identity', in *Philosophy, Activism, Nature*, vol 3, pp41–55

Albrecht, G. (2006) 'Environmental distress as solastalgia', *Alternatives*, vol 32, no 4/5, pp34–35, www.for.gov.bc.ca/HRE/topics/mpb.htm, accessed October 2009

Albrecht, G., Higginbotham, N., Connor, L. and Freeman, S. (2008) 'Human health and ecosystem health: A social perspective', in K. Heggenhougen (ed) *Encyclopaedia of Public Health*, Elsevier, San Diego, California

Albrecht, G., Sartore, G., Connor, L., Higginbotham, N., Freeman, S., Kelly, B., Stain, H., Tonna, A. and Pollard, G. (2007) 'Solastalgia: The distress caused by environmental change', *Australasian Psychiatry*, vol 15, special supplement, pp95–98

Bateson, G. (1972) *Steps to an Ecology of Mind: Collected Essays in Anthropology, Psychiatry, Evolution, and Epistemology*, Chandler Publishing Company, New York

BC Government (2009) 'Mountain pine beetle research', www.for.gov.bc.ca/HRE/topics/mpb.htm

Berkes, F. (2008) *Sacred Ecology* (2nd edition), Routledge, New York

Bird Rose, D. (1996), *Nourishing Terrains: Australian Aboriginal Views of Landscape and Wilderness*, Australian Heritage Commission, Canberra

Casey, E. (1993) *Getting Back Into Place: Toward a Renewed Understanding of the Place-World*, Indiana University Press, Bloomington

Connor, L., Albrecht, G., Higginbotham, N., Smith W. and Freeman, S. (2004) 'Environmental change and human health in upper Hunter communities of New South Wales, Australia', *EcoHealth*, vol 1, supplement 2, pp47–58

Fiennes, R. (2002) *The Snow Geese*, Picador, London

Fromm, E. (1965) *The Heart of Man: Its Genius for Good and Evil*, Routledge and Kegan Paul, London

Fromm, E. (1994) *On Being Human*, Continuum, New York

Gunnoe, M. (2009) 'Stop mountaintop removal', www.earthjustice.org/our_work/campaigns/stop-mountaintop-removal.html

Higginbotham, N., Connor, L., Albrecht, G., Freeman, S. and Agho, K. (2006) 'Validation of an environmental distress scale (EDS)', *EcoHealth*, vol 3, no 4, pp245–254

Johnston, F., Jacups, S., Vickery, A. J. and Bowman, D. (2007) 'Ecohealth and aboriginal testimony of the nexus between human health and place', *EcoHealth*, vol 4, no 7, pp489–499

Knudtson, P. and Suzuki, D. (1992), *Wisdom of the Elders: Native and Scientific Ways of Knowing About Nature*, Allen & Unwin, Sydney

Lear, J. (2006) *Radical Hope: Ethics in the Face of Cultural Devastation*, Harvard University Press, Cambridge, Massachusetts

Leopold, A. (1949) *A Sand County Almanac and Sketches Here and There*, Oxford University Press, London and New York

Lowenthal, D. (1985), *The Past is a Foreign Country*, Cambridge, Cambridge University Press

Matherne, B. (2002) 'Bees by Rudolf Steiner', Book Review, www.doyletics.com/arj/beesrvw.htm, accessed 13 October 2009

Mitchell, E. (1946) *Soil and Civilization*, Halstead Press, Sydney

Rapport, D. and Whitford, W. G. (1999), 'How ecosystems respond to stress', *Bioscience*, vol 49, p193–203

Real Dirt (2008) 'How coal companies are stealing our rivers', http://realdirt.com.au/2008/11/21/riverstalgia-how-coal-companies-are-stealing-our-rivers, accessed October 2009

Soyinka, W. (2004) 'The Reith Lectures: Climate of fear', www.bbc.co.uk/radio4/reith2004/lecture3.shtml

Stanner, W. E. H. (2009) *The Dreaming and Other Essays*, Black Inc. Agenda, Melbourne

Steiner, R. (1998) *Bees*, Anthroposophic Press, New York

Torsten (2007) 'Scariest consequence of climate change (global warming)', Bautforum, www.bautforum.com/science-technology/65319-scariest-consequence-climate-change-global-warming-3.html, accessed October 2009

Tuan, Y. F. (1974) *Topophilia: A Study of Environmental Perception, Attitudes, and Values*, Prentice-Hall, New Jersey

Wilson, E. O. (1984) *Biophilia*, Harvard University Press, Cambridge, Massachusetts

13
Ecocultural Revitalization: Replenishing Community Connections to the Land

Sarah Pilgrim, Colin Samson and Jules Pretty

The holistic reconnection of people to each other and to the land, affording reserve-based and urban populations the opportunity to engage with each other and their homelands, will be the foundation of individual psychophysical health and community resurgence (Alfred, 2009).

Introduction

Humans have evolved in natural environments over a period of several million years. Subsequently, their cultures have shaped, and in turn been shaped by, local ecosystems and their constituent parts (Balée, 1994; Norgaard, 1994; Denevan, 2001; Maffi, 2001; Toledo, 2001; Gunderson and Holling, 2002; Harmon, 2002). In industrialized countries[1] today, many communities live their daily lives in urban environments, with only intermittent exposure to green space. However, in indigenous[2] and non-industrial communities (such as subsistence hunter-gatherers, agriculturalists and pastoralists), most people retain stronger links with the natural environment through resource use and management (Milton, 1998; Berkes, 2004, 2008). Consequently, the knowledge and practices of these communities are being drawn upon across the world today for use in community conservation (Cinner et al, 2005; Smith et al, 2007; Robson and Berkes, 2010, this volume), demonstrating why distinctions made between social and natural systems are considered by most indigenous peoples and some recent commentators to be artificial (Berkes et al, 2003; Pretty et al, 2008).

While learning how to live off the land, human communities developed and refined knowledge, skills and tenure systems that still persist in many non-industrial communities (Turner and Berkes, 2006; Pilgrim et al, 2008). This extrinsic physical dependency has evolved in line with a more intrinsic dependency. For instance, many indigenous societies have evolved spiritual beliefs, ceremonial traditions, sacred designations and worldviews based on their own lands. Where this is the case, both personal and cultural identity is intertwined with the physical landscape and nature as a whole (Basso, 1996; Milton, 1998; Berkes, 2004). However, this interconnection is increasingly being lost in communities across the world as the daily lives of a growing number of peoples are becoming separated from nature (Kellert and Wilson, 1993; Nabhan and St Antoine, 1993; Pyle, 2003; Pretty, 2004, 2007; Pretty et al, 2005, 2006, 2007).

Disconnection from nature can be caused by physical or psychological separation. The former can be brought about through the physical dislocation of an entire community away from their homelands and to a different environment. Forced resettlement follows from state-directed policies to assimilate indigenous and other marginal groups or from environmental destruction caused by industrial projects. Local communities are rarely consulted and displacement is often against their will (Cernea, 1988, 1997; Cernea and Schmidt-Soltau, 2006). Forced resettlement has occurred through the policies of a large number of colonizing state entities, settler states and dominant populations that have expanded into the territories of indigenous and native peoples over a long period of time (Colson, 1971; Turnbull, 1973; Brody, 1981; Marcus, 1995; Gall, 2002; McKnight, 2002; Samson, 2003; McGrath, 2006; Alfred, 2009). Examples can be seen in Canada, the US, Australia and New Zealand, parts of Asia and Africa including the Kalahari, and throughout Amazonia. Dislocations, both small- and large-scale, stem from natural resource policies on the one hand, advocating the reclamation and expropriation of particular lands and resources, and assimilation campaigns on the other. Assimilation does not necessarily separate communities from their lands physically, but strives to diminish and erode the intrinsic connection with land, spiritually, mentally and emotionally, leading to a form of psychological separation (Samson, 2003; Samson and Pretty, 2006; Pretty, 2007; Alfred, 2009; Albrecht, 2010, this volume).

Disconnection from the land has the capacity to damage and even destroy cultures that remain closely tied to their environments. The effects of dispossession often create political chaos and social discord within indigenous communities, leading to psychological, physical and financial dependency on the state (Alfred, 2009). This, in turn, can lead to mental and psychological ills which often also manifest as physical ailments and social pathologies, particularly if disconnection is rapid, for instance, across one or two generations. Ills include increasing prevalence of modern day conditions such as obesity, type II diabetes, hypertension and coronary heart disease, as inactivity increases and diets shift from wild to store-bought foods. Depression and anxiety, as well as suicide and alcohol abuse, have also been shown to be associated with reduced time in the natural environment (Shkilnyk, 1985; Basso, 1996; Samson, 2003; Samson and Pretty, 2006; Johnston et al, 2007; Pretty, 2007; Alfred, 2009;

Gracey and King, 2009; King et al, 2009). Thus a broken connection can lead to a broken community.

The Emergence of Revitalization Projects

Rapid disconnection is most significantly felt today by indigenous and non-industrial societies marginalized by limited wealth, power and status and suffering from associated social pathologies (Milton, 1998). However, many such communities that have fallen victim to disconnection in the past are now striving to reinvigorate their traditional cultures and reconnect with their homelands, in spite of continuing pressures such as globalization and commodification of resources (Berkes, 2001; Pilgrim et al, 2008, Gigoux and Samson, 2009). Recognizing the health and societal repercussions of being disconnected from nature (Johnston et al, 2007), non-industrial communities in a number of locations are taking action to reclaim or maintain their unique beliefs and practices through ecocultural revitalization projects (so termed for their focus on reconnecting cultural systems with the ecosystems upon which they are based). Like the cultures they seek to rejuvenate, revitalization projects are very diverse, ranging from hunter-support schemes and local food policies to language initiatives and ecotourism projects.

All revitalization projects share a similar objective: to maintain or reclaim the culture of local peoples and reconnect them to the land for long-term individual and societal health. Cultures exist in many different contexts today, for instance, in science, policy and business, and by no means are all cultural ideas and practices good for nature. However, this chapter focuses on cultural arrangements (largely of resource dependent communities) that do have positive synergies with nature, and how these synergies can be actively fostered for the future. Pretty et al (2008) use four intrinsic components to assess these changes; (i) beliefs, meanings and worldviews, (ii) livelihoods, practices and resource management systems, (iii) knowledge bases and languages, and (iv) institutions, norms and regulations. All four must be sustained if cultural continuity is to be successfully attained.[3]

Today, revitalization projects are evolving independently of one another among communities and groups of communities across the world, at a time when international policy-makers are only just starting to acknowledge the interdependence between human and environmental health – e.g. the Man and the Biosphere Programme from the United Nations Educational, Scientific and Cultural Organization (UNESCO), and the 2007 flagship report *Global Environment Outlook* from the United Nations Environment Programme (UNEP). Efforts currently span North America, parts of the Arctic and sub-Arctic, as well as locations in Asia and Africa. Traditional food, healthcare, language and culture projects focus on reviving specific elements of community culture, such as local diets, medicines, languages, ceremonial traditions and land-based practices (e.g. specialized hunting techniques). Ecotourism projects have a similar objective, but utilize these cultural elements as an income-generating strategy to encourage tourist activity. Education projects focus on developing

Table 13.1 *Typology of Revitalization Projects*

Project type	Objectives of project type
1 Traditional Foods	To increase the consumption of traditional local foods and revive food collection and preparation practices
2 Traditional Healthcare	To revive knowledge of traditional healthcare practices including the preparation and ethnobotanical skills they are based upon
3 Ecotourism	To revive traditional cultural practices and ceremonies as part of an income generating strategy
4 Education	To provide a more balanced, culturally-appropriate education system either separate from or as part of a state education system
5 Language	To protect or enhance the competency of speakers of endangered languages and open communication channels between community elders and young people
6 Cultural	To revive particular aspects of a way of life that may have been neglected
7 Rights	To campaign for the recognition of the human rights and land rights of indigenous cultures with a view to ensuring cultural continuity and diversity into the future

Source: Pilgrim et al, 2009

culturally-appropriate education schemes and transferring traditional knowledge and practices to younger generations. Rights revitalization efforts are based on the renewal and strengthening of traditional rights, most commonly land rights. These project types (see Table 13.1) will be discussed in turn in the forthcoming sections, along with selected reviews of specific projects.

Traditional Foods Revitalization Projects

Foods play a role above and beyond nutrition in human societies. They help to define identity and shape social structure, and are often used in communication, group activities and religious observances. As a result, it is not uncommon for traditional foods to be a major defining characteristic of society. Local diets epitomise the ways in which a culture uses, classifies and thinks about its natural resources, and strengthens the connection between a society, its landscape and its ancestral roots (Pars et al 2001; Tansey, 2004; Raine, 2005; Willows, 2005). Moreover, there is a growing evidence base which suggests that reverting to a traditional diet offers physical health benefits too (Samson and Pretty, 2006; Johnston et al, 2007; Pretty, 2007).

Compared with the low saturated fat, low sugar and low salt diets most non-industrial communities are used to, highly processed store bought foods, combined with the lack of physical exercise needed to acquire them, has led to substantial health costs, including obesity and related diseases such as hypertension and heart disease (Cordain et al, 2000; Kozlov and Zdor, 2003; Waldram et al, 2006). Recognizing this, a number of revitalization projects have been established to reintroduce traditional foods into modern diets (Marquardt and Caulfield, 1996; Nuttall, 1998; Kishigami, 2000; IITC, 2003a,b; Sonjica, 2004; Bersamin and Simpson, 2005; OHEP, 2008).

A range of incentives from local organizations and national governments are being used to promote this shift from modern to traditional foods, including the creation of markets for local foods (Marquardt and Caulfield, 1996; Nuttall, 1998; Kishigami, 2000). However, reintroduction comprises more than just renewed consumption. In many cases, communities are reviving the livelihood skills, practices and knowledge needed to find, collect and prepare traditional foods. In the Inuit communities of Akulivik in Quebec, Canada, a hunter-support programme has been established to provide economic support to hunters while ensuring the distribution of traditional foods. Market channels have been created so that hunters in local communities can sell the meat and fish they catch to village councils for redistribution. Each village councillor is given a portion of the project funds to pay hunters and to buy and repair equipment. Every village that participates in the project is entitled to a community hunting boat and a communal cold storage house. If project funds are not used locally for hunter and fisher wages, then monies may be used to buy fish and meat from nearby villages to distribute among community members, in particular widows, elders or full-time wage earners who are unable to hunt. This provides local hunters with an income source and ensures continued consumption of traditional foods (Kishigami, 2000).

The Greenland government has created a similar market but on a national scale. The Home Rule Government, established in 1979, has made the promotion and expansion of traditional country food markets a priority. The government distributes licences (prioritizing commercial hunters and fishers, and nationals unemployed for more than 125 days per year) and territories to communities and, in doing so, ensures livelihoods and healthy diets even in isolated communities and settlements. At the same time, residents of more industrial towns who do not have the time to hunt (e.g. on the west coast) are able to purchase healthier traditional foods (Nuttall, 1998). To ensure price competition does not drive down populations, all country foods are sold at fixed prices agreed upon by the local hunters' and fishers' association (Marquardt and Caulfield, 1996; Pars et al, 2001). Hunters are encouraged to sell their surplus catch to Royal Greenland, the national meat and fish processing and distribution company (Nuttall, 1998). By providing a source of full time income, or even just supplementary income for households, the marketization of country foods in Greenland has increased the self sufficiency and cultural continuity of Greenlandic communities (Marquardt and Caulfield, 1996; Nuttall, 1998; Pars et al, 2001). The success of the Greenlandic government supersedes the notion that traditional hunter-gatherer economies exist outside of the modern capitalist one, and instead demonstrates that an indigenous hybrid economy exists based on a three sector model: private, public and customary. In this economy, customary activities are integrated into the global capitalist economy (Altman, 2005).

In South Africa, the Department of Science and Technology, the Council for Scientific and Industrial Research, and groups set up to promote indigenous foods launched Indiza Foods in 2004. Indiza Foods is a community-owned company aimed at developing sustainable community enterprise while promoting traditional foods. It encourages the consumption of country foods by publishing local recipe books and holding local, district and provincial food fairs. It has also set up processing centres to supply supermarkets and businesses with indigenous foods (Sonjica, 2004).

Traditional foods revitalization projects, from local to national in scale, have succeeded in market creation, the establishment of support programmes, the creation of new businesses, in particular the emergence of micro-enterprise, and incentivizing traditional food collection and consumption. Their success lies in their ability to reaffirm cultural identity while ensuring livelihood security.

Traditional Healthcare Revitalization Projects

Traditional healthcare is a vital dimension of indigenous knowledge, and is also central to cultural continuity and connectedness to the land. Many indigenous and non-industrial groups developed systems of medicine long before modern medicine evolved, for instance, Ayurvedic medicine in India. These systems, and versions of them, still persist today in communities across the world where Western healthcare is physically and/or financially inaccessible. However, there is now emerging a sense that these healthcare systems should be protected even where biomedicine is available. Apart from time-tested efficacy for common ailments, the reason is the cultural continuity and historical identity they represent. Some are based on ethnobotanical species, others on spiritual belief systems centred on landscape features, and some contain elements of both. However, all represent a specific form of cultural identity, knowledge and belief systems that modern healthcare systems threaten to displace.

The Foundation for Revitalisation of Local Health Traditions (FRLHT) in Bangalore is a local health NGO seeking to revive local healthcare traditions. Its efforts include publishing books in the local language, encouraging the establishment of kitchen herbal gardens which contain up to 20 ethnobotanical species, disseminating information using videos and slides, and training village leaders in traditional healthcare practices. Similarly, the TRAMIL Program of Applied Research into Popular Medicine in the Caribbean is focused upon encouraging villagers to create homegardens containing medicinal species enabling home healthcare. Ethnobotanical information is promoted and transferred through the use of pamphlets, videos, music, dance, puppet shows and community meetings (Burford and Ngila, 2002).

Rukararwe Partnership Workshop for Rural Development, an NGO based in western Bushenyi District, Uganda, has established a fully functional health clinic which combines modern medicine with traditional healthcare practices. To facilitate this, a cooperative of local healers have created a herbarium with more than 100 medicinal species. Herbariums have also been created by schoolchildren in Colombia through the New Schools of Colombia Project, which encourages children's participation in community research. Children are involved in creating kitchen gardens and documenting local ethnobotanical knowledge. Similarly, in Morocco, schoolchildren have created a school yard herbarium through Dar Taliba Educational Project. Ethnobotanical species grown in this garden can be used for community healthcare, and ethnobotanical teachings are integrated into the school curriculum (Burford and Ngila, 2002). In Tanzania, the NGO Aang Serian recruits international volunteers each summer to train up Noonkodin Secondary School students (16–18 year olds) in anthropological and ethnobotanical research skills in order to help them to undertake their own research projects on an aspect of Maasai traditional healthcare (Burford, pers comm).

Traditional healthcare systems depend upon open communication channels between elder and younger generations. Their revival has the capacity to improve local living conditions, both in terms of community health and economic status. Furthermore, there is the potential to develop micro-enterprise opportunities from the traditional products for marketing locally, nationally or indeed internationally, providing the Intellectual Property Rights of communities are protected and there is equitable benefit sharing among knowledge holders.

Ecotourism Revitalization Projects

Ecotourism is becoming increasingly popular as wealthy people seek to learn and experience non-industrial lifestyles and the 'back to nature' image they represent. Tourists are willing to pay to enjoy this experience. By reviving traditional skills for the purpose of income generation, ecotourism projects permit an integrative approach to revitalization by providing an economic market for traditional cultures and the knowledge and skills they hold. Making full use of this emerging market, many Sámi reindeer herders have begun to open their reindeer farms to the public. The Inari Reindeer Farm in Finland is one such example. Visitors to Inari enjoy traditional game dishes, singing traditional *joiks* and experiencing reindeer safaris by sled (Inari Event, 2008). The Purnumukka Reindeer Farm, Finland, offers a similar package (Purnumukan Porofarmi, 2008).

Reindeer farms are often family owned businesses, whereas other ecotourism facilities are community owned or at least community centred, with jobs and other benefits filtering back to local community members. Examples include Bathurst Inlet Lodge and Elu Inlet Lodge, both in Canada, which offer accommodation and participation in traditional local activities (Bathurst Inlet Lodge, 2008; Elu Inlet Lodge, 2008). Other companies offer outfitting services. With the growing ecotourism industry, many communities are working on developing their infrastructure and training the workforce in order to develop their tourism potential based on local culture and traditional lifestyles, for instance, Alaska Chukotka Development Program (ACDP) in Chukotka, Russia (ACDP, 2003). By developing tourism infrastructure locally and rejuvenating cultural traditions, non-industrial communities can utilize their traditional practices, foods and skills to generate income and compete in today's capitalist economic markets.

Ecotourism projects incentivize cultural revival by combining it with modern day economic markets. Thus, some would say, offering an integrative approach to revitalization and sustainable community development. However, ecotourism can also bring challenges by incentivizing traditional skills and practices through economic means rather than traditional values, creating a divide between elder and younger generations. An increasing number of external visitors with their own values and materialistic wealth can also shift local values and worldviews. Furthermore, there is the potential that the money generated from ecotourism schemes does not always benefit the community as a whole and, instead, is only captured by a few with the most power.

Education Revitalization Projects

A recent report from the United Nations Children's Fund (UNICEF) on the *State of the World Minorities and Indigenous Peoples 2009* highlights the pressing need for culturally-inclusive education systems, in order to ensure that formal education does not eradicate minority cultures and their way of life (UNICEF, 2009). Education revitalization efforts take place in one of two forms; some occur within the formal school curriculum and others are more informal, promoting a rounded education in a child's own cultural environment. The latter often employ a combination of cultural tools (such as talking sticks as mnemonic devices), narratives and direct experience. The content of teaching varies, although most projects attempt to combine the modern day curriculum with a cultural education in land skills, ancestral stories and belief systems (Crockatt and Smythe, 2002; Ball, 2004; Lepani, 2004; Takano, 2004; Pember, 2007).

One such project was developed in Igloolik, Canada, in response to Iglulingmiut elders' concerns that young people were failing to learn the knowledge and skills vital to the future survival of their culture. This project was organized by the Inullariit Society and works to expand young people's education, teaching them the knowledge and skills that would have traditionally been taught in the family and are lacking from modern education systems. Through this project, young people are learning how to 'be and become an Inuk', acquiring the land skills, knowledge, values and beliefs that enable them to establish a traditional connection with the land (Takano, 2004).

In Russian Mission, a Yup'ik village in Alaska, the local school developed a similar initiative as a consequence of the physical isolation of the village and its dependency on a subsistence economy. The project was set up in 2000 and was incorporated into the formal school curriculum from the 6th to the 12th grade. The school sought to provide education appropriate for the Yup'ik way of life. The initiative incorporates seasonal subsistence activities throughout the school year, outdoor experiences integrated into other subjects including reading and writing, and a three-week field trip teaching students hunting and fishing skills. As well as equipping students intellectually, Russian Mission school works to equip young people practically, emotionally and physically, enabling them to become contributing members of the Yup'ik community (Takano, 2004).

The Nunavut Literacy Council's Family and Community Literacy Project was set up to teach literacy and oral-history in a culturally appropriate setting. In Cambridge Bay, for example, children put on their *atigis* (traditional coats worn by Inuit) after school and attend the local library's after-school home-work club. As well as doing their homework, children learn traditional arts and crafts, and read stories, many of which take place in the reading tent painted by local artists (Crockatt and Smythe, 2002).

In Tibet, the Junyong Community Foundation (established in 2004) in coop-eration with the Rigdzin Foundation have established an education initiative for local communities. This aims to combine literacy teaching with appropriate live-lihood and life skills training, cultural renewal, and values education, particularly targeting nomadic families in eastern Tibet (Lepani, 2004; Rigdzin Foundation, 2006). In Tanzania, Aang Serian's Noonkodin Secondary School runs an indige-nous knowledge programme targeting Maasai children. However, in recognition of the other ethnic groups in Tanzania, Aang Serian is in the process of developing an intercultural education framework using 'study circle' teaching methods that could be adapted to any indigenous group in the world (Burford, pers comm). In Hawaii, entire schools have been set up to teach local culture, language and practical skills alongside the modern day curriculum, termed Immersion Schools (Boyea, pers comm). Efforts are even being made to target pre-school children with culturally-sustaining childcare, such as Cree-ative Daycare (Ball, 2004).

The failures of modern day curricula for marginalized populations have been highlighted over past decades, often with little effect on formal education systems. However, education revitalization projects offer an integrative approach that has long been absent from the school curriculum. Despite this, efforts remain limited in scope and capacity. The most common problem is the mismatch in the timing of school breaks and when people would traditionally be on the land. In Alaskan communities, the timing and length of school breaks should be tied in with spring whaling; in Labrador, the spring and autumn caribou migrations occur during school time, while the long summer break occurs at a time when people prefer not to be in the forests. In Ngāi Tahu communities of southern New Zealand, children are regularly punished for missing school during the muttonbird harvest in spring. There is a vital need to ensure that formal educa-tion comes at no cost to cultural continuity and intergenerational teachings but many obstacles still remain embedded in state-run education systems which emphasize standardization of curricula and timetabling.

Language Revitalization Projects

Language revitalization projects are geographically widespread among non-industrial communities and are a key tool to replenishing beliefs and practices by opening communication channels between generations both now and in the future (Crockatt and Smythe, 2002; Martin and Tagalik, 2004). Indigenous languages often contain vocabularies for specific places, topographies and climatic conditions which assist in the understanding of the local environment more so than an imported language such as English. Languages also contain within them assumptions about the relationship of humans to the world that cannot always be replicated or made commensurate with imported languages (Maffi, 1998, 2001). For this reason, language revitalization projects are numerous and highly diverse – from formally taught modules in schools, to informal sessions held in a community setting with local village elders.

In Oklahoma and Florida, Native American language teachers are being trained in the skills of language revival. These teachers are then being encouraged to take these skills back to their communities and remind people how to speak, read and write in their native languages (Hirata-Edds et al, 2003). Teaching tools include stories, songs and written language in visible places (e.g. cafeterias, traffic signs and clothing) (Hirata-Edds et al, 2003). The Inuit language, Inuktitut, is being strengthened through more formal channels using the modern day curriculum in Nunavut (Martin and Tagalik, 2004). At the same time the Nunavut Literacy Council is working on publishing an array of Inuktitut reading materials including newspapers, magazines and elders' stories (Crockatt and Smythe, 2002).

The importance of language to a culture, and the threat of its loss, has mobilized entire communities and cultures to establish local projects and reform school curricula. Some linguists are directing their efforts at documenting local languages now only spoken by community elders before they die out altogether. However, the only assured way of protecting a language from extinction is to ensure continued use among its holders in the environmental context where it is most meaningful.

Cultural Revitalization Projects

A culture can be defined as:

> *the system of shared symbols, behaviours, beliefs, values, norms, artefacts and institutions that the members of a society use to cope with their world and with one another, and that are transmitted from generation to generation through learning* (Brey, 2007).

However, a culture holds more than just a utilitarian function, as highlighted by Smith (2001) who describes cultural change as:

> *a form of coevolution between cultural information and the social and natural environment.*

This depicts a culture to be a complex and intrinsic system of interlinked components that contributes to individuals' identity by representing relationships with the surrounding environment (Milton, 1998; Berkes, 2004, 2008). While peoples may voluntarily change, psychological ills often arise where communities retain their values and belief systems but are involuntarily separated from the landscape features upon which they are based, such as ancient burial grounds or sacred sites, and traditional plants and animals used for ceremonial purposes. Spiritual beliefs, religious ceremonies and practices can lose their meaning outside of a society's traditional territory, threatening a community's sense of identity (Samson, 2003; Samson and Pretty, 2006).

Cultures comprise many components, however, most cultural revitalization projects only focus on one or two of these, for instance, reintroducing a particular event or ceremony, recreating a set of spiritual beliefs or reviving traditional stories and arts (Crockatt and Smythe, 2002; Waldram et al, 2006). Some of the earliest cultural revitalization efforts were among the Sámi of Norway in the 1950s and 1960s. After centuries of cultural repression and forced assimilation, the Nordic Sámi Council was formed in 1956 and Sámi rights to preserve and protect their own culture were recognized in the 1960s. Early small-scale revitalization efforts included the publication of Sámi newspapers and magazines, Sámi television programmes, storytelling sessions, poetry readings and culture days. The Sámi language was taught in schools and universities and finally, after 300 years of repression, Sámi Parliaments were set up in Finland, Sweden and Norway to protect the needs and traditions of its culture (Brenna, 1997; Nuttall, 1998).

Smaller scale cultural revitalization efforts are more common today, and include the revival of the Spirit Dance ceremony in British Columbia. In this project, spiritual leaders are working to teach the traditions of the ceremony to young people and, in doing so, renew a local sense of Aboriginal identity and cultural pride (Waldram et al, 2006). Similarly, in Kugaaruk, Canada, efforts are being made to record elders' stories, read traditional narratives aloud over the radio and publish the stories in local languages and English (Crockatt and Smythe, 2002). In Hokkaidō, Japan, communities are striving to expand the ethnic base of Ainu cultural traditions despite rapidly modernizing Japanese culture. Efforts include seasonal camps established at important Ainu sites, the making and reinterpretation of traditional arts and crafts, sewing and weaving, boat building, storytelling and salmon fishing (Ohtsuka, 1999; Kayano, pers comm; Keira, pers comm).

Across the US, culture camps are being established by and within First Nation communities. These enable participants to experience total immersion into their culture (Takano, 2004; ANKN, 2007; APRN, 2008; Aqqaluk Trust, 2008; KCAW, 2008). Most target young people and are non-profit, organized and funded by regional tribal associations, school districts, other local non-profit organizations or a combination of these (Droulias, pers comm). For instance, Gaalee'ya Spirit Camp near Fairbanks, Alaska, runs youth camps, spirit camps and elder camps, as well as educational workshops and language preservation schemes. Camp facilities include a smoke house and a cook shelter. Activities vary from season to season and include berry picking and fish wheel building in the summer, and dog sledding and trapping in the winter. Participants learn traditional skills such as tanning hides, drum making, food preservation, beading and story-telling (ANKN, 2007). Similarly, Dog Point Fish Camp, Camp Sivu (from *sivunniigvik* meaning 'the planning place') and Urban Unangax, all in Alaska, teach seasonal food collection techniques and other local skills such as drum making, kayak construction and basket weaving (APRN, 2008; Aqqaluk Trust, 2008; KCAW, 2008). At Camp Sivu, all hunting and food preparation must be carried out according to Inupiaq values (Aqqaluk Trust, 2008).

The Canadian First Nation, Tr'ondëk Hwëch'in, is situated on the upper Yukon River valley. Cultural revitalization efforts for the Tr'ondëk Hwëch'in began with the establishment of a summer fish camp at an important heritage site in Moosehide in the 1970s. Since that time, the Tr'ondëk Hwëch'in have successfully established a number of cultural revitalization projects including public heritage research and activities, revival of traditional artistic expression, relearning traditional songs and dances, and community storytelling contests, all of which relate to the land. In doing so, they claim to have improved community health through the revival of cultural pride (Neufeld, 2004).

Cultural revitalization focuses on renewing non-income generating activities and inspiring the younger generation about their cultural identity and their ancestor's way of life. Cultural immersion schemes, however, are still often limited by funding and resources. Participation, therefore, is usually sporadic and short term, limiting their potential impact. If the benefits of these more holistic approaches can be systematically analysed and demonstrated, then economic funding may be secured to enable longer periods of cultural immersion and greater community respect.

Rights Revitalization Projects

Some revitalization efforts focus on altering policy or promoting community rights to enhance cultural revival and continuity, for instance, by legalizing access to traditional lands for hunting and gathering. Many such schemes are, directly or indirectly, based upon legitimizing free access to and use of land enabling continuity of cultural activities. These projects often reach beyond the level of community to alter regional or even national policy. One of the most important policies passed in this respect was the designation of Nunavut, meaning 'our land' in Inuit. This 200 million hectare territory was designated in 1999 and is seen by many as a step towards resolving Inuit native land claims and harvesting areas. In Nunavut, situated in the Canadian eastern Arctic, Inuit have free access to hunt and fish. Approximately 30,000 people live in Nunavut. Although Nunavut has its own government, it is still strongly led by the Canadian Constitution and the terms of the agreement required Inuit to extinguish their legal aboriginal title to the land. Despite this, the territory itself, both land and sea, is under permanent Inuit management (Nuttall, 1998; Jull, 2001).

An international social movement for the promotion of the rights of indigenous peoples can be dated back to the early 20th century, but the political agitation for the promotion and protection of indigenous distinctiveness has mushroomed into a global political force since the 1980s. Around the world, indigenous peoples have become ever more conscious of their right to an indigenous identity. Much of this has been fashioned against the growing number of threats from development projects, such as logging, mining, hydroelectric power generation and farming, which often compromise existing ways of life. The movement also combats assimilation measures such as the North American boarding schools and the removal of Australian Aboriginal children from their families, as well as more general policies to diminish indigenous beliefs and practices, and to culturally blend indigenous peoples into the dominant ethnic population.

Global indigenous political organization has centred on the promotion of international legal and human rights instruments. The International Labour Organization (ILO) led the way with its Resolutions 107 and 169. Additional standards are provided in the International Covenant on Economic, Social and Cultural Rights and the Inter American Commission on Human Rights. After more than a decade of debate, the UN finally adopted the Declaration on the Rights of Indigenous Peoples in 2007. This sets standards for nation states, many of which had constituted indigenous populations as having inferior rights to other citizens. The declaration gives indigenous peoples collective rights to their lands, languages, religions and laws, as well as rights to determine their own political status (although it does not condone the break-up of nation states).

The UN declaration would not have been possible without the social movement of indigenous representatives who travelled several times a year to Geneva and New York (Niezen, 2003; Morgan, 2004). The goal of these representatives focused around the promotion of their right to their own distinct practices, all of which are based on active engagement with nature. While it remains to be seen how well states will respect and enforce these human rights standards (Samson, 2008a), the adoption of the declaration is widely seen as a triumph for those indigenous groups channelling their efforts into cultural preservation through legal protection.

The Impacts of Revitalization Projects

Up until recently, efforts to deal with the range of ill-health and social pathologies arising from environmental disconnection and loss of cultural continuity have been Western in their approach. Examples include equipping communities with modern health clinics and teams of mental health counsellors (Samson, 2003, 2008b). However, these approaches have been limited in their success because they are unable to deal with the root cause of the problem (Alfred, 2009). They remain external to the local culture and therefore community, and in doing so have the capacity to contribute further to a community's sense of dislocation and loss of identity. Featuring prominently here are medical and psychological approaches emphasizing individual sickness and removing the problem from the historical and contemporary experiences of entire communities suffering cultural dispossession. Revitalization projects offer an alternative to these extrinsic and externally-imposed projects. They are often established by or with communities. Rather than targeting the symptoms of the illness, revitalization projects target the cause by attempting to revive community cultures and reconnect people with their lands.

Factors likely to affect the success of revitalization projects include longevity, available resources, policy frameworks and organizational capacity. As many revitalization projects are either recent to emerge or not widely reported, their ability to promote cultural continuity and alleviate local health problems has not been formally evaluated or assessed, but there exist localized reports of success stories. For instance, students from Russian Mission school in Alaska, although unable to fluently speak the Yup'ik language, still share traditional Yup'ik belief systems and values with their elders. An Inullariit Society informal education project in Igloolik, Canada, had similar success. When interviewed, young participants expressed the same value of 'being on the land' as their elders. Participants' also based their identity on the land and held an intrinsic respect for their home environments, although they admit that they would struggle to live off the land for long periods (Takano, 2004). Table 13.2 summarizes some of the characteristics and potential outcomes of the different categories of revitalization project.

Table 13.2 *Some outcomes of different categories of revitalization project*

Project type	Characteristics of project type
1 Traditional Foods	• Capacity to benefit both mental and physical health • Potential to generate income and jobs where markets are successfully created • Effective in the transfer of new skills and knowledge associated with traditional foods collection and preparation • Likely to strengthen community bonds, particularly where hunting groups are established, and intergenerational knowledge transfer • Target young people and community elders alike • Likely to positively impact local people, ecosystems (where combined with effective resource management) and economies (where traditional food markets exist)
2 Traditional Healthcare	• Likely to offer physical and mental/spiritual health benefits depending on the system itself • Potential to generate micro-enterprise opportunities where locally-developed medicinal products can be marketed • Based on the revival of local skills and knowledge • Often restricted to community elders and leaders, but sometimes knowledge is shared among all community members • Likely to benefit health and economic status, and ecosystem biodiversity where combined with sustainable management of ethnobotanical species
3 Ecotourism	• Where activities are physically demanding, projects are likely to benefit the physical health of participants • If established in line with values and not just with income generation in mind, ecotourism schemes are likely to reinforce cultural identity and therefore offer mental health benefits • Successful at job creation and income generation, although measures need to be taken to ensure equitable benefit sharing within communities • Effective at teaching/reviving traditional knowledge and skills • Ecotourism schemes often target the younger generations for training and employment opportunities • Unlikely to strengthen community bonds and may weaken traditional values • Likely to benefit local people and economies, and ecosystems only where land-based activities are revived and sustainably managed
4 Education	• Likely to offer mental health benefits and physical health benefits where projects incorporate the teaching of physical activities

Table 13.2 *(Continued)*

Project type	Characteristics of project type
	• Heavily focused on knowledge transmission and the revival of traditional skills and knowledge • Where elders are involved in projects, education schemes are likely to strengthen community bonds • Primarily target younger generations • Offer very few employment or income-generating opportunities • Potential to benefit local people, but no direct impact likely on local economies and ecosystem
5 Language	• Through the reinforcement of cultural identity, successful language projects are likely to benefit community mental health, but not physical health • Capacity to create jobs for local linguists but unlikely to increase income generating opportunities for the society as a whole • Successful at transferring new knowledge, although practical skills are rarely emphasized • Likely to strengthen social bonds, in particular by opening communication channels between older generations and young people • Can target adults and young people, but many focus on younger generations • Potential to benefit local people but impacts on economy or ecosystem unlikely
6 Cultural	• Capacity to benefit community health mentally, and also physically if physical activities are taught, depending on the longevity of the project • Effective in the transfer of new knowledge and skills • Likely to strengthen community bonds, particularly where community elders are recruited to teach young people • Have the potential to benefit children and adults alike, although most established schemes target young people • With the exception of project organizers, cultural projects are unlikely to increase jobs and income-generating activities • Likely to benefit local people and capacity to benefit local ecosystems if traditional land management regimes are revived but unlikely to benefit local economy
7 Rights	• Capacity to offer both mental and physical health benefits to community • Although traditional skills and knowledge are not directly taught, where land rights are re-established communities have the capacity and resources to relearn these traditional skills and practices • Potential to strongly enhance community bonds, particularly where hunting groups are established and intergenerational lines of knowledge transfer reopened through land-based activities • Benefits accrued by community elders and young people alike • Unlikely to directly increase full-time employment options, although renewed access to traditional lands will increase income-generating opportunities for many, particularly where markets for traditional foods exist • Likely to benefit local people, local economy (where traditional foods collection and practices are revived on the land) and ecosystems (where traditional resource management is reintroduced and sustainable in the long-term)

Source: Pilgrim et al, 2009

Revitalization for the Future

Ecocultural revitalization projects established by indigenous and marginalized groups offer insight into elements that may be used to reconnect industrial communities undergoing long-term disconnections from nature. For instance, green exercise and green care initiatives are an emerging trend, particularly in the UK and across Europe. Such projects offer health benefits to participants, have the capacity to create and strengthen social relationships, and are open to all community cohorts. Bushcraft and foraging courses have also increased in popularity, teaching participants new skills and practices, strengthening bonds and benefiting human health. Thus, many of the principles that revitalization projects centre around can be applied to efforts to reconnect modern industrial communities with the local environment, for instance, new business creation, a need to incentivize reconnection, the establishment of support networks and local knowledge transfer.

Revitalization projects, although highly diverse, are currently being developed independently of one another in communities around the world. Their emergence is in response to shared concerns about disconnection from nature and motivations to revive traditional ways of living through reconnection with the land. They are commonly initiated by elders who perceive younger generations to be disconnected from both their culture and homelands (Johnston et al, 2007). It is important to note that the categories of project described in this chapter are by no means mutually-exclusive, and in fact there are a great many revitalization efforts that traverse these type boundaries, with the mixture of objectives reflecting the various different community concerns at play. For instance, Chamorro cultures in Guam, US, employ a range of different approaches, including culture camps, craft workshops, language projects, and the revival of traditional fishing practices. In Tyva in southern Siberia, ecotourism efforts are being combined with livestock camps dedicated to teaching the skills of nomadic pastoralism to young Tuvinians (Pretty, 2009). The Tshikapisk Foundation is another organization of indigenous peoples attempting to engage outsiders such as ecotourists, biologists, artists, students and wildlife enthusiasts as a way of perpetuating the Innu hunting life in northern Labrador, Canada. Tshikapisk have also established independently funded initiatives to assist with the inter-generational transmission of Innu knowledge through canoe trips, walks and hunting expeditions (see Samson and Pretty, 2006). Furthermore, the organization plays an important role in using land-based activities as an 'alternative school' and a means of improving the mental health of young Innu, ravaged by assimilation policies (Samson, 2008b).

It could be suggested that cultural revitalization formally began with the recognition of the outstation movement of the early 1970s, when the failure of assimilation policies resulted in rural exodus by indigenous communities, for instance, Aborigines in Australia. People chose to return to traditional ways of living, including livelihoods, diets and connections with the land (Altman, 2002; Johnston et al, 2007). Today, there are a number of international organizations working towards ecocultural revitalization, for instance the Global

Initiative for Traditional Systems of Health and Terralingua. Such organizations are dedicated to ensuring that cultural connections to indigenous homelands persist, albeit in a dynamic state, in today's rapidly changing world.

Policy-makers dealing with disconnected communities should look towards revitalization projects for long-term solutions to cultural continuity. Government funding would facilitate the establishment of such projects. As well as assisting with compliance to the 2007 UN Declaration on the Rights of Indigenous Peoples, this is likely to result in reduced healthcare costs and reduced loss of earnings caused by social pathologies and ill-health in the long term. Revitalization projects offer a new community-centred approach to dealing with the problems of disconnection. In some cases, communities have developed their own political, economic and social organization that could not exist without strong connections to the land and cultural identity (Alfred, 2009). By being locally-driven, projects are more likely to encourage long-term support and participation, and have the capacity to empower non-industrial peoples, thus reinvigorating communities, cultures and connections with the land.

Notes

1   In this paper, we use the terms industrialized and non-industrialized to refer to differing social contexts variously referred to as developed, modernized, developing, traditional, native, pre-industrial, in-transition, Western, Third World or indigenous. None of these terms are without difficulty, as they seem to suggest that patterns of development or cultural change are linear and similar. Some indigenous communities, for example, participate in resource extraction activities consistent with industrialization; others have never actively industrialized their own economies but may use technologies, manufactured clothing, processed foods and other products of industrialization; some exist in developing countries, and others in industrialized countries. We do not imply an inevitable and singular pathway for change by the use of any of these terms.
2   The definition of indigenous peoples used here is taken from Jose R. Martinez Cobo's working definition adopted by the UN: 'Indigenous communities, peoples and nations are those which, having a historical continuity with pre-invasion and pre-colonial societies that developed on their territories, consider themselves distinct from other sectors of the societies now prevailing on those territories, or parts of them. They form at present non-dominant sectors of society and are determined to preserve, develop and transmit to future generations their ancestral territories, and their ethnic identity, as the basis of their continued existence as peoples, in accordance with their own cultural patterns, social institutions and legal system' (UN, 2004).
3   Cultural continuity does not imply no cultural change; rather it suggests the need to maintain core components of cultures in light of cultural evolution caused by externally-driven change.

References

ACDP (2003) 'Economic development', Alaska Chukotka Development Program, www.chukotka.uaa.alaska.edu/Economic%20Development/ econdev.htm

Alfred, T. (2009) 'Colonialism and state dependency', National Aboriginal Health Organization Project 'Communities in Crisis', http://web.uvic.ca/igov/uploads/pdf/Colonialism%20and%20State%20Dependency%20NAHO%20(Alfred).pdf

Altman, J. (2002) 'Indigenous hunter-gatherers in the 21st century: Beyond the limits of universalism in Australian social policy', in T. Eardley and B. Bradbury (eds) *Competing Visions: Refereed Proceedings of the National Social Policy Conference 2001*, SPRC Report 1/02, Social Policy Research Centre, University of New South Wales, Sydney, pp35–44

Altman, J. (2005) 'The indigenous hybrid economy, realistic sustainable option for remote communities: Radical approach or plain common sense?' Address to the Australian Fabian Society, 26 October 2005

ANKN (2007) 'Gaalee'ya Spirit Camp', Alaska Native Knowledge Network, www.ankn.uaf.edu/ANCR/Athabascan.html

APRN (2008) 'Learning to cook the traditional way', Alaska Public Radio Network, http://aprn.org/2008 /06/26/alaska-news-nightly-june-26-2008

Aqqaluk Trust (2008) 'Camp Sivunniigvik', www.aqqaluktrust.com/pages/camp_sivu.html

Balée, W. (1994) *Footprints of the Forest. Ka' apor Ethnobotany*, Columbia University Press, New York

Ball, J. (2004) 'First Nations Partnerships Programs: Incorporating culture in ECE training', *The Early Childhood Educator*, Spring, pp1–5

Basso, K. (1996) *Wisdom Sits in Places*, University of New Mexico Press, Albuquerque

Bathurst Inlet Lodge (2008) 'Welcome to Bathurst Inlet Lodge', www.bathurstinlet-lodge.com /index.htm

Berkes, F. (2001) 'Religious traditions and biodiversity', *Encyclopaedia of Biodiversity*, vol 5, pp109–120

Berkes, F. (2004) 'Rethinking community-based conservation', *Conservation Biology*, vol 18, no 3, pp621–630

Berkes, F. (2008) *Sacred Ecology* (2nd edition), Routledge, New York

Berkes, F., Colding, J. and Folke, C. (eds) (2003) *Navigating Social-Ecological Systems: Building Resilience for Complexity and Change*, Cambridge University Press, Cambridge

Bersamin, A. and Simpson, A. (2005) 'Reindeer meat is a healthy food', UAF Newsroom, www.uaf.edu/news/featured/05/reindeer/meat.html

Brenna, W. (1997) 'The Sami of Norway', Ministry of Foreign Affairs, Norway, http://odin.dep. no/odin/engelsk/norway/history/032005-990463/index-dok000- b-n-a.html

Brey, P. (2007) 'Theorizing the cultural quality of new media', *Technè*, vol 11, no 1, http://scholar.lib.vt.edu/ejournals/SPT/v11n1/pdf/v11n1.pdf

Brody, H. (1981) *Maps and Dreams*, Pantheon, New York

Burford, G. and Ngila, L. O. (2002) 'Beyond ethnobotanical inventory: Community-level strategies for documenting and developing traditional health care services', Unpublished

Cernea, M. (1988) 'Involuntary resettlement in development projects: Policy guidelines in World Bank-financed projects', World Bank Technical Paper, no 80

Cernea, M. (1997) 'The risks and reconstruction model for resettling displaced populations', *World Development*, vol 25, no 10, pp1569–1587

Cernea, M. and Schmidt-Soltau, K. (2006) 'Poverty risks and national parks: Policy issues in conservation and resettlement', *World Development*, vol 34, no 10, pp1808–1830

Cinner, J. E., Marnane, M. J. and McClanahan, T. R. (2005) 'Conservation and community benefits from traditional coral reef management at Ahus Island, Papau New Guinea', *Conservation Biology*, vol 19, pp1714–1723

Colson, E. (1971) *The Social Consequences of Resettlement: The Impact of the Kariba Resettlement on the Gwembe Tonga*, Manchester University Press, Manchester

Cordain, L., Miller, J. B., Eaton, S. B., Mann, N., Holt, S. H. A. and Speth, J. D. (2000) 'Plant-animal subsistence ratios and macronutrient energy estimations in worldwide hunter-gatherer diets', *American Journal of Clinical Nutrition*, vol 71, pp682–692

Crockatt, K. and Smythe, S. (2002) 'Building culture and community: Family and community literacy partnerships in Canada's north', Nunavut Literacy Council, www.nunavutliteracy.ca/english/resource/reports/building/ building.pdf

Denevan, W. M. (2001) *Cultivated Landscapes of Native Amazonia and the Andes*, Oxford University Press, Oxford

Elu Inlet Lodge (2008) 'Welcome to Elu Inlet Lodge', www.elulodge.com

Gall, S. (2002) *The Bushmen of Southern Africa: Slaughter of the Innocent*, Pimlico, London

Gigoux, C. and Samson, C. (2009) 'Globalization and indigenous peoples: New old patterns', in B. Turner (ed) *Handbook of Globalization Studies*, Routledge, London

Gracey, M. and King, M. (2009) 'Indigenous health part 1: Determinants and disease patterns', *The Lancet*, vol 374, no 9683, pp65–75

Gunderson, L. H. and Holling, C. S. (eds) (2002) *Panarchy: Understanding Transformations in Human and Natural Systems*, Island Press, Washington, DC

Harmon, D. (2002) *In Light of Our Differences*, Smithsonian Institution Press, Washington, DC

Hirata-Edds, T., Linn, M., Peter, L. and Yamamoto, A. (2003) 'Language training in Oklahoma and Florida', *Cultural Survival Quarterly*, Winter, vol 27, no.4

IITC (2003a) 'Questionnaire on indigenous peoples' traditional foods and cultures: Results', Distributed by the International Indian Treaty Council (IITC) and submitted to the United Nations Food and Agriculture Organization (FAO) Rural Development Division (SDAR), www.treatycouncil.org/QRE%20RESULTS.pdf

IITC (2003b) 'Final report on an indigenous peoples' initiative to establish cultural indicators for SARD: Questionnaire on indigenous peoples' traditional food and cultures', Distributed by the International Indian Treaty Council (IITC) and submitted to the United Nations Food and Agriculture Organization (FAO) Rural Development Division (SDAR), www.foodsovereignty.org/public/documenti/CultIndicQre%20REPORT_FIN.doc

Inari Event (2008) 'Winter activities in Inari', www.saariselka.fi/inarievent/winterac-tivities.html

Johnston, F. H., Jacups, S. P., Vickery, A. J. and Bowman, D. M. J. S. (2007) 'Ecohealth and aboriginal testimony of the nexus between human health and place', *EcoHealth*, vol 4, pp489–499

Jull, P. (2001) 'Nunavut: The still small voice of indigenous governance', *Indigenous Affairs*, vol 3, no 1, pp42–51

KCAW (2008) 'Fish camp residents learn the hard way', http://kcaw.org/modules/local_news/index.php?op=sideBlock&ID=189

Kellert, S. R. and Wilson, E. O. (eds) (1993) *The Biophilia Hypothesis*, Island Press, Washington, DC

King, M., Smith, A. and Gracey, M. (2009) 'Indigenous health part 2: The underlying causes of the health gap', *The Lancet*, vol 374, no 9683, pp76–84

Kishigami, N. (2000) 'Contemporary Inuit food sharing and hunter support program of Nunavik, Canada', in G. W. Wenzel, G. Hovelsrud-Broda and N. Kishigami (eds) *The Social Economy of Sharing: Resource Allocation and Modern Hunter-Gatherers*, Senri Ethnological Studies, vol 53, National Museum of Ethnology, Osaka, p171–192

Kozlov, A. I. and Zdor, E. V. (2003) 'Whaling products as an element of indigenous diet in Chukotka', The Anthropology of East Europe Review: Central Europe, Eastern Europe and Eurasia, *Food and Foodways*, vol 21, no 1, pp127–137

Lepani, B. (2004) 'Tibet: An educational strategy for cultural revitalisation and overcoming economic marginalisation', International Development Studies Conference, Auckland University, 3–5 December

Maffi, L. (1998) 'Language: A resource for nature', *Nature and Resources: The UNESCO Journal on Environment and Natural Resources Research*, vol 34, no 4, pp12–21

Maffi, L. (ed) (2001) *On Biocultural Diversity*, Smithsonian Institution Press, Washington, DC

Marcus, A. R. (1995) *Relocating Eden: The Image and Politics of Inuit Exile in the Canadian Arctic*, University Press of New England, Hanover, New Hampshire

Marquardt, O. and Caulfield, R. A. (1996) 'Development of west Greenlandic markets for country foods since the 18th century', *Arctic*, vol 49, no 2, pp107–119

Martin, I. and Tagalik, S. (2004) 'Aajiiqatigiingniq: Lesson's learned from Nunavut's language if instruction research project', Department of Education, Government of Nunavut, http://pubs. aina.ucalgary.ca/aina/14thISCProceedings.pdf#page=185

McGrath, M. (2006) *The Long Exile: A True Story of Deception and Survival in the Canadian Arctic*, Harper Perennial, London

McKnight, D. (2002) *From Hunting to Drinking: The Devastating Effects of Alcohol on an Australian Aboriginal Community*, Routledge, London

Milton, K. (1998) 'Nature and the environment in indigenous and traditional cultures', in D. E. Cooper and J. A. Palmer (eds) *Spirit of the Environment: Religion, Value and Environmental Concern*, Routledge, London and New York, pp86–99

Morgan, R. (2004) 'Advancing indigenous rights at the United Nations: Strategic framing and its impact on the normative development of international law', *Social and Legal Studies*, vol 13, no 4, pp481–500

Nabhan, G. P. and St. Antoine, S. (1993) 'The loss of floral and faunal story: The extinction of experience', in S. R. Kellert and E. O. Wilson (eds) *The Biophilia Hypothesis*, Island Press, Washington, DC, pp229–250

Neufeld, D. (2004) 'Nomination of the Government of Tr'ondëk Hwëch'in For The Robert Kelly Memorial Award of the National Council on Public History', Unpublished

Niezen, R. (2003) *The Origins of Indigenism: Human Rights and the Politics of Identity*, University of California Press, Berkeley

Norgaard, R. B. (1994) *Development Betrayed*, Routledge, London

Nuttall, M. (1998) *Protecting the Arctic: Indigenous Peoples and Cultural Survival*, Harwood Academic Publishers, Amsterdam

OHEP (2008) 'Ontario's Hunter Education Program', www.ohep.net

Ohtsuka, K. (1999) 'Itaomachip: Reviving a boat building and trading tradition', in W. Fitzhugh and C. Dubreuil (eds) *Ainu: Spirit of a Northern People*, Smithsonian Institution, Washington, DC, pp374–376

Pars, T., Osler, M. and Bjerregaard, P. (2001) 'Contemporary use of traditional and imported food among Greenlandic Inuit', *Arctic*, vol 54, no 1, pp22–31

Pember, M. A. (2007) 'A mandate for native history: The Montana Indian Education for All Act', *Diverse Issues in Higher Education*, www.highbeam.com/doc/1G1-164421874.html

Pilgrim, S., Cullen, L., Smith, D. J. and Pretty, J. (2008) 'Ecological knowledge is lost in wealthier communities and countries', *Environmental Science and Technology*, vol 42, no 4, pp1004–1009

Pilgrim, S., Samson, C. and Pretty, J. (2009) 'Rebuilding lost connections: How revitalisation projects contribute to cultural continuity and improve the environment', Interdisciplinary Centre for Environment and Society Occasional Paper 2009-01, University of Essex, www.essex.ac.uk/ces/esu/occ-papers.shtm

Pretty, J. (2004) 'How nature contributes to mental and physical health', *Spirituality Health International*, vol 5, no 2, pp68–78

Pretty, J. (2007) *The Earth Only Endures: On Reconnecting with Nature and Our Place in it*, Earthscan, London

Pretty, J. (2009) 'Integrated community development and biodiversity conservation in the Republic of Tyva', Report of a Mission to Tyva commissioned by Oxfam GB, Interdisciplinary Centre for Environment and Society, University of Essex

Pretty, J., Adams, B., Berkes, F., de Athayde, S., Dudley, N., Hunn, E., Maffi, L., Milton, K., Rapport, D., Robbins, P., Samson, C., Sterling, E., Stolton, S., Takeuchi, K., Tsing, A., Vintinner, E. and Pilgrim, S. (2008) 'How do nature and culture intersect?' Plenary paper for Conference 'Sustaining Cultural and Biological Diversity in a Rapidly Changing World: Lessons for Global Policy', Organized by AMNH, IUCN-The World Conservation Union/Theme on Culture and Conservation, and Terralingua, New York, 2–5 April

Pretty, J., Hine, R. and Peacock, J. (2006) 'Green exercise: The benefits of activities in green places', *The Biologist*, vol 53, no 3, pp143–148

Pretty, J., Peacock, J., Hine, R., Sellens, M., South., N. and Griffin, M. (2007) 'Green exercise in the UK countryside: Effects on health and psychological well-being, and implications for policy and planning', *Journal of Environmental Planning and Management*, vol 50, no 2, pp211–231

Pretty, J., Peacock, J., Sellens, M. and Griffin, M. (2005) 'The mental and physical health outcomes of green exercise', *International Journal of Environmental Health Research*, vol 15, no 5, pp319–337

Purnumukan Porofarmi (2008) 'Purnumukka Reindeer Farm', www.porofarmi.com/en/porofarmi.php

Pyle, R. M. (2003) 'Nature matrix: Reconnecting people and nature', *Oryx*, vol 37, pp206–214

Raine, K. D. (2005) 'Determinants of healthy eating in Canada: An overview and synthesis', *Canadian Journal of Public Health*, vol 96, supplement 3, ppS8–S14

Rigdzin Foundation (2006) 'Leadership and literacy: Cultural revitalisation in East Tibet', www.rigdzinfoundation.org/projects/leadership_and_literacy/images/Leadership%20and%20Literacy.pdf

Samson, C. (2003) *A Way of Life that Does Not Exist*, ISER, Canada

Samson, C. (2008a) 'The rule of Terra Nullius and the impotence of international human rights for indigenous peoples', *Essex Human Rights Review*, vol 5, no 1, pp69–82

Samson, C. (2008b) 'A colonial double bind: The social and historical contexts of Innu mental health', in L. Kirmayer and G. Valaskakis (eds) *The Mental Health of Canadian Aboriginal Peoples: Transformations of Identity and Community*, University of British Columbia Press, Vancouver, pp195–243

Samson, C. and Pretty, J. (2006) 'Environmental and health benefits of hunting diets and lifestyles for the Innu of Labrador', *Food Policy*, vol 31, no 6, pp528–553

Shkilnyk, A. (1985) *A Poison Stronger than Love: The Destruction of an Ojibwa Community*, Yale University Press, New Haven

Smith. D. J., Pilgrim, S. E. and Cullen, L. C. (2007) 'Coral reefs and people', in J. Pretty, A. Ball, T. Benton, J. Guivant, D. Lee, D. Orr, M. Pfeffer and H. Ward (eds) *Sage Handbook on Environment and Society*, Sage Publications, London, pp1081–1117

Smith, E. A. (2001) 'On the coevolution of cultural, linguistic, and biological diversity', in L. Maffi (ed) *On Biocultural Diversity*, Smithsonian Institution Press, Washington, DC, pp95–117

Sonjica, B. (2004) 'Speech by Ms Buyelwa Sonjica, the Deputy Minister of Arts, Culture, Science and Technology, at the launch of Indiza Food and Indigenous Food Fair', Port St Johns, Eastern Cape, 12 March, Department of Science and Technology and the Council for Scientific and Industrial Research of South Africa, www.polity.org. za/article.php?a_id=48176

Takano, T. (2004) 'Meanings of "connection" with the environment: Findings from outdoor educational programs in Scotland, Alaska and Nunavut', www.latrobe.edu. au/oent/OE_conference_ 2004/papers/takano.pdf

Tansey, G. (2004) 'A food system overview', in S. Twarog and P. Kapoor (eds) *Protecting and Promoting Traditional Knowledge: Systems, National Experiences and International Dimensions*, UN, Geneva, pp41–58

Toledo, V. M. (2001) 'Biodiversity and indigenous peoples', *Encyclopedia of Biodiversity*, vol 3, pp451–463

Turnbull, C. (1973) *The Mountain People*, Picador, London

Turner, N. and Berkes, F. (2006) 'Coming to understanding: Developing conservation through incremental learning in the Pacific Northwest', *Human Ecology*, vol 34, pp495–513

UN (2004) 'The concept of indigenous peoples', Background paper prepared by the Secretariat of the Permanent Forum on Indigenous Issues, Department of Economic and Social Affairs, www.un.org/esa/socdev/unpfii/documents/workshop_data_background.doc

UNICEF (2009) *State of the World's Minorities and Indigenous Peoples 2009*, file:/// C:/96ca7b39421f67a28ee517cadf4cfb/My%20Documents/Revitalisation%20 projects/state-of-the-worlds-minorities-and-indigenous-peoples-2009.html

Waldram, J. B., Herring, A. and Young, T. K. (2006) *Aboriginal Health in Canada: Historical, Cultural, and Epidemiological Perspectives* (2nd edition), University of Toronto Press, London

Willows, N. D. (2005) 'Determinants of healthy eating in Aboriginal peoples in Canada: The current state of knowledge and research gaps', *Canadian Journal of Public Health*, vol 96, supplement 3, ppS32–S36

14
Nature and Culture: Looking to the Future for Human-Environment Systems

Jules Pretty and Sarah Pilgrim

Growing Pressures

An unprecedented combination of pressures is emerging to threaten the health of human and ecological systems across the world. Continued population growth, rapidly changing consumption patterns and the signals of climate change are driving limited resources of food, energy, water and materials towards and beyond critical thresholds (Pretty, 2007; Brown, 2008; Jackson, 2009). Modern life has brought astonishing technological advances, but modernization is also a story of traditional and place-based cultures eroding beneath swift currents of change. However, vulnerabilities now reach into the global arena, as exemplified by the latest economic crisis and the rolling consequences of oil depletion.

It is evident that human cultures and environment systems are intimately linked in ways that are only just beginning to be appreciated (Pretty et al, 2007, 2008; Escobar, 2008; Pretty and Pilgrim, 2008; Hulme and Ong, 2009). We do know, however, that there are certain sets of circumstances by which a society is resilient to perturbations in a system, and others by which a society is so vulnerable that it will be unable to be sustained. Resilience refers to the capacity of a system to absorb or even benefit from changes to the system, and so persist without a qualitative change in structure (Holling, 1973; Costanza et al, 2007). Vulnerability, on the other hand, refers to instances when they lack resilience and robustness, so being driven to rapid change, chaos or collapse (Berkes et al, 2005; Diamond, 2005; Costanza et al, 2007; Pretty, 2007).

It is increasingly clear that certain cultural and ecological components build system resilience: natural capital that delivers a flow of ecosystem goods and services; social capital in the form of relations of trust, norms, obligations and institutions fundamental for collective action; human capital that provides knowledge, skills and capabilities to produce the technologies for well-being; and physical and financial capital that provide infrastructure and financial resources (Pretty, 2003; MEA, 2005). Our current period, termed the 'anthropocene' for humanity's intense modification of the environment, has resulted in dramatic worldwide declines in these renewable capital assets (MEA, 2005; Pilgrim et al, 2008). These concerns are not new, emerging first from the Club of Rome in the 1970s, then the measured tones of the Brundtland Commission in the late 1980s, the growing evidence base of the Intergovernmental Panel on Climate Change (IPCC) and Millennium Ecosystem Assessment of the 1990s and 2000s, and finally the as-yet unmet aspirations of the Millennium Development Goals (MDGs) of the 2000s.

What is new, though, is the growing recognition that human-environment systems are more vulnerable than formerly predicted and that further global change is inevitable. A key question is whether human systems will act quickly enough to avoid severe non-linearities, or whether they will simply be in responsive mode (MEA, 2005; Folke et al, 2007; Brown, 2008). The immense challenge and diverse consequences mean it will take a novel combination of economic, social, political, legal and management expertise to analyse and develop solutions to address these challenges. Such changes are going to be required over large geographic areas – communities, towns or cities, landscapes or watersheds, mountains or marshes, and whole continents.

David Ehrenfeld (2005) has pointed out that 'globalization is creating an environment that will prove hostile to its own survival'. We think we have control too (another delusion), but this is a chimera: 'Our ability to manage global systems, which depends on our being able to predict the results of the things we do, or even to understand the systems we have created, has been greatly exaggerated. Much of our alleged control is science fiction.' The crisis is, as he says, here now.

David Orr (2005) has further stated that, 'no broadly informed scientist can be optimistic about the long-term future of human-kind without assuming we will soon recalibrate human numbers, wants, needs and actions ... within a finite biosphere'. He goes on to say, 'the time for reason and reasonableness is running short'. Curiously, if we do see a collapse, it will be the citizens of many cultures disconnected to the global political economy who will be best placed to survive, as they are comparatively self-sufficient and the least dependent on the technologies and interconnected markets and institutions of the industrial world.

Despite the prevailing gloom about global prospects, there are communities or ecocultures, as termed here, in a variety of political and governance contexts in both industrial and non-industrial regions who are living in ways that build natural, social and human capital, maintain well-being and happiness, and contribute to the sustainable use of resources, as a result of various projects, programmes and cultural institutions (Samson and Pretty, 2006; Pilgrim et al, 2009). Some of these have been deliberately designed as cultural revitalization projects (Pilgrim et al, 2009), others are historical relics; some are distinct communities, such as indigenous or religious groups, others are indistinct from the surrounding or neighbouring human systems.

Consumption and Convergence

Choices matter, and how the many cultures and peoples of the world choose to live their lives will affect the whole world (Ehrenfield, 1981; McNeill, 2000; Diamond, 2005; Lovelock, 2005). If there is increasing convergence towards high consumption patterns of physical resources, then the finite limits of the world will be pressured and broken. The consumption figures are deeply worrying. Nearly a quarter of the world's population has no access to water; average per capita consumption in Europe is 150 litres per day, in North America it is 430 litres per day; in Las Vegas it is 1600 litres per day. In North America, 308kg of paper is consumed by each person annually; in Europe 125kg, in China 34kg, and in India and across the African continent just 4kg (Pretty, 2007).

In North America, there are 80 motor vehicles for every 100 people, in Japan 57, in Europe 24, and in China, India and Africa so far just 0.6 to 0.9. World-wide, some 400,000ha of cropland are paved per year for roads and parking lots (the US's 16 million hectares of land under asphalt will soon pass their total area under wheat) (Pretty, 2007). The world motor vehicle fleet grows alarmingly, as does the consumption of fossil fuels. By almost every measure of resource consumption or proxy for waste production, industrial countries lead the way. And what do people in other cultures and contexts increasingly wish to do? What model is being held up as the one to follow? Many in the world now aspire to the same levels of consumption and same ways of living as those in North America. After all, that is what we imply is a pinnacle of economic achievement.

The new consumers, as Myers and Kent (2004) call them, have already entered the global economy and are aspiring to lead the lifestyles currently enjoyed by the wealthy. A number of formerly poor countries are seeing the growing influence of affluence, as the middle classes of China, India, Indonesia, Pakistan, Philippines, South Korea, Thailand, Argentina, Brazil, Colombia and Mexico engage in greater conspicuous consumption. The side-effects are already being felt – the average car in Bangkok spends 44 days a year stuck in traffic. But there is still a long way to go. The car fleet of the whole of India is still smaller than that of Chicago, and that of China comprises half the number of cars in greater Los Angeles.

What, then, should we make of the current drivers of consumption? There is a simple and stark problem. If everyone consumes at the levels and patterns of industrial country citizens, then modern civilization is seriously threatened. Yet our conspicuous consumption behaviour, as Thorstein Veblen first called it in 1899, is making things worse day by day. Most people think that having more money will make them happier. But the more we have, the more we seem to want. It is of course true that the poorest who lack basic resources of food, water and security and the human dignity they bring are certainly made happier as they get wealthier, but after a certain threshold is passed, well-being becomes independent of consumption. The problem, it seems, is that we appear to be somewhat hard-wired to assess our position relative to that of others, and if others are getting more, we want more too. We have bigger houses and larger cars, spend less time with family and friends, work harder to get more money, take shorter holidays, and all for what? To have something larger or brighter or newer than someone else. It is true that we can choose what we spend and so appear to have free will, but we cannot choose what someone else spends, and that makes a difference to our own behaviour.

The problem, it seems, is that humans are remarkably good at adaptation. We get used to things, habitualized, and then take them for granted. Adaptation is very useful in a world of misery – it allows people to cope, but as Schwartz (2004) says, 'if you live in a world of plenty then adaptation defeats your attempts to enjoy good fortune'. Yet if we could choose, most people would think that winning the lottery was clearly a desirable thing. Robert Frank (1999) points out in his *Luxury Fever* that we do get a rush of satisfaction when we get a new TV, fridge or mobile phone, but unfortunately it soon wears off. He states: 'The US remains by far the richest nation on Earth. Yet we are currently squandering much of our wealth on fruitless mine-is-bigger consumption arms races.' A seemingly inevitable outcome is when faced with a huge array of options, we end up regretting the fact that our choice might have been the wrong one. And these regrets are dangerously corrosive.

But this dominant idea about the inevitable material benefits of progress would appear to be something of a modern invention. Indigenous peoples do not believe that their current community is any better than those in the past. To them, past and future are the same as current time. Their ancestors, and those of animals too, constantly remind them to be humble as they move about their landscapes. But the myth of progress permits the losses of both species and special places, as we believe we can offset any losses by doing something else that is better. The myth permits belief in technological fixes, which are indeed effective in many ways, but rarely seem to make us happier in the long term. Our environmental problems are, after all, human problems. And this myth is built on memory loss too – those people before today surely could not have been as happy, as they had less material wealth, and anyway did not survive to this day, so must be less clever too.

To believe in the inevitability of progress is also to believe in a freedom from the constraints of the world. It is to have faith in an idea that we are masters of our own destiny. Gray (2004) observes, 'we have inherited the faith that as the world becomes more modern, it will become more reasonable, more enlightened and more balanced. But such faith is not more than superstition'. We can and will, of course, do so much with intentionality – think of the advances in medical technology made in recent years. None of us would want to give these up. But will these advances be enough to make us more happy? Those people in the richer parts of the world have a wealth of material goods that entertain and reward them for all the hard work spent in obtaining them. Yet, extraordinarily, they have not made us any happier.

Contrast this situation again with that of hunter-gatherer communities, the way of life for all but a fifth of one per cent of our time on this Earth. Marshall

Sahlins famously called hunter-gatherers the original affluent society, and they typically do spend long periods resting, talking, telling stories and eating, with short periods of intense activity for hunting or gathering. They work two to five hours per day, live well and happily, while the rest of us are given a life sentence of hard labour (Lee and Daly, 1999). For moderns brought up to believe we must not waste a minute, this feels like anathema. But it can be a release.

Divergence

The authors of this book have shown that there have been unparalleled losses in biological and cultural diversity in recent decades (see Chapter 3 by Harmon, Woodley and Loh). These arise from common drivers that include the modernization of services such as healthcare and education which can lead to language erosion, decrease in cultural knowledge transfer (including resource management strategies), and less time spent experiencing nature with community elders and family members. Privatization of lands, exclusion policies and urban migration create a shift away from traditional resource management often causing a decline in biodiversity. In turn, this leads to the erosion of cultures as they are physically separated from the lands that their beliefs and worldviews are based upon. The health of both people and ecosystems suffers, as David Rapport and Luisa Maffi show in Chapter 6.

At the same time, the globalization of traditional food systems is leading to biodiversity decline as monocultures are favoured over on farm and wild diversity (see Chapters 9 and 10 by Patricia L. Howard and E. N. Anderson). This leads to the loss of traditional knowledge, skills and practices central to many non-industrial cultures. Livelihood diversification and resource commodification are threatening global diversity causing a departure from traditional resource use and management practices, the loss of traditional livelihoods and the local knowledge they are based upon. Such shifts may result from aspirations for consumer lifestyles or, indeed, new commercial resource uses being introduced, for instance, in first or second generation biofuels.

Helen Newing and W. M. Adams (Chapters 2 and 4) have shown the importance of shaping and self-shaping. Nature is a social construction (in the way that we relate to it), but so is society. Life on Earth is an emergent property of life itself, and thus cultures grow up in specific places and relate to these specificities of ecology and circumstance. Terms like wilderness need to be revised, as all that appears wild is now in some way influenced by humans. But different groups of people understand resources and systems in different ways – whether distinct disciplines from natural and social sciences, or whether hunters, conservationists, farmers and scientists. The chapters by Tirso Gonzales and Maria Gonzalez, and Martina Tyrrell (Chapters 5 and 7), show what happens when assumptions are made about the primacy of one knowledge system over another. The outcomes are often perverse for both cultures and biodiversity. We live in a world of capitalist landscapes, as Bill Adams suggests, and the fight for the future will be partly centred upon creating the space for divergence from this model. Recognition of the validity of different worldviews is a start (Gonzales and Gonzalez, Chapter 5). As a number of chapters show, many indigenous peoples have pluricultural views of the world – they accept that there are, and indeed should be, differences.

Loss of knowledge about local circumstances, both ecological and social, is another threat to the re-establishment of ways of living tied more closely to the land (see Howard, Tyrrell, Pilgrim et al). Nonetheless, there are signs of success, as James P. Robson and Fikret Berkes (Chapter 11) show with the emergence of community conserved areas. Building customary governance around local knowledge, monitoring, norms and rules can contribute to standard protected areas targets. Yet many indigenous and rural groups still associate National Parks with dispossession. Once again, professionals in the form of conservationists are going to need to be more inclusive and pluralistic. Some have doubts as to whether this can happen. Tyrrell (Chapter 7) points to the spiral of destructive relationships arising from resource regulations and restrictions that undermine local cultures and result in perverse biological outcomes too. The normal hunt is a relaxed affair, social, affirming; the hunt with quotas is a race. In Chapter 8, Garry Marvin focuses on the nature of recreational hunting in industrial countries, how it is practiced and experienced from the points of view and the perspectives of hunters themselves. Once again, activities on the land are part end in themselves, such as obtaining food, and part about the means to a social end. People hunt to tell stories, and stories underpin cultures. Hunting is not one thing either – it is underpinned by a rich spectrum of attitudes and activities ranging from treating animals as mere objects, targets for shooting, to the most respectful of pursuits. As Marvin says, 'hunters have a passionate interest in wild animals and much of their lives are orientated to such animals and the natural world'.

It is now known that time spent directly experiencing and interacting with nature improves physical and mental health (Pretty et al, 2005, 2006, 2007). On the other hand, spending less time in nature results in an intrinsic disconnection with nature, which not only undermines health but perhaps whole cultures too. Solastalgia is a term coined by Glenn Albrecht (Chapter 12) to describe the homesickness or nostalgia people can feel in their own home environment, usually brought on by environmental change such as development, devastation or disconnection. The term solastalgia derives from 'nostalgia' meaning homesickness, 'solace' meaning to provide comfort and relieve stress, 'desolation' meaning a personal feeling of abandonment (in humans) or devastation and ruin (of the environment), and 'algia' meaning anguish or pain. The pain or sickness that solastalgia describes is caused by the loss of, or inability to, derive solace from a person's home environment due to its change in state, usually resulting from a negative transformation of the physical environment. It can lead to feelings of bereavement and more serious physical and psychological ills, including symptoms of depression and illness. Thus instead of a spatial or temporal dislocation, solastalgia causes a person to feel homesick while at home due to environmental desolation. On the other hand, soliphilia can be the foundation of expressions of new ways of being that reunite us, as people, and with the foundations of life.

This, then, is the topic of the final chapter on revitalization (Pilgrim et al). Recognizing the health and societal repercussions of being disconnected from nature, communities in a number of locations have taken action to reclaim or maintain their unique beliefs and practices through ecocultural revitalization (so termed for their focus on reconnecting cultural systems with the ecosystems upon which they are based). Like the cultures they seek to rejuvenate, revitalization projects are diverse, ranging from hunter-support schemes and local food policies to language initiatives and ecotourism projects. All revitalization projects share a similar objective: to maintain or reclaim the culture of local peoples and reconnect them to the land for long-term individual and societal health. As their success is demonstrated, through changes to beliefs, meanings and worldviews, to livelihoods, practices and resource management systems, to knowledge bases and languages, and to institutions, norms and regulations, so there emerge the possibilities for wider divergence. Policy-makers should not only look towards revitalization projects for long-term solutions to cultural continuity, but also assess how these principles can be internalized across industrializing societies.

Integrated Policies and Options

The role of and need for effective policies in biodiversity protection has long been understood, but the importance of cultural protection policies as they relate to the environment is only just emerging (Maffi, 2001; MEA, 2005). Since many common drivers exist between biological and cultural diversity and their existence is so inherently linked, policy responses should target both in a new integrative conservation approach. The need for a parallel approach to the conservation of biological and cultural diversity has been acknowledged in the MDGs. However, policy responses to this paradigm have been slow to emerge. Responses to date include local recovery projects and revitalization schemes such as outpost and hunter-support programmes, culturally appropriate education schemes, ecotourism projects and language revitalization initiatives (see Chapter 13). Other revitalization efforts include the revival of culturally-appropriate healthcare systems, the protection and careful commercialization of traditional food systems, and the greening of businesses.

Larger-scale movements that have contributed to the dual protection of biological and cultural diversity include the fair-trade movement and other certification programmes. The land rights of indigenous and other rural people are being recognized in some locations, for instance, in the designation of the Nunavut Inuit territory in Canada. Investment into community-based conservation and the dissemination of power to grass-roots initiatives and institutions has increased, strengthening the mechanisms that favour ecocultural system sustainability. Entrepreneurship-based conservation development projects are also emerging (UNEP, 2007). However, despite their diversity, many efforts remain fragmented, localized and on a small scale.

Nonetheless, most of these initiatives focus primarily on biological diversity, perceiving cultural diversity as a secondary objective or a stepping stone to protecting biodiversity. Therefore, a great deal still needs to be done in the international arena to strengthen this movement and to ensure that integrative policies are applied widely. A paradigm shift is needed to transform the way people think about all global diversity, whereby biological and cultural diversity are thought of as parts of the same whole. This is emerging in the literature but has yet to emerge in protection policies, for instance, with the implementation of Ecocultural Protection Plans (EPPs) and the designation of Sites of Ecocultural Importance (SEIs).

One important development has been a dramatic reshaping of the way in which protected areas are conceived. There is increasing recognition of the importance of Indigenous and Community Conserved Areas (Robson and Berkes, Chapter 11). These are places managed by local communities in ways that support high levels of biodiversity but which often have no official protected status (Borrini-Feyerabend et al, 2004). There is also growing agreement that cultural landscapes are worthy of protection (under IUCN Category V Protected Areas) where the interaction of humans and nature over time has produced a particular set of natural and cultural conditions (Phillips, 2002). Emerging partnerships between faith groups and conservation organizations present another powerful opportunity.

Thus a range of policy options do exist to pave the way towards the joint protection of nature and culture and the development of fully functional ecocultural systems. However, to conserve global diversity effectively, policy efforts need to be both locally and internationally driven, large-scale, multi-level and inclusive. For instance, policies emphasizing political empowerment, self-governance and territorial control at the grass-roots level have the potential to provide a solid platform from which communities can play a central role in biodiversity conservation at the same time as retaining their own cultural distinctiveness and connectedness to the land (Colchester, 2000; Schwartzman et al, 2000; Peres and Zimmerman, 2001; Heckenberger, 2004; Athayde et al, 2007).

The degree to which the diversity of the world's ecosystems is linked to the diversity of its cultures is only beginning to be understood, and there is a great deal about this connection that we have yet to learn. There is always a need for more research. But it is precisely as our knowledge of this linkage is advancing that these cultures are receding. In the absence of an extensive and sensitive accounting of the mutual influences and effective policies targeting these issues, endangered species, threatened habitats, dying languages and knowledge bases are being lost at rates that are orders of magnitude higher than natural extinction rates. While conserving nature alongside human cultures presents unique challenges, any hope of saving biological diversity, or even recreating lost environments through restoration ecology, is predicated on a concomitant effort to appreciate, protect and respect cultural diversity. The future of the world depends on it.

References

Athayde, S. F., Kaiabi, A., Ono, K. Y. and Alexiades, M. N. (2007) 'Weaving power: Displacement and the dynamics of artistic knowledge amongst the Kaiabi in the Brazilian Amazon', in M. N. Alexiades (eds) *Mobility and Migration in Indigenous Amazonia: Contemporary Ethnoecological Perspectives*, Berghahn Books, Oxford

Berkes, F., Huebert, R., Fast, H., Manseau, M. and Diduck, A. (eds) (2005) *Breaking Ice: Renewable Resource and Ocean Management in the Canadian North*, University of Calgary Press, Calgary, pp225–247

Borrini-Feyerabend, G., Kothari, A. and Oviedo, G. (2004) *Indigenous and Local Communities and Protected Areas*, Cardiff University and IUCN, Cardiff and Cambridge

Brown, L. R. (2008) *Plan B 3.0: Mobilizing to Save Civilization*, W. W. Norton, New York

Colchester, M. (2000) 'Self-determination or environmental determinism for indigenous peoples in tropical forest conservation', *Conservation Biology*, vol 14, no 5, pp1365–1367

Costanza, R., Graumlich, L. J. and Steffen, W. (eds) (2007) *Sustainability or Collapse?* MIT Press, Cambridge, Massachusetts

Diamond, J. (2005) *Collapse: How Societies Choose to Fail or Survive*, Penguin, London

Ehrenfield, D. (1981) *The Arrogance of Humanism*, Oxford University Press, New York

Ehrenfield, D. (2005) 'The environmental limits to globalization', *Conservation Biology*, vol 19, no 2, pp318–326

Escobar, A. (2008) *Territories of Difference: Place, Movements, Life*, Duke University Pess, Redes

Folke, C., Colding, J., Olsson, P. and Hahn, T. (2007) 'Interdependent social-ecological systems', in J. Pretty, A. Ball, T. Benton, J. Guivant, D. Lee, D. Orr, M. Pfeffer and H. Ward (eds) *Sage Handbook on Environment and Society*, Sage, London, pp536–552

Frank, R. H. (1999) *Luxury Fever*, Princeton University Press, Princeton

Gray, J. (2004) *Heresies: Against Progress and Other Illusions*, Granta, London

Holling, C. S. (1973) 'Resilience and stability of ecological systems', *Annual Review of Ecology Systematics*, vol 4, pp1–23

Heckenberger, M. J. (2004) 'Archeology as indigenous advocacy in Amazonia', *Practicing Anthropology*, vol 26, pp34–38

Hulme, K. and Ong, D. (2009) 'The challenge of global environmental change for international law', in P. S. Low (ed) *Global Change and Sustainable Development*, Cambridge University Press, Cambridge

Jackson, T. (2009) *Prosperity Without Growth*, Earthscan, London

Lee, R. B. and Daly, R. (1999) *The Cambridge Encyclopedia of Hunters and Gatherers*, Cambridge University Press, Cambridge

Lovelock, J. (2005) *The Revenge of Gaia*, Allen Lane, London

Maffi, L. (ed) (2001) *On Biocultural Diversity*, Smithsonian Institution Press, Washington, DC

McNeill, J. (2000) *Something New Under the Sun*, Penguin, London

MEA (2005) *Ecosystems and Human Well-being: Current State and Trends*, Millennium Ecosystem Assessment, Island Press, Washington, DC

Myers, N. and Kent, J. (2004) *The New Consumers*, Island Press, Washington, DC

Orr, D. W. (2005) *The Right to Life*, Oberlin College, Oberlin

Peres, C. and Zimmerman, B. (2001) 'Perils in parks or parks in peril? Reconciling conservation in Amazonian reserves with and without use', *Conservation Biology*, vol 15, no 3, pp793–797

Phillips, A. (2002) 'Management guidelines for IUCN category V Protected Areas: Protected landscapes and seascapes', IUCN and the University of Cardiff, Cambridge and Cardiff

Pilgrim, S., Cullen, L., Smith, D. J. and Pretty, J. (2008) 'Ecological knowledge is lost in wealthier communities and countries', *Environmental Science and Technology*, vol 42, no 4, pp1004–1009

Pilgrim, S., Samson, C. and Pretty, J. (2009) 'Rebuilding lost connections: How revitalisation projects contribute to cultural continuity and improve the environment', iCES Occasional Paper, University of Essex

Pretty, J. (2003) 'Social capital and the collective management of resources', *Science*, vol 302, pp1912–1915

Pretty, J. (2007) *The Earth Only Endures: On Reconnecting with Nature and Our Place in it*, Earthscan, London

Pretty, J., Adams, B., Berkes, F., de Athayde, S., Dudley, N., Hunn, E., Maffi, L., Milton, K., Rapport, D., Robbins, P., Samson, C., Sterling, E., Stolton, S., Takeuchi, K., Tsing, A., Vintinner, E. and Pilgrim, S. (2008) How do nature and culture intersect? Plenary paper for Conference *Sustaining Cultural and Biological Diversity in a Rapidly Changing World: Lessons for Global Policy*, Organized by AMNH, IUCN-The World Conservation Union/Theme on Culture and Conservation, and Terralingua, New York, 2–5 April, 2008

Pretty, J., Ball, A. S., Benton, T., Guivant, J., Lee, D., Orr, D., Pfeffer, M. and Ward, H. (eds) (2007) *Sage Handbook on Environment and Society*, Sage, London

Pretty, J., Noble, A. D., Bossio, D., Dixon, J., Hine, R. E., Penning de Vries, F. W. T. and Morison, J. I. L. (2006) 'Resource-conserving agriculture increases yields in developing countries', *Environment, Science and Technology*, vol 3, no 1, pp24–43

Pretty, J., Peacock, J., Sellens, M. and Griffin, M. (2005) 'The mental and physical health outcomes of green exercise', *International Journal of Environmental Health Research*, vol 15, no 5, pp319–337

Pretty, J. and Pilgrim, S. (2008) 'Nature and culture', *Resurgence*, vol 250, September/October

Samson, C. and Pretty, J. (2006) 'Environmental and health benefits of hunting lifestyles and diets for the Innu of Labrador', *Food Policy*, vol 31, no 6, pp528–553

Schwartz, B. (2004) *The Paradox of Choice*, Harper, New York

Schwartzman, S., Moreira, A. and Nepstad, D. (2000) 'Rethinking tropical forest conservation: Perils in parks', *Conservation Biology*, vol 14, no 5, pp1351–1357

UNEP (2007) *Global Environment Outlook 4 (GEO-4)*, UNEP, Nairobi, Kenya

Index

Significant information in notes is indexed as 123n1, ie. note 1 on page 123

Printed in the USA/Agawam, MA
January 11, 2012

563480.055